W9-CEW-484

CRACKING THE CODES

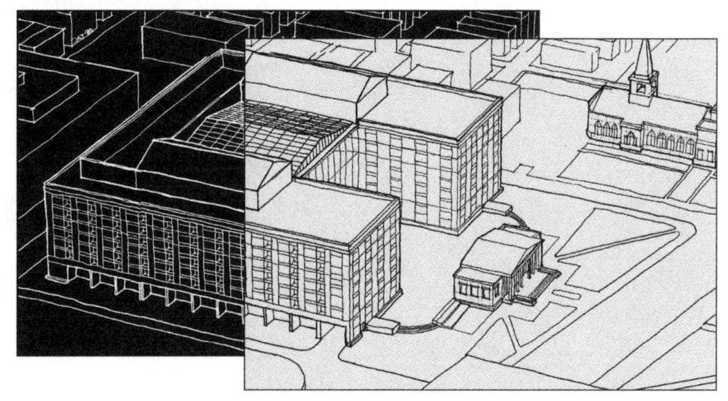

CRACKING THE CODES

An Architect's Guide to Building Regulations

BARRY D. YATT

Associate Professor

School of Architecture and Planning
The Catholic University of America

John Wiley & Sons, Inc

New York • Chichester • Weinheim • Brisbane • Singapore • Toronto

This book is printed on acid-free paper.

Published by John Wiley & Sons, Inc.

Published simultaneously in Canada.

Library of Congress Cataloging in Publication Data:

Yatt, Barry D.
Cracking the Codes : an architect's guide to building regulations / Barry D. Yatt
p. cm.
Includes index.
ISBN 0-471-16967-6 (acid-free paper)
1. Building laws—United States. 2. Architects—United States—Handbooks, manuals, etc. I. Title.
KF5701.Y38' 1998
343.73'07869—dc21 97-49806
CIP

Printed in the United States of America.

10 9 8 7 6 5 4 3 2 1

This book is dedicated in grateful memory to
Joseph P. Vanecko
(1931–1985),
a mentor of unparalled integrity.

CONTENTS

ACKNOWLEDGMENTS

I would like to express my deepest thanks to the many people who helped bring me to this point.

For material contributions: A very special thanks to Roger Lewis for allowing me complete access to and use of his witty and wonderful cartoons. I am indebted to John Michael Day for his beautiful fantasy city renderings and Carla Graeff for her painstaking compilation of codes into the Appendix matrices, both very significant efforts. Thanks to Kevin Wyllie and Quito Banogan for their photographs, Rebecca Gordon and Tom Andrews for their reviews and research assistance, and to Patrick Haggerty, Zoe Rivera, Diana Engles, Richard Loosle, and Caren Brown for their efforts. To the students of "Issues in Practice" at The Catholic University of America for whom this book was initially developed, thanks for their help in trying to teach me how to explain the right things at the right time in the right way.

For technical input and checking: Thanks to Doryan Winkelman (Subdivision Regulations), Judith Meany (Planning), Tony O'Neill (Fire Suppression), Quy Do (Getting Local), Edward Balter and Robert Miller (Geotechnical and Environmental), Roy Eugene Graham (Preservation), John O'Neill (Life Safety), Jonathan Binder (Environmental protection), Carol Stockton (Accessibility), George Marcou (Urban Design), Michael Pantano (Hygiene), Thomas Rohrbaugh (Mechanical), George Colomiris (Politics, Preservation), and to Stephanie Vierra, Simin Naaseh, and Chris Arnold (Seismic). A special thanks to Sara O'Neil-Manion, Jim Howe, and Pelin Atasoy for their general reviews and comments.

For providing information: Thanks to the officials of building departments and planning offices, the model code staffs, and others who patiently answered questions and mailed information. This includes Steve Jacobson of Reston Association; Ed Nunley, District of Columbia Zoning Review Branch; The Arlington County, Virginia, Department of Public Works; and Sophia Zager of Fairfax County.

For technical support: Thanks to David Sassian at John Wiley & Sons. Thanks to Donna Levine, and to Walter Schwarz of Figaro, Inc., whose editing and production were a comforting and very professional cap to a frenetic effort. To Lenny Mayor and Baran Kilical, who gave so generously of their time maintaining and rescuing my computer.

For professional support: Thanks to Stanley Hallet and Stephen Kendall, without whose encouragement I would never have moved from practice to teaching. To the faculty at The Catholic University of America, who gave me complete independence to develop courses that took their cues from the students, with a special thanks to James O'Hear, III, Milton Shinberg, and Donald Little. To Joseph Bilello, Dale Ellickson, Charles Heuer, Charles Amado, Ronald Gribble, Steve Hill, and Joseph Scalabrin for their years of professional encouragement, mentorship, and friendship. To David Herstein, Emily Lipsitz, and the Baltimore City Public Schools, who taught me how to think.

For very special support: Thanks to Forrest Wilson, who got me into the writing racket, the mental sparring partner with whom I've had the rare pleasure of sharing an office since I began teaching. To James Howe, whose work with students inspired me to attempt to teach codes, whose energy and input kept me on track, and whose in-progress reviews of the manuscript were instrumental. To the proposal reviewers, Edward Allen and Maryrose McGowan, for their criticism, insight, and support. And to my editor at John Wiley & Sons, Amanda Miller, who extended a rare level of guidance, perspective, and friendship.

For personal support: Most of all, thanks to my amazing wife, Sandy, for her encouragement, support, faith, sacrifice, and patience, while I poured so many hours into this and so many other efforts. To my son, Benjamin, for his patience during my hours at the keyboard. To my parents, Harold and Bonnie, for always believing, and giving so much for so many years. To my in-laws, Saul and Lina, for their unfailing encouragement and help. And to Jerry, Mark, Minda, and Dory: you already know.

To any I may have inadvertently left off this list, I apologize deeply, thank you sincerely, and hope you know that your efforts and support are appreciated.

CRACKING THE CODES

INTRODUCTION

WHAT?

This book is about developing and designing better projects, aided by the accumulated knowledge that regulations represent. It posits the radical notion that, far from being a hindrance, regulations strengthen design when understood and used properly. It shows them as lessons learned through centuries of skilled craftsmanship and tragic building failures, lessons far too valuable to be ignored or marginalized. It makes codes accessible by establishing the project planning process as the Rosetta stone of building regulation, translating the wide range of rules through the perspective of project development, design, and construction. After all, this process is the common target of all codes, ordinances, regulations, standards, and guidelines, so why not use it as a lingua franca? In effect, this book unbinds the individual regulations, throws them in the air, and lets them fall back as a unified, integrated, understandable package reflecting the thinking of owners, builders, and designers rather than the necessities of code writing.

WHY?

Buildings are bigger, more complex, and contain more people than ever before. They are therefore more expensive and prone to bigger failures and greater loss than ever before. Society depends on their continued functionality and financial stability. Besides, the cost of error is too high. None of us can learn enough to make all the right decisions for every aspect of our projects. Thankfully, help is available for producing excellent projects, in the form of the organized body of knowledge represented by statutes, regulations, ordinances, codes, standards, and guidelines. This is a body we might generically refer to as rules.

These rules contain carefully tailored directives aimed at maintaining the value and safety of buildings and the communities they form. But the rules have also become somewhat inaccessible. There is simply too much uncontrolled information, with too many overlaps and too much confusion. Perhaps that's the inevitable result of any attempt to describe such complex and critically important issues using words. So what can one do?

Any member of a project team—owner, designer, or builder—can call on regulatory consultants to review project plans and report back on what is acceptable and what isn't. These consultants are specialists who spend a majority of their time familiarizing themselves with current requirements and proposed changes to them. For other members of a project team, doing this while simultaneously trying to fulfill so many other roles would be daunting.

But this may not solve the problem either. In this situation, project teams are often forced to accept their consultants' conclusions, because they rarely have enough time to check the facts themselves or to find out if the consultants' recommended solutions are the best. They're held hostage by their own unfamiliarity with rules, and by their consultants' unfamiliarity with the project team's vision for the project. These conditions are, unfortunately, far too understandable in light of the complexities and pressures of the contemporary real estate, design, and construction businesses.

Great design work requires a thorough understanding of the issues that regulators worry about. Why? Because one can't be "cutting edge" without knowing where the edge is. For example, without understanding the difference between a "project," a "building," and a "fire area," project teams would likely be forced, through misinterpretation, to design with unnecessarily reduced floor areas. Without understanding the differences between "fire walls" and "fire area separations," teams might locate walls in areas that could otherwise be opened through use of windows and railings.

This book is intended to strike a middle ground, bringing a reasonably deep yet conceptual understanding of regulatory issues without getting caught up in the minutiae. It is intended to identify and consolidate the basic requirements that must be addressed when planning building projects. The point of all this is to empower members of project teams to be able to

- recognize issues that call for regulatory research,
- work effectively with appropriate consultants as needed,
- suggest strategic directions in which to move the project effort as a result of regulatory factors, and
- make informed project decisions in response to their consultants' advice, all without necessitating continuous and exhaustive study of rules.

This book is not intended to eliminate the need to read rules in the original. The choice of words found there aims for accuracy over digestibility. With this book, the priorities are reversed. Precision and completeness are somewhat compromised to promote understanding. Readers involved in projects who can't or won't read rules in the original should find someone on their teams who can and will.

It is also not intended to replace the handbooks published by each of the organizations that write model building codes. Such books go through the codes, paragraph by paragraph, explaining nuances and offering usage examples. While such commentaries are extremely useful to those who already have a working knowledge of the codes, they are not as accessible to those who don't understand the codes' structure.

HOW MUCH?

This book explores the body of knowledge related to program, site, and value issues codified in four types of rules:

- Zoning ordinances and covenants
- Building codes and standards
- Accessibility statutes, regulations, and guidelines
- Historic preservation standards and guidelines

However, it does not do so in an all-inclusive way. Rules exist for every aspect of our lives and human endeavor. There are millions of rules in thousands of regulations in the United States alone. Almost all these rules contain enormous numbers of exceptions and qualifications.

The extent to which issues are explored in this book is limited by the realities of readership and authorship. In perhaps no case was the author able to review all the available options, or even address all the issues involved in any of the topics explored. This was deliberate. Such an exhaustive study would not be terribly useful, since no single book or reference could ever contain, or even reference, all the rules in existence. This book attempts to move readers onto the path of doing their own explorations. To that end, it seeks to provide a glimpse of the richness available to those with the time, interest, and motivation to look.

WHY NOW?

With a unified building code (the International Building Code, or IBC) being developed to replace the multiplicity of model building codes currently in use, the information in this book is becoming more valuable than ever. Once the current codes are withdrawn, only the IBC's logic, definitions of design problems, and recommendations of possible solutions will be common knowledge. As with any reduction in diversity, the effect of this consolidation will necessarily reduce the number of alternative solutions embodied by or suggested by the codes. As other codes have, the IBC will encourage design professionals to pursue any and all effective and appropriate design solutions, but it will contain fewer hints than the three model codes that preceded it. This is particularly true since its orientation is performance based, in contrast to the prescriptive orientation of current codes. While being less restrictive, this approach will also require more original thinking.

This book starts the process of recording the diverse approaches previously used for dealing with diverse design issues. Further archival research, analysis, and dissemination would be most appropriate. Still, and regardless of the changes made through consolidation of the building codes, concepts and basic approaches will remain fairly constant with those embodied in the old model codes.

Last, while building codes are undergoing sweeping and fairly long-term changes, they are not the only rules referenced in developing construction projects. Zoning ordinances are always locally based and therefore very diversified. Historic preservation guidelines and accessibility rules are in constant flux. Now is as good a time as any to observe them critically.

HOW?

Methodology

The planning and design of a building is a very complex thing. Every decision is affected by every other decision. Unless issues are considered *en masse* rather than individually, the design solutions achieved may not be truly optimal even if they do fulfill many goals. Successful businesses run on slim margins, and optimal solutions are not only demanded but expected.

Such tasks demand nonlinear thinking. Project teams must see many issues at once. Although the information in this book is presented sequentially, chapter by chapter, it is also extensively cross-referenced. To reduce redundancy of content, each issue is discussed only once, but liberal use of citations will help readers find everything that relates to any particular issue, regardless of where they start their search. Readers may find that the structure of this book approaches that of a matrix.

Process

In an effort to reduce the inevitable confusion that plagues regulatory issues, a concerted attempt has been made to compare apples with apples. All regulatory issues analyzed in this book are investigated using the same methodology. Issues are grouped by the design question that would elicit their consideration. The issues are first discussed in general, and then broken down into specific regulatory responses related to particular issues. This methodology contains the following components, in order:

Context

This section provides a context for understanding the rules and includes the following parts:

Why the issue is important

How does consideration of the issue make for better built environments? What are the functional, societal, and legal implications of the issue?

Why regulatory bodies care

What is the thinking behind the issue? What are the rules getting at? Why do they get involved? What outcomes are they intending to avoid or promote, and why?

Where to find the requirements

Code books can be unwieldy and intimidating. Where does one get information related to a particular issue? Is it all gathered in one part of a rule, scattered among several related topics, or even addressed by several different regulatory bodies? Are there sources one should consult before or after turning to the rules?

Analysis of specific rules related to the issue

The hope of any project team in doing a project is to optimally fulfill its goals. The factors that constitute optimization vary from project to project. It may mean maximization of one characteristic or another. It may mean achieving a certain level of balance between critical factors. Whatever it means, in an ideal project, rules supplement, rather than stand in the way of, the fight for such optimization.

This section responds to optimization with systematic analyses of each of the specific issues related to the topic by looking at three aspects.

Background

What, specifically, does the rule mean? Why is it relevant?

What considerations pertain to the planning question raised, regardless of the specific methodology used by rules to address it?

Requirements

The thinking underlying specific regulatory requirements is examined. The requirements themselves are listed only where they are useful to illuminate the underlying thinking. Those that are listed are generalized where possible, to transcend any particular edition or model rule. Therefore, readers who need specific requirements must read the source rules.

The discussion of implications may be one of the most valuable components of this book, not only for project teams but for those who draft rules. In it, the inherent but unwritten and often unintended biases contained in regulatory requirements are discussed, and the kinds of designs such requirements are likely to foster are made explicit. Regulatory requirements are written by human beings, after all, who are trying, through their requirements, to prevent problematic design solutions and encourage desired outcomes. They aren't given crystal balls and can't necessarily predict the kinds of projects that will result from their regulatory requirements. This is one reason rules need to be revised periodically. In the meantime, project teams would do well to understand these hidden implications.

Optimizing the rules

When rules appear to block some aspect of a preferred design, savvy project teams don't abandon the design, they look for ways to reconcile the conflict. This part of the book reviews strategies for creatively applying rules to projects, explaining how to get a desired design without having to violate rules. It explores allowable exceptions and alternatives to the base rules. Particular effort has been made to point out those aspects of each issue that are rule-free. Difficult requirements can often be optimized without skirting or illegally ignoring them. Conversely, avoidance tactics are often noticed and rejected by regulatory officials who have seen them attempted by others.

When optimizing rules, open-minded project teams attempt to fulfill code goals while maintaining or even broadening design options. Those that stretch regulatory prescriptions simply for the sake of aesthetic goals too easily wind up compromising the overall value of their projects. Such value depends on solutions balanced in terms of Vitruvius's famous trio, "firmness, commodity, and delight," or their contemporary equivalents, durability, functionality, and appeal. The issue is, therefore, not one of finding ways to justify inferior design solutions for the sake of aesthetics but rather one of fulfilling regulatory goals while maintaining, or even expanding, design options.

Related case studies

These are examples of actual projects that have found ways to successfully optimize rules, and that managed to get approval for aspects that might otherwise have been considered noncompliant.

Integrated Strategies

In this section, we return to the planning issue, the design question, that the specific regulatory mechanisms are intending to address. Are there broad strategies that can be used to deal with several specific regulatory requirements at once, in an integrated, systemic way? Relevant examples and case studies are presented.

Sequence

Organized by Design Issue

Designers work by identifying their projects' design issues and then working through different potential responses to those issues to find the most elegant solutions. The issues come first.

With this book, the issues also come first. Its table of contents is organized by design issues rather than by the table of contents of any existing regulation, ordinance, or code. Consequently, readers will be able to find regulatory issues related to their design issues by looking in the sections that explore those design issues. Rather than trying to remember where "means of egress" issues are, they can simply look under "circulation." More important, once there they will find a discussion not only of building code rules but also of accessibility rules, historic preservation rules, and other rules relevant to the same design issue.

Ordered by a Traditional Design Sequence

Many building projects are too big and complex to be planned or designed without some kind of process or sequence. For many, teams work in a macro to micro sequence, considering first the issues that will make or break the project, then those that have the broadest, most public impact, and only later the finer details, those that make a basic building into a supremely functional and delightful environment. This sequence is familiar to all project teams, notwithstanding the validity of other sequences and the cyclical rather than strictly linear reality of the design process.

This book therefore follows the macro to micro sequence. This sequence considers the project team's ability

- to get permission to put projects on the intended sites with the intended program,
- to determine where on the intended sites to place the projects,
- to determine the massing (size and shape) of the buildings,
- to determine facade and roof designs,
- to determine the layouts of the buildings,
- to choose materials and components,
- to determine the way that materials will be combined into assemblies.

No previous experience in real estate or construction issues is necessary to understand this book. It does, however, go from this zero point to a fairly sophisticated level of investigation. It gives readers a basic, conceptual, and systemic understanding and appreciation of the regulatory issues to which successful projects must respond.

CONVENTIONS

Throughout this book, the term *rule* has been used to refer to any of a number of classes of directives intended to regulate the design and construction industries. The term may refer to codes, ordinances, standards, guidelines, statutes, or regulations, or all of them. There is no industry or legal consensus on a single word that refers to so many types of documents. *Rule* was chosen partly for its simplicity—it isn't obscured by jargon or legalistic baggage. We start understanding the importance of "rules" in earliest childhood.

Citations are shown in brackets. For example, [BOCA 96 2304.1] refers readers to Section 2304.1 of the 1996 edition of the BOCA National Building Code. References are from a variety of rules and a variety of publication dates, since the author tried to select citations that would best illustrate each particular topic of exploration. The code referenced may therefore not pertain to any specific project currently being developed by readers. Also, although all three American building codes converted to a common sequence of chapters in 1993, section references within chapters are somewhat inconsistent. The Appendix contains a matrix correlating the issues discussed in this book with article citations in each of the major codes.

Code citations are indicated as follows:

- BOCA*yy*, SBC*yy*, UBCyy, CABO*yy*: Refers to the four model building codes. The *yy* number indicates the year of publication.
- *xx* CFR *xx*: Code of Federal Regulations (ADAAG, etc.). The *xx* numbers refer to Title and Part.
- IBC: The International Building Code, Working Draft, May 1997. Although this model code is still a long way from finished, it is a valid reference for the issues described in this book because it simply combines language found in the current model building codes. As noted above, upon final release, its language will be modified and will quite likely be less reflective of the diversity embodied in its constituent model codes.
- IMC*yy*, IPC*yy*: International Mechanical Code, International Plumbing Code.
- LSC*yy*: Life Safety Code (NFPA 101), 1994 edition, unless noted otherwise.
- LSCH: Life Safety Code Handbook, © 1994.
- NEC*yy*: National Electric Code, NFPA 70.
- NFPA*xx*: Refers to any of the publications of the National Fire Protection Association, except the Life Safety Code (NFPA 101) and the National Electric Code (NFPA 70). In this case, the *xx* number specifies the particular publication. Thus NFPA13 refers to NFPA publication number 13, Installation of Sprinkler Systems.
- *xx* Pub. L. *xx*: Public Laws, U.S. Statutes at Large. The *xx* numbers refer to Title and Section.
- *xx* USC *xx*: United States Code. The *xx* numbers refer to Title and Section.

ATTITUDE

Try as I might, I wasn't able to limit the discussion in this book strictly to rules and regulations. Rules are inseparably bound up in design and development issues. So, you may ask, is this a book about construction technology, conceptual ideas, pragmatic advice, law, behavior, politics, science, or art? Yes.

My perspective is that of a practicing architect who has learned something about education. It is not that of the code official or consultant. The advantage of this is that

- I am not focused on only one code or type of code,
- this book may be better oriented toward the concerns of readers in any of the construction professions or industries,
- I have learned about codes by using them, and learned about relating what I've learned by teaching them.

While I have done my best to research and confirm everything explored in this book, the experience forces me to paraphrase Mae West: There are so many rules and so little time. I hope that this book's logic structure triggers insights that lead to greater vision and creativity. For some topics in this book, implications and optimizations remain to be explored. Readers are invited and encouraged to join me in furthering this work by sending comments, corrections, insights, case studies, and favorite optimization techniques to me in care of the publisher.

DISCLAIMER

The information and analyses in this book have been derived from sources that include zoning ordinances, building codes, government regulations, private covenants, industry standards, articles from periodicals, the Internet, and personal contacts with many individuals. Although it is presented in good faith, the author and publisher do not warrant, and assume no liability for, its accuracy or completeness or its fitness for any particular purpose, even though every reasonable effort has been made to make this book accurate and authoritative.

Readers should note especially that this is a book of concepts, ideas that are not intended to be used for final design of any building or structure. It is the responsibility of readers to apply their professional knowledge in the use of information contained in this book, and where needed, to consult original sources for more detailed information and to seek expert advice.

PART I

BACKGROUND

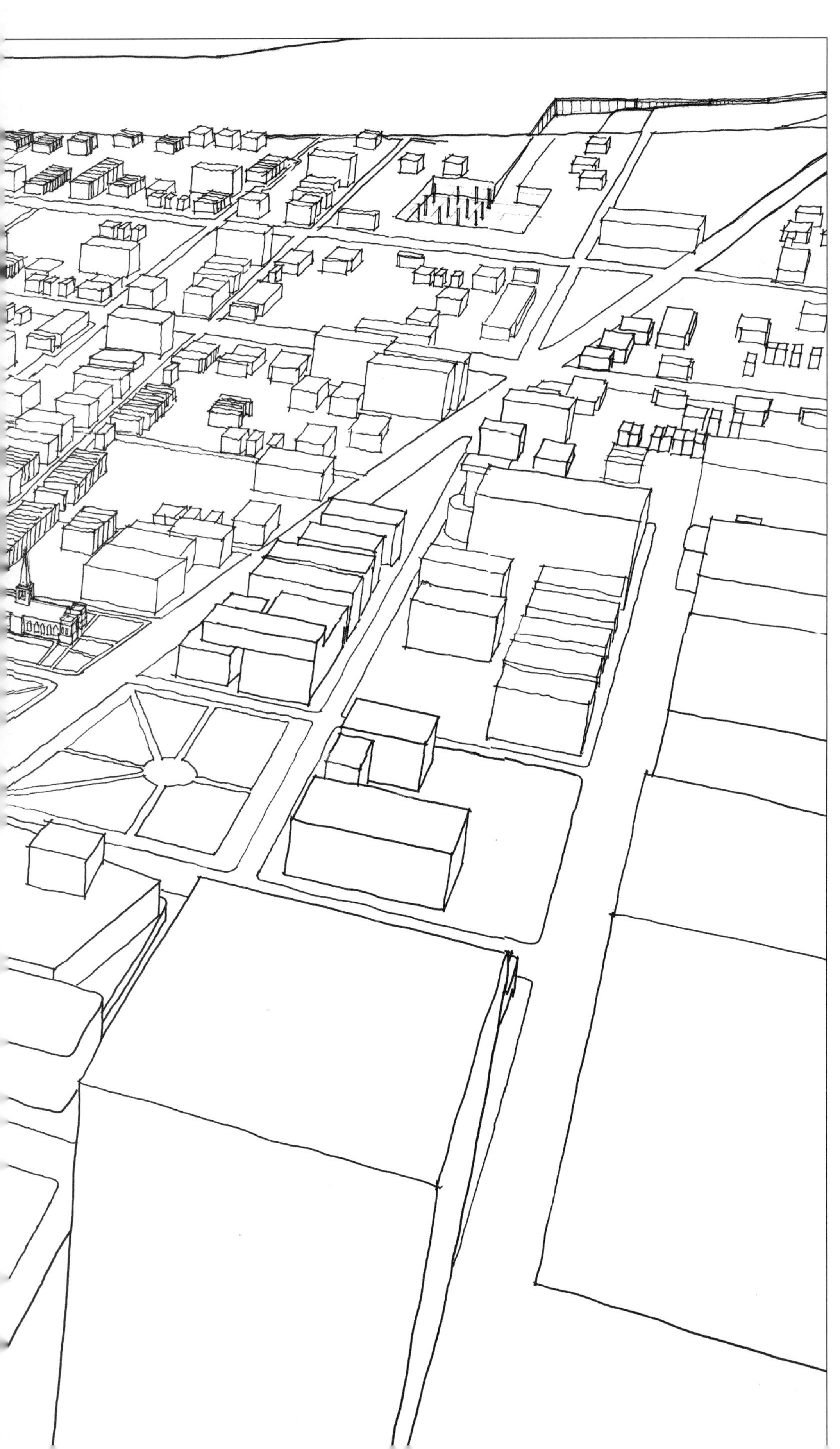

Before jumping into design issues, one must understand what rules are and why we have them. Part I explores:

- Why we choose to have rules in the building industry.
- The types and functions of building industry rules.
- Who has what rights in the process of drafting, adopting, revising, applying, and adjusting rules, and why.

1 PEOPLE

The Building Industry

CONTEXT

The thought of dealing with codes is enough to strike if not terror, then at least resentment and boredom in the hearts of many a developer, architect, and builder. Building, zoning, historic preservation, and accessibility rules are usually seen as the stick that prods them, grudgingly, to design and build things that weren't on their agendas. Rarely are the rules understood as either the societal assurance or as the designers' resource that they are.

RULES AS SOCIETAL ASSURANCE

Why are project teams required to deal with externally imposed rules? Who gave governments the authority to impose such restrictions on projects built by private groups? The answer can be understood if one considers the nature of market economies and contractual obligations.

In relationships between a project's three primary players—owners, builders, and design professionals—the owners exert direct control over their projects.[1] Traditionally, the three-party relationship has been expressed by the "three-legged stool" diagram:

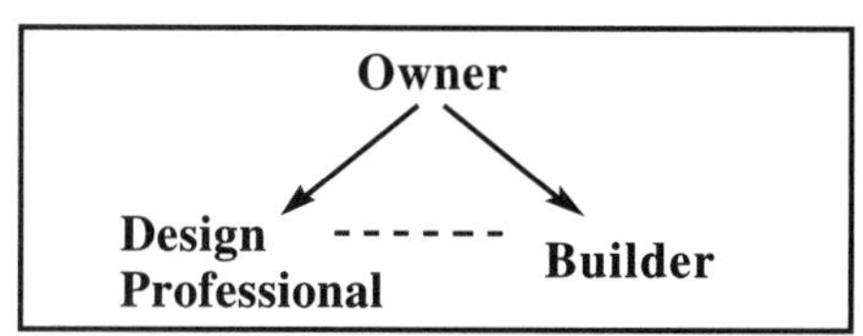

The twin arrows represent the separate contracts governing the owner's relationships with the design professional and with the builder. The dashed line between the design professional and the builder expresses the design professional's duties administering the contract for construction.

Real estate relationships—simplified.

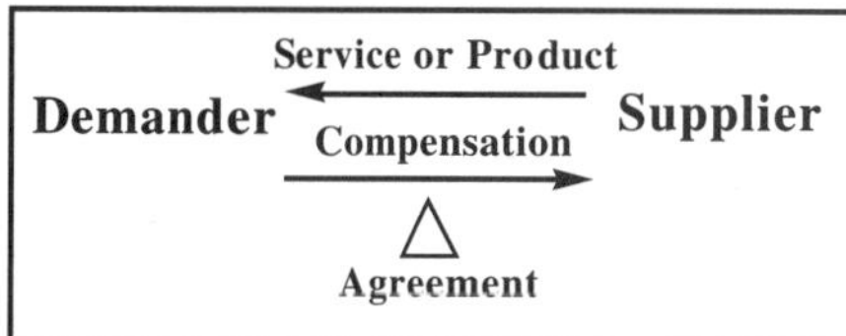

This scenario works adequately when the Demander is able to clearly and completely explain to the Supplier the service or product he wants supplied.

Typical market economy business relationships.

This diagram expresses the two-part nature of the relationship, whereby the owner has separate contracts with the design professional and with the builder. The main problem with this diagram is in its simplicity. It does not express the exchange inherent in relationships that are based on supply and demand. The following diagram is a general expression of this exchange.

These two diagrams can be combined to more specifically and accurately express real estate relationships, as shown in the next diagram.

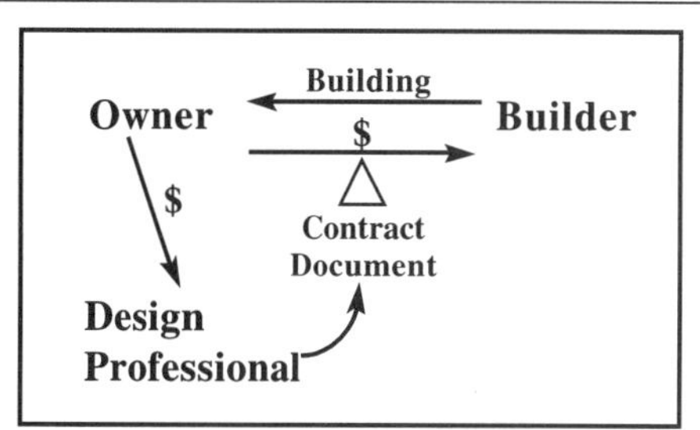

Before owners ask builders to "deliver" buildings in exchange for payment, they ask design professionals to prepare contract documents in exchange for fees. These documents then structure their relationships with their builders.

Real estate relationships—expanded.

While perhaps more accurate, this diagram still ignores rules and governments. To include them, we must recognize that owners generally act in their own interest, and that their interests do not necessarily coincide with the interests of others affected by the project. These others like to know that the buildings they visit, work in, live in, own, or that are adjacent to their real estate investments won't fall on their heads, make them sick, or compromise their investments.

So how do they get a voice? People pay taxes, in part, so that someone (the government) will look after their interests. This is expressed in the basic civics relationship.

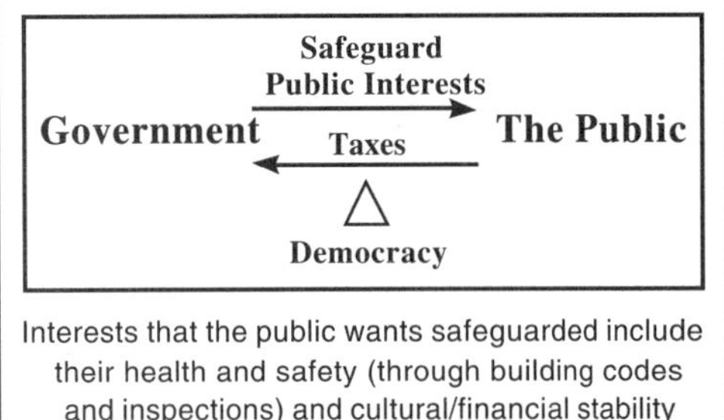

Interests that the public wants safeguarded include their health and safety (through building codes and inspections) and cultural/financial stability (through planning laws).

Basic civics relationship.

All that's left is to link the expanded real estate relationship to the civics relationship. Such a link would allow governments to act on behalf of the public. As shown in the diagram below, licensing laws provide this link, by which the public is able to exert indirect control over real estate projects.

Licensing laws are much more stringent for design professionals than for builders, because builders already have a direct contractual obligation with the owner to build what the contract documents describe. So long as the public finds the contract documents acceptable, they don't have much to worry about with builders. The public therefore is more careful about licensing design professionals, which it does through practice and title statutes. In exchange for their code-compliant designs, the professionals are given the marketing advantage of not having to compete with unlicensed designers.

The public has the right to impose regulatory limits, not only on public projects for which, ultimately, they are the client, but also on private construction projects. The government is a necessary mediator in this process.

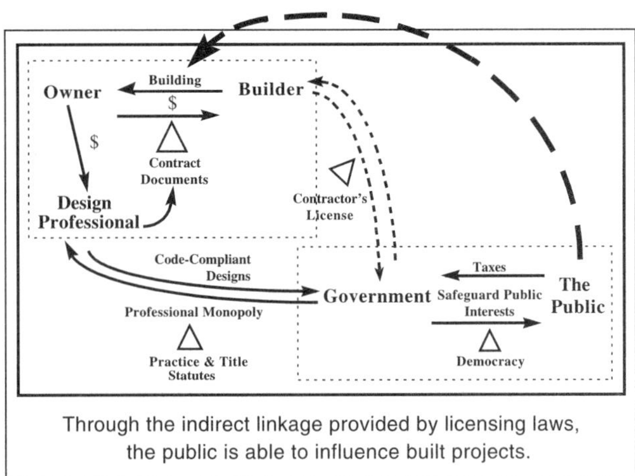

Through the indirect linkage provided by licensing laws, the public is able to influence built projects.

The complete situation—linking the real estate relationship with the civics relationship.

Perhaps these diagrams help convey the vital necessity of public attendance and participation when project teams petition for regulatory relief, and even of public involvement in the drafting and revising of rules, particularly at the local level.

Rules as Designers' Resource

Rules are information resources for designers who want their projects to be adequately functional, appropriate, strong, and safe. Such rules bring a public relations benefit to designers, associating them with serving the larger public interest.

People used to say "You are what you eat" and "Acorns don't fall far from their trees." Now they say "Garbage in, garbage out." Either way, they're expressing a concept that applies very directly to the relationship of buildings to their developers and designers. Anyone involved in the production of the built environment needs to understand the basic elements that affect and determine what is built. Deciding what should be produced, which one might call design content, is every bit as important as deciding how it should be produced—its form. But how does one get an understanding of such things? Where is such information to be found? While some "whats" are taught in accredited programs of architecture and engineering, many are learned through work experience. Some information must be developed from scratch for each new project as a response to unique conditions. Other information is more broadly applicable, and compiled into codes, standards, and guidelines. The information in such documents is organized, methodical, and relevant to many design issues, and therefore useful to anyone involved in producing buildings.

Buildings, like people, reflect their heritage and formative influences, whether they directly express them or not. Before construction begins, and even before the design process begins, proper research (predesign or feasibility work) allows the project team to discover and explore a project's "roots." It also provides a basis for testing the limits, extents, and validity of those roots.

The wealth of knowledge contained in rules is compiled by experts who have studied the causes and consequences of building failures and by local legislators who have worked to prevent urban planning mistakes. Even when such rules fail, the control they provide can turn failures into learning experiences. In current jargon, we could say that the use of rules compiled by experts allows project teams to "outsource" some of the research that precedes building design.

These rules help the industry establish and maintain a standard of care, a baseline that sets minimum acceptable levels of quality. In a profession and an industry that are compelled to constant invention (since every project introduces a new site, new goals, and new user issues), where products often do not have the luxury of being refined in successive production runs, and where feedback after completion rarely makes it back to those who create the projects, rules can rightly be viewed as providing a relatively stable and sorely needed knowledge base. The more we understand each project's formative influences, the more responsive and appropriate (and therefore successful) our building designs can be. This is a major factor in design excellence.

This is the carrot in the codes, the upside of rules that intimately affect the way the building industry works. The public is willing to allow greater design freedoms to projects that are inherently safer, and greater concessions to developers of projects that include plazas or other public domain benefits. Practitioners and industry leaders who recognize the value of this carrot find ways to take advantage of the incentives, exceptions, and bonuses that rules allow for responsive projects.

PARTICIPANTS

Before exploring building industry rules, it is important to understand the participants, for the following reasons:

- Knowledge of those who influence particular issues helps in understanding why some voices are more influential than others in the shaping of building industry rules.
- An understanding of the participants helps explain why only some of them are subject to regulatory oversight.

The "building delivery" industry is a combination of the real estate, finance, and investment industries; the design professions; and the construction industry. For every project, these industries and others have vested interests. Further, these vested parties can be arranged into many different relationships based on whether the issue at hand is social, political, economic, or something else. All their agendas are relevant to the consideration and fulfillment of building industry rules.

Participants can be categorized as part of the investment team, design team, construction team, or users group.

Participants may each have their own agendas. © Roger K. Lewis.

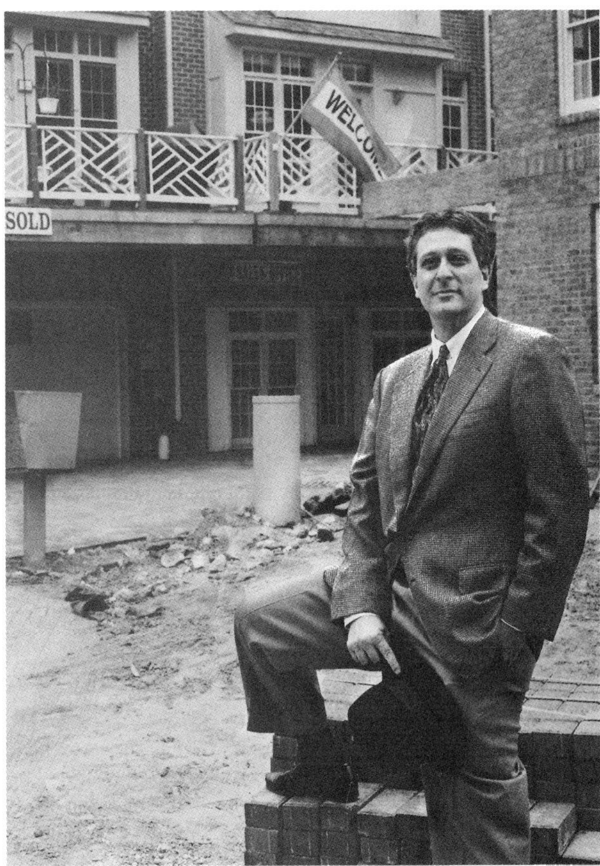

Photo by the author.

The Investment Team

This group initiates, produces, and directs the project in addition to taking the financial risk for it. They deal with the building long after the design and construction teams have left. Further, as project managers, owners, investors, and users, they often carry several votes. What elements from their agenda can be aided through regulatory responses?

Developers commit their reputation to the project's ability to generate a profit. And to ensure that the project will be able to generate that profit, they bring a very definite agenda to the process.

Investors have a say in what the project will be, but they generally don't care much as long as their investment pays off.

One might think of **lenders** as temporary investors. They most immediately and directly pay for the project, through construction draws. Even though investors generally guarantee any losses with collateral, lenders want to be sure projects will be able to pay them back.

Users ultimately pay for the project, through rents and use fees. They must find the project sufficiently desirable to want to occupy it.

Until **designers and builders** get paid, they may have some of the legal rights of investors. Designers have invested time, and builders have invested time and materials. Builders can enforce their claims to payment by filing a "mechanic's lien." In a growing number of states, designers are also being given the right to "lien" a property.[2] Doing so allows builders and designers to claim partial ownership so long as they haven't been fully paid. The most effective action they can take is to refuse to agree when the rest of the investment team is ready to sell or refinance the property, generally stopping the deal. This act tends to get the attention of the other investors. These two groups are fundamentally different, though, from other investors. Their financial interest in the project lasts only until they are paid, and their goals are not generally relevant to long-term planning.

Neighbors might be thought of as investors even though they don't directly pay for a project. Their concern over being able to pay for their own projects makes them concerned about the value of other properties in their area. This concern is sometimes strong enough to motivate the neighbors to action, action that can affect a project's financial viability. Zoning ordinances have their roots in the concerns of neighbors.

The Design Team

Before construction begins, everyone, including the public (through the permitting process), must approve proposed projects. This necessitates making most construction decisions before construction actually starts. The process of making, and then documenting, these decisions is the process of design. Who handles it, and what are their agendas?

Design is primarily the purview of the investment teams' design consultants. These consultants compose the project, in the musical sense of that word, based on direction from the investment team. They develop several alternative ways to meet their projects' design goals, and then present them, with recommendations, to investment teams for decisions and approvals. This is as true of conceptual design work as it is of the design of roofing details. The main difference between these two widely different scales is that investment teams rely much more on trust for recommendations in solving the more detailed, technically oriented issues. They often approve thousands of very small decisions as a package, and they sometimes do so with a relatively abbreviated review process. By contrast, for conceptual design decisions, they often prefer to make more of the design decisions themselves.

Regarding sequence, investment teams may allow design teams to investigate the consequences of proposed designs before proceeding with construction. Sometimes, however, investment teams find greater value in compressed schedules (those that eliminate design or make it simultaneous with execution) such as fast-track delivery than in those that allow design to occur before construction such as the traditional design-procure-build (also known as design-bid-build).

Courtesy of the Association of Collegiate Schools of Architecture.

Imposed rules address many of the potentially troublesome consequences of design. For the reasons noted above, design teams are the teams primarily, if not exclusively, responsible for ensuring that planned projects fulfill regulatory requirements. Licensed designers generally cannot legally pass responsibility for regulatory fulfillment on to others on the project team, including owners and builders. They are regulated and legally sanctioned by the public to make design decisions. The buck stops there.

CASE STUDY

An interesting case illustrating the relationships between owner, architect, builder, and users involved blocked sight lines from designated accessible seats at the MCI Arena in Washington, D.C. As the arena was designed, sports fans who leap to their feet during exciting plays block the views of those in wheelchairs. The Paralyzed Veterans of America (PVA) sued both the owner and the architect, Ellerbe Becket, alleging that the latter had been in a position to prevent such discrimination resulting from architectural barriers. The architecture firm made two defenses:

- That it had done what the rules required. Even the judge came down hard on the U.S. Justice Department for not "providing concrete guidance for architects and builders." He noted that, under the current rules, design professionals could make good-faith efforts to comply with regulations yet fail to comply with statutes.
- That it was not directly liable for any discrimination caused by the seating arrangement, claiming that only the owner had direct responsibility for providing an accessible facility. The judge "asserted that the statute explicitly places responsibility for compliance with ADA [Americans with Disabilities Act] on owners, operators, lessors, and lessees. 'It would make sense for the Congress to have included architects and engineers and designers under the design and construction section of the statute...but the simple language strictly limits the individuals who are responsible.'"

Ellerbe Becket was dropped from the list of defendants in this suit. Of course, this ruling does not ultimately address the performance of the architecture firm. Ellerbe Becket remains liable to its client, the owner, for the adequacy of the design services it rendered. So long as MCI is satisfied, the architecture firm will not have to defend itself.[3]

Licensed Designers

Licensed designers are responsible for regulatory compliance.

The "Prime" Designer (Who May Be an Architect or an Engineer)

Whoever is party to the design contract with an investment team is considered the prime designer, responsible for providing complete design services. When the project is primarily an occupiable building, the prime designer is generally an architect. When it is a road, a bridge, or primarily occupied by machinery or equipment, the prime designer is generally an engineer. State practice statutes regulate this issue, which is also influenced by local business custom. Evolving customs are manifest in the twentieth-century shift of primary design responsibility for bridges from architects to engineers.

Generally, in standard construction industry contracts, design teams agree to provide project planning (design) activities for all building components, including architectural, structural, mechanical, and electrical work. Prime designers may hire others to help them fulfill this obligation, but in the final analysis, they are the ones directly responsible to the investment team for such work.

Team Designers

While **architects** are legally allowed to provide many engineering and all interior design services, they are uniquely responsible for services related to the planning of weather protection—designing the roof and enclosure walls, with their complement of windows and doors, insulating materials and waterproofing membranes. These are areas that architects generally do not subcontract, although they may retain specialist consultants.

Civil engineers are uniquely responsible for services related to the planning of roads and exterior utilities. With respect to the latter, they design connections between buildings and available sewer, water, gas, power, telephone, and data lines.

Structural engineers are uniquely responsible for services related to the planning of loads imposed on the project by gravity, wind, and earthquake—designing the structure. Design includes establishing quality levels for materials, determining how much of each material to use for each structural member ("sizing" the members), and determining how to connect them all.

Mechanical and plumbing engineers are uniquely responsible for services related to planning the control of the interior environment—designing the ventilating, heating, and cooling systems. In addition, they plan water systems related to sanitation, drinking, and storm runoff. Their services pick up where the civil engineer's services leave off, a point accepted in the industry as five feet beyond a building's perimeter.

Fire protection engineers specialize in the design of suppression systems and the application of many of the concepts explored in this book, including compartmentation, fire-resistive assemblies and protectives, timed exiting, and egress in general. While some mechanical engineering firms have in-house fire protection engineers, many do not. Interested prime designers should ask.

Electrical engineers are uniquely responsible for services related to the planning of power, lighting, and communications. Like mechanical engineers, they also pick up at the five-foot mark.

Designers Who May Be Unlicensed

Each state decides whether to license the designers listed below. Where they choose not to, governments impose no or minimal educational or testing requirements. In these cases, such unlicensed designers are generally not responsible for regulatory compliance.

Interior designers are generally brought on to the project team to select finishes and furnishings, neither of which were traditionally critical to the issues addressed by rules except with respect to flammability of the materials and hygiene. Now accessibility has also become an issue.

Although many states are considering requests from the interior design profession to require licensing of their practitioners, most as yet do not regulate interior designers. Decisions will no doubt turn on whether regulatory requirements apply to the portions of a project that they design. Since many interior design decisions affect regulatory performance, such as those involving fire ratings of partitions and positioning of exit access corridors, the public has generally tended toward reserving such work for licensed professionals.

Landscape architects design spaces that do not involve weather protection, such as plazas, gardens, gazebos, and band shells. In addition, they may prepare plans for regrading of soils. When projects include structural elements or geotechnical work, an engineer's stamp is generally required. Where states regulate landscape architects, it is often in recognition of the conse-

CASE STUDY

Ambiguity in the command chain over this particular issue is what led to the collapse of a paired set of suspended walkways at the Kansas City Hyatt Regency Hotel and the deaths of 130 people.

According to the architect's design, the second and fourth floor walkways were to be hung from the roof structure above using a set of one-piece rods extending from the second floor to the roof. The fabricator's detailer thought it was too unwieldy, and changed the design to two separate lengths of rod that transferred load from one to another through the box beams at the fourth floor walkways. This revised design was documented in a shop drawing and sent to the architects and their engineers, none of whom questioned it. Likewise, the city's inspector didn't question it. The engineer who was eventually found liable, and subsequently lost his license, had never seen the fabricator's shop drawings.[4]

quences their work has for health, safety, and welfare issues such as:

- zoning issues, environmental impact, and accessibility;
- grading, including the risks posed to people and buildings by soils instability when graded too steeply;
- erosion concerns, not only of open land, but of soils adjacent to foundations.

Miscellaneous Design Consultants

Design consultants may also include advisers in many other areas. Three key ones are:

- regulatory consultants, who advise on design issues related to any and all of the areas discussed in this book,
- equipment consultants, who advise on design issues related to elevator, medical, audio, kitchen, and other special equipment needs,
- programming or "design criteria" consultants, who advise on design issues related to security, geriatric, and other special program needs, and help establish the design goals for a project.

Others

Ideally, **investors** rely on their legally sanctioned design consultants to develop the "hows" of the design. This is the clearest way to delegate design responsibility. This is not, however, the way it always happens. Often owners help generate design ideas. This can yield excellent results so long as the licensed design consultants thoroughly investigate the appropriateness and implications of the ideas. Either way, owners might always be said to ultimately design their projects, since they decide which design ideas will be built.

The role of **builders** as designers is a debate that is awaiting resolution. In any construction project there are design problems that are not addressed before construction starts, either because they aren't discovered or because a deliberate decision is made to postpone their resolution until after a contract for construction has been signed. In such cases, builders may make significant design decisions in the course of executing a set of contract documents (blueprints and specifications). This commonly happens when the contractor is required to make decisions when preparing shop drawings. Many manufacturers employ registered architects and engineers to make the detailed design decisions that architect and engineer consultants contractually delegate to them. The trend in construction is toward more extensive use of, and larger, prefabricated elements, making coordination and communication that much more critical.

Of key importance is the fact that contractors are not regulated as designers. As explored above, their licenses allow the public, via the government, to oversee their execution of design decisions made by consulting architects and engineers. Contractor licenses do not permit them to generate design decisions. Of course, when licensed design professionals work for construction companies, they have the general rights and responsibilities appropriate for any licensed design professional. In the case of design-build delivery, where a single company provides both design and construction services to an investment team, the design portion of the work must be provided by licensed design professionals within the design-build firm. Of course, this doesn't apply to very small projects or projects otherwise exempted from practice statutes.

THE CONSTRUCTION TEAM

The real task of this group is to execute the designers' documented interpretation of the investors' project goals. For the sake of clarity, the discussion below narrowly defines the construction team members in their role as builders, recognizing that members of this team frequently serve as designers and even investors, as noted above.

What is their agenda?

Contractors generally control construction sites. They handle the front and back ends of projects. At the front end, as a project begins, they control the delegation of work among on-site construction workers, manufacturers, and distributors. At the back end, as it nears completion, they ensure that components are placed in their proper positions and that work that can only be done after all components are assembled is, in fact, done. Important questions include:

- What are the skill levels and quantities of the workers?
- With what kinds of materials are they experienced?
- What kinds of equipment do they have?

Answers can be solicited during the project procurement (bidding) process.

One might think of **manufacturers** as contractors who just happen to do their work away from the construction site. Many of the issues that apply to contractors also apply to manufacturers, with a few additional ones thrown in.

- How will the factory ship the components to the site?
- What kinds of processes can their equipment handle?
- What kinds of conditions exist in their factories?

Manufacturers are often concerned primarily with product development and manufacturing, and market to a limited number of parties called distributors. This is more efficient than marketing to the thousands of individual builders or coordinating customer services such as storage, staging, and delivery. Because of this, some manufacturers prefer to delegate their contact with builders to distributors, including many who retain contact with designers. This contact ensures that the designers are properly informed of the availability, performance, and use of manufactured products.

Distributors are the middle men who help contractors work with manufacturers. They handle the transactions and help facilitate the flow of products from the manufacturers to the contractors. For interior furnishings, interior designers may serve in this capacity.

Others

Some **investors,** and occasionally some **users** (such as tanants), like to participate in building the project particularly when the project is small. Before they do so, some reflection may be in order. Do they have sufficient knowledge of the building industry? Will they know what their requirements are? Do they realize that their involvement could increase costs and reduce protections?

Photo by the author.

The Users Group

Finally, there are the users, acting in their role as the project's occupants. This is the group for whom construction industry rules are primarily written. While not coordinated as a team, and perhaps lacking a common goal, this group is critical in its role as the project's market. Without it, there would be no reason to do the project. It is also the group that will be living with the building long after design and construction teams have left.

The interests of these often silent users are championed on the project team by representative agents whose job is to ensure a minimum level of project quality on the users' behalf. These agents do this by developing, and then enforcing, building industry rules.

This process and these rules are the focus of this book.

Who Are the Users?

Ultimately, to establish what the users want, one must establish who constitutes the users. One might define a user as anyone who occupies, or is affected in any way by the project. The wants and needs of many are considered, from the property insurer to the end user. Those who live and/or have a financial stake in an area get to influence what gets built in their area, by hiring, directly (electing) or indirectly (appointing), many people to represent their concerns. Users include property owners, renters, tenants, customers, visitors, and passersby. The only common link among them is that they spend time, even if only fleeting, in physical proximity to the project.

What Are Their Agendas?

Users expect, and are entitled to, a lot. They want property values to be retained, want safe and convenient designs to be ensured, and want accessibility to be a given. In some cases, they want the places they inhabit to evoke certain feelings. Some of these expectations might be met as a function of market forces. Others must be mandated by rules.

Maintaining Property Values

Achieving value stability or growth involves many issues other than the business climate. Key among them is focusing and maintaining community identity. This issue is championed primarily by zoning boards, design review boards, historic preservation committees, and homeowners' associations.

Establishing a Mood

Not infrequently, users want their built environments to evoke certain feelings in them. They might want to feel nostalgic, or patriotic, or welcome. Although coming from a different perspective, this agenda is not substantively different from maintaining property values. Properties have the values they do in large part because of the way people react to them. In addressing property values, the civic organizations listed above also address mood.

Establishing Safety

Safety is basic to the idea of shelter. People want to know that their environments won't hurt them or make them sick. When doctors and the health care industry make mistakes, individuals who were already sick get sicker and some die. When design professionals and the construction industry make mistakes, groups of healthy people can get sick and hundreds can die.

Establishing Quality

Poor quality breeds inconvenience. A poor-quality roof may leak, and leaks interfere with the lives of users, who are inconvenienced by having to arrange and pay for the repair of failed materials and the collateral damage caused by the failure. As Richard Ross, a facilities manager at The Catholic University of America has said, "I may not remember the designer's name ten years later, but I'll remember the problems [arising from his design]."

Establishing Access

Society decided, as recently as 1990, that disabled people have the right to participate fully in public activities. The ADA (the Americans with Disabilities Act) has disabilities as its middle name, and its provisions protect those it defines as disabled. Designing for accessibility has limits. Not every facility can be designed for access by anyone at any time. The ADA defines disabled as those with real or perceived physical or mental impairments that limit major life activities [42 U.S.C. 12102(2)]. Clearly this does not help the agendas of those pushing strollers or coming home with arms full of groceries. (See pp 205–206.)

Head	Agency	Rules	Control	Jurisdiction	
				Issues	Location
Planning Commissioner	Dept. of Parks & Planning	Comprehensive plans	Public	Planning policies (land use, transportation)	All land
Commissioner of Public Works	Dept. of Public Works	Public facilities manuals	Public	Construction and maintenance of public property	Public ways (streets, parks)
Zoning Commissioner	Dept. of Zoning	Zoning ordinances and subdivision regulations	Public	Urban design (the design formed by groups of buildings)	Properties
Building Commissioner	Building Dept.	Building codes	Public	Health, safety, and welfare of occupants	Properties
Fire Marshal	Fire Dept. Office	Life safety codes	Public	Life safety and fire resistance	All land
Director	Historic Preserv. Office	Historic preserv. standards	Public	Historic Preservation	Properties
President	Developer	Development guidelines	Private	Design of new units	Properties
President	Homeowner Association	Covenants, conditions, and restrictions	Private	Maintenance, alterations, and activities	Properties

Some of the groups representing users.

Who Represents User Interests?

A diverse collection of public and private agencies and organizations develop and enforce rules on behalf of the general public.

Those Who Draft and Interpret Policy

Legislative Bodies and Agencies

These include the elected representatives who write laws and their staffs. They also include the employees of government agencies, who write regulations that define fulfillment of the legislators' laws. The intent of either group is to set and control minimum quality standards for the design of built projects. (See p. 25.)

Design Review Boards

Design Review Boards (DRBs), including such well known bodies as the National Capitol Planning Commission in Washing-ton, D.C., and the California Coastal Commission, tend to be commissioned for districts with unusual characteristics that community members feel are worth preserving. For example, planned communities usually have them, and they are quite popular with historic districts. They allow communities some degree of control over design.

Boards have their own standards, called guidelines. They serve to provide some control over more subjective matters than those handled by plans reviewers and inspectors. (See pp. 28–29). They are usually made up of design professionals, leaning toward architects, landscape architects, and urban designers, who are appointed by mayors, county councils, or homeowners' associations. Many of those serving are retired, since active practitioners too often find themselves facing conflicts of interest by virtue of having worked on a project or with one of the parties involved in it.

As a process, design review has its share of champions and detractors. Brenda C. Lightner, a professor of planning at the University of Cincinnati studying the issue, says, "Design review rewards ordinary performance and discourages extraordinary performance. Designers adhere to the range of acceptability held by particular reviewers, and therefore rarely waste their clients' time proposing something original or exceptional."[6]

Conservation Boards

These boards are empowered to review projects for compliance with historic and environmental rules. They seek to retain "historically contributory" properties through preservation (see pp. 59–60) and to conserve open space, sometimes through mitigation of wetlands, forests, and habitats. (See pp. 111–114.)

Boards of Appeal or Adjustment

No set of rules can ever anticipate every situation that might occur. At federal, state, and local levels, hearings are held to deal with special cases. Boards of appeal and boards of adjustment provide a mechanism for reasonable exceptions to be granted. Their members are generally appointed from members of the local professional community. (See also pp. 66–69 for more on the process.)

CASE STUDY

The following appeal was made by the state of Indiana to its local governments requesting input on a model floodplain ordinance it had prepared. It is a very direct demonstration of the work that governments do as servants of the users group, as well as the interdependence of federal, state, and local governments.[5]

"It's commonly known that every community participating in the National Flood Insurance Program [NFIP] has a floodplain ordinance that contains all current federal and state regulations. What may not be as well known is how these ordinances are developed.

"Pursuant to federal and state regulations, [the Indiana] Division of Water staff review [local] floodplain ordinances for compliance. For many years ordinances were received that were incomplete. Community officials told us it was difficult to write an ordinance that includes both federal and state regulations. As a result, our office developed the Indiana Model Ordinance for Flood Hazard Areas. Much effort was put into the development of this prototype to ensure that it contained all necessary language. Finally, after many months of reviews and revisions, the Federal Emergency Management Agency [FEMA] gave the model its blessing.

"At last! Something we could provide the communities to relieve them of the burden of developing an ordinance on their own. Many communities were very receptive to this idea. It took all the guesswork out of 'building the perfect floodplain ordinance.'

"Unfortunately, others seem to be very displeased with our endeavor to make life easier for NFIP participating communities and are quick to scrutinize the document without actually offering any concrete suggestions to improve it. We would like to point out that the model ordinance for flood hazard areas is just that, a model by which a community can develop and adopt a compliant ordinance to remain eligible to participate in the National Flood Insurance Program. Many communities choose to adopt the model verbatim, simply inserting the appropriate name, community number and map dates where necessary.

"Those communities choosing not to utilize the model may develop their own format. It is not a mandatory requirement to use the model ordinance in order to have compliant floodplain regulations. It is, however, necessary that an NFIP participating community adopt a floodplain ordinance that complies with both federal and state regulations pertaining to floodplain management. In addition, we are open to any constructive suggestions which will improve the model ordinance and make it more user friendly."

Those Who Enforce Policy (Ministerial)

These people enforce rules on behalf of those who hired them, the public. Enforcers include plan reviewers and inspectors.

Among the members of organizations that write model rules are many reviewers and inspectors. Much of the drafting and revising of the model rules is based on comments made by these members.

PLAN REVIEWERS

Reviewers compare building plans with regulatory requirements before construction begins. If the design complies, a building permit is issued. They are often willing to review in-progress plans for the project team, giving preliminary assessments of compliance and helping to identify areas that need further consideration.

INSPECTORS

Inspectors compare construction in place with the plans approved by the reviewer. Their job is to confirm, as well as they can, that no rule-infringing modifications have occurred along the way. They also provide a second level of review, since the reviewers' approvals do not relieve projects of their obligation to fulfill rules. Violations found must be corrected, even if missed by the plan reviewer.

TRENDS IN CODE REVIEW

Recently the relationship between design teams and code officials representing the users group has been the subject of some experimentation. Alternatives are being explored for their ability to reduce administrative costs to taxpayers and reduce the liability exposure of code officials and the government.

A potentially significant experiment now being tried in several jurisdictions is design certification. With certification, statements made by design professionals confirming regulatory compliance of project designs are accepted in lieu of the traditional plans review process. This reduces the number of projects that governments need to review and approve prior to issuing permits. For qualifying projects, statements may be accepted from one of two sources:

- the designing architect or engineer
- deputized licensed design professionals who may or may not work in the office of the designing professional

While it doesn't really change the design professional's liability, it can potentially eliminate one opportunity to catch problems before they are built. This is particularly true considering the conflict of interest design professionals would have in certifying their own projects.

Design through compromise. © Roger K. Lewis.

2 RULES

An Overview

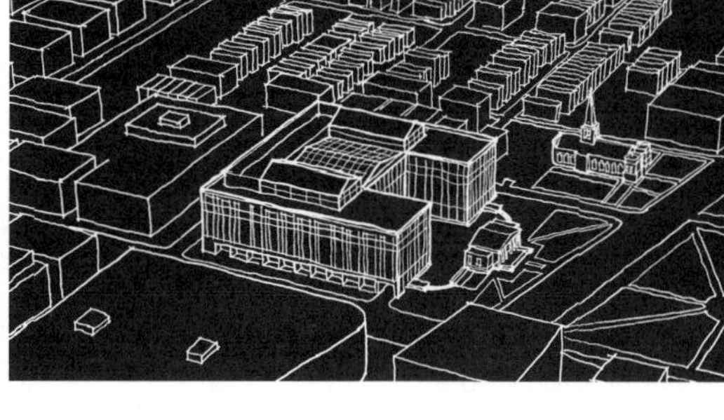

There is a distinction between nature's laws and our rules. We work by rules, but we employ nature's laws to make something. The rule is made to be changed, but nature cannot change its laws. If it did, there would be no Order whatsoever. . . . A rule is a conscious act needing circumstances to prove its validity or its need for change.

Any rule you have is really there on trial. The greatest moment of a rule is change: when that rule comes to a higher level of realization, that leads to a new rule. To discover a new rule is to discover a new avenue of expression.

LOUIS I. KAHN[7]

THE CONCEPT

Rules as a public endeavor.

JUSTIFICATIONS

What Are Rules?

Buildings must respond to the emotional/perceptual, social/behavioral, physical, functional, and economic goals they are intended to fulfill. The quality of their designs reflects the care with which their goals are established and then fulfilled.

Establishing and fulfilling design goals appropriately depends on how well preparatory research is handled, how far it is pushed, the direction in which it is taken, and how well it is addressed. The people developing, designing, and building projects must understand the applicable issues and respond appropriately.

Understanding and responding require knowledge. Knowledge requires access to information (a "knowledge base") combined with the judgment to apply it appropriately. Knowledge can be used to more easily generate appropriate responses when it is codified, meaning organized and distilled into concrete requirements. Packages of codified knowledge enforced by government are called by such names as statutes, regulations, ordinances, codes, guidelines, and standards. This book refers to them collectively as rules.

If rules are followed, the task of building planning and design may be greatly simplified. But is simplicity more important than thoroughness or freedom of choice? Although this question may frequently be asked, it isn't generally germane. Regarding thoroughness, rules do not prohibit project teams from conducting their own research. Regarding freedom of choice, rules are not responses, but simply tools to be used in generating responses. For some issues they do establish minimum acceptable levels of response quality, but they never prohibit responses of a higher quality.

Who Needs Rules?

You decide. Rules are a form of societal control, with some qualifying as law. Ava Abramovitz, Esq., a lawyer specializing in professional liability, defines law as "a mechanism for increasing the predictability of life." We'd all like to predict that the buildings we occupy won't collapse on our heads, that we won't wind up with nasty respiratory diseases from microbes in our heating systems, that our commutes won't be made miserable by levels of development that exceed the capacities of our roads and mass transit systems, or that the beloved old train depot won't be all but destroyed by an insensitive renovation. In stark contrast, we also don't want anybody telling us what to do with buildings we happen to own.

What happens if we decide that we don't need rules? Buildings are less likely to be capable of responding to conditions of use and environment. In the case of building codes, an illustration is provided by the hazard insurance carriers in the 1990s. Several pulled out of that business after paying millions of dollars in claims for properties damaged or destroyed in West Coast earthquakes, Midwest floods, and East Coast hurricanes. These carriers decided that continuing to insure properties was not in their financial interest. The disasters motivating them were roughly predictable—weather related rather than acts of terrorism - and therefore could have been managed through planning. Although it is difficult to justify designing for acts of terrorism, buildings can reasonably be designed to withstand weather, even when extreme. Proper use of codes figures prominently among the following actions, any one of which might have changed the insurance situation in the 1990s:

- Property owners could have agreed to pay the higher taxes needed to support increased enforcement of existing building codes.
- Residents could have pressed their legislators to develop or adopt better rules to improve the odds that buildings would withstand the increased severity of recent weather trends. Of course, this plan would involve higher costs for new construction and for retrofitting existing properties.
- Those with insurance coverage could have agreed to pay higher premiums to their carriers to handle the increased risk.
- Insurance companies could have decided to reduce their profit margins, assuming there were any, with whatever implications that might have had.

In the end, some insurance carriers felt that the market would not support their risk. When GEICO turned its insureds over to Aetna in 1996, some of those who had carried GEICO homeowner's policies for thirty years were upset and confused. They did not understand the interrelationship between construction quality, design, rules, enforcement, market forces, and financial health. Let's look into this.

Rules (Perhaps) Help Establish Professional Liability

By nature of their intent, rules establish a minimum quality of design, and therefore a "standard of care" for architects and engineers. At the same time, by nature of their drafting processes and overlapping jurisdictions, they can be ambiguous and contradictory. When properly considered, they can make the lives of design professionals easier by taking the burden of meeting basic needs and desires from the design team's shoulders, allowing them to concentrate their efforts on goals that exceed this base level.

In favor of rules, look at what happened to doctors because they have no codes. Their standard of care is set only by the competence level of their peers. This would be fine except that it is hard to establish. Court cases tend to rely exclusively on expert witnesses. The same is true for professionals whose actions are regulated more directly, but the rules provide a more stable reference point for standard of care. The standard procedures doctors follow are not the medical equivalent of construction industry codes. This complicates efforts to defend against charges of negligence, since it doesn't let doctors claim adherence to procedures publicly recognized as prudent. They therefore require patients to undergo extensive tests that may be medically unnecessary but that establish a legally defensible position. It certainly causes expenses to rise.

On the other hand, codes can be ambiguous and can also get project teams in trouble. This is especially true for teams that mistakenly assume that a permit in hand is evidence of compliance. With accessibility issues in particular, local code officials may grant permits to noncompliant projects that they believe to be acceptable, without understanding the primacy of federal statutes and regulations. Approval by local code officials does not absolve project teams of responsibility for compliance.

In summary, although they can present problems, rules provide a great deal of value in establishing professional liability.

The Price of Ignorance

Governments are under no obligation to give people building permits. In fact, governments are empowered to take existing properties away from owners who don't fulfill regulatory requirements. Whether you agree with them or not, you ignore these guidelines at your own peril. (See p. 84 for further discussion of this issue.)

LEGAL STRUCTURE

The Behavioral Basis of Rules

Laws of nature, such as those of gravity and thermodynamics, have been extremely effective at getting buildings to obey. On the other hand, laws passed by a community or a society tend to be ignored by buildings. Therefore, for a community to get the kinds of buildings it desires, it must adopt and enforce laws that target those in the construction (including design) and real estate industries. When a law says that "windows shall resist a wind load of 25 psf," it isn't intending to intimidate a rectangular assembly of wood and glass into compliance. Rather, it is telling the owner of the building that failure to use a window of such quality would be considered criminal behavior. Conceptually, this is an important reinforcement of the idea that at the heart of all building projects are people.

Regulatory Hierarchy

The basic categories of public rule have a "pecking order."

1. This being a democracy, all public rules originate in the individual. Think about it. Individuals, ultimately, decide what is public and what is private. In the public sector, they retain the right to approve constitutional amendments, which thereafter control all governmental action. In the private sector, individuals retain the option of organizing themselves outside of government into such citizen groups as labor unions, religious organizations, and homeowners' and tenants' associations.
2. All public rules grow from these constitutions. State and federal constitutions set forth the guiding principles of society, against which new laws are measured. State constitutions derive their authority from the federal constitution.
3. Statutes (legislated laws) apply constitutions to daily life. They are strategic and establish intentions for responding to problems affecting society in general. Local governments derive their authority from state constitutions.

   On a track parallel with statutes, local legislatures develop ordinances. These are often strategic and tactical, but always locally focused. Like statutes, ordinances can incorporate standards and guidelines by adoption.
4. Regulations: Regulations are tactical and are drafted at the direction of statutes. They spell out specific behaviors that when controlled through incentives and penalties are expected to achieve the statutes' intentions.
5 Standards and guidelines: For specific issues, regulations may mandate supplemental codes, standards and guidelines. These may be developed within government agencies or outsourced to private organizations.
   - Standards are supplements that have been officially adopted following a public comment and review period in which public consensus is sought.
   - Guidelines are supplements that have not yet been subjected to public comment or review. They are developed by committees or private organizations and can be given regulatory power, despite their lack of public input, if legislative bodies decide to adopt them.

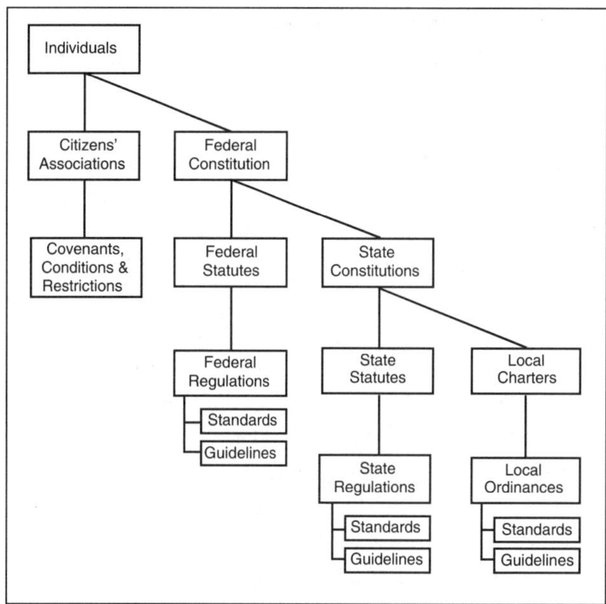

A chart of authorities for the American system of government.

(Codes are compilations of individual statutes or regulations, or standards, but are not of themselves a unique form of rule.)

Integral to this hierarchy is the general precedence of federal laws over state or local laws. Many issues are subject to laws at several levels of government. When there are federal, state, and local laws affecting a particular issue, local laws must comply with state laws, and state laws must comply with federal ones. In a limit to this overlap, the U.S. Constitution reserves some issues for the states, prohibiting the federal government from imposing its own laws.

Practically speaking, federal laws are often relatively global in nature, dealing with issues that pertain to groups of states. Local laws more commonly deal with specifics as a response to the unique characteristics that make one part of the country different from another. In many cases, a project team that follows local laws is inherently following state and local laws. Many is not all, however, and as they say, ignorance of the law is no excuse. It makes sense to check.

Statutes and Regulations

So what are these statutes and regulations?

	Statutes	Regulations
Branch	Legislative	Executive
Written by	Representative legislatures	Government agencies
Empowered by	Constitution	Statute
Establishes	Strategic intent	Tactics for fulfilling statutes
Where found (federal)	Public Law (Pub. L.) and United States Code (USC)	Code of Federal Regulations (CFR)

Summary of statute and regulation characteristics.

Statutes come from the legislative branch of government, regulations from the executive branch. A statute is a law drafted and adopted by a representative legislature. A regulation is an order with the power of law, drafted by a government agency (nonrepresentative) in response to a statute. Statutes establish strategic intent; regulations establish tactics for achieving statutory intentions. Violations of either statutes or regulations may be viewed as criminal or civil, so penalties may include imprisonment, censure, or fines.

To find any of these laws, one must understand their classification systems. The complexity of our government is reflected in the complexity of these systems. Statutes and regulations exist at federal, state, and local levels of government, and projects must be designed to comply with any that are applicable. At the local level, laws are generally classified as ordinances, as described in the next section. State classification systems vary, but the organizational structure of many of them parallel the federal systems explained below. The need to comply with all laws underscores the value of legal counsel familiar with the law of the place of jurisdiction for the project.

Since federal laws apply to all locations and have a structure similar to many of the state classification systems, understanding the system for federal laws is of broad value.

Classification of Federal Statutes

At the federal level, statutes are referred to by four numbering systems, two of which are useful to the construction industry. The first is a number assigned before new ideas become law. Bills pending before the Senate are given "S" numbers; bills pending before the House of Representatives are given "HR" numbers. When new construction-related legislation is being considered, concerned citizens can send their comments to senators and congressmen involved in the vote, using the S and HR numbers for reference.

After a bill is signed into law, it receives three more numbers, only one of which is particularly useful. Since Congress considers new laws whenever senators and congressmen feel they are needed, the chronology of passage has nothing to do with the subject matter of the proposed law. New laws first receive Public Law (Pub. L.) numbers indicating the order of their passage in Congress.

Since they are organized chronologically rather than by subject matter, statutes are not readily accessible until they receive United States Code (USC) numbers. The U.S. Code is a compilation of laws organized by topic. By assigning USC numbers to new laws, the government makes it easier for citizens to find new laws by looking in the same place as existing laws on the same subject.

Primary headings in the U.S. Code are called titles. There are at least two systems for subdividing titles, as indicated in the matrix below. Some titles use one system, some another, with no apparent pattern. Thankfully, USC designations list title and section numbers only, omitting intermediate numbers. Thus 42 USC 12181 would be found in Section 12181 of Title 42.

As for finding statutes and regulations, libraries and the Internet are likely candidates. Statutes are published in the *United States Statutes at Large* and the *United States Congressional and Administrative News,* one or both of which are available through many public library systems. The complete U.S. Code is published by the Government Printing Office as the *United States Code* (USC). It can be found in most public library systems and several places on the Internet, including www.law.cornell.edu/uscode/.

Classification of Federal Regulations

Federal regulations developed to supplement statutes receive Code of Federal Regulations (CFR) numbers referencing titles and parts, such as 28 CFR Part 36. Copies of individual titles can be bought through the Government Printing Office or are available for reference in many public and law school libraries and on the Internet at www.access.gpo.gov/nara/cfr/cfr-table-search.html. Any relationship between CFR numbers and USC numbers is strictly coincidental.

Regulations related to building design and construction are scattered throughout the CFR. The following sections give some sense of the scope and reach of these rules for project teams:

- Title 10: Energy (regulating mechanical systems, passive solar, sustainability)
- Title 14: Aeronautics (regulating structures built within a few miles of airports)
- Title 23: Highways (regulating parking)
- Title 24: Health, Education, and Welfare (regulating public housing)
- Title 36: Parks, Forests, and Public Property (historic preservation)
- Title 41: Public Contracts and Property Management, Subtitle C: Federal Property Management Regulations (regulating work done for the government as might be provided by design professionals and construction firms)
    - Part 105: GSA (General Services Administration, the group that manages federal construction projects)
    - Part 109: DOE (Department of Energy)
    - Part 114: DOI (Department of Interior)
    - Part 115: EPA (Environmental Protection Agency)
    - Part 128: DOJ (Department of Justice)
- Title 44: Emergency Management and Assistance (regulating reconstruction after disasters)
    - Subtitle I: FEMA (Federal Emergency Management Agency)
    - Subtitle IV: Departments of Commerce and Transportation
- Title 45: Public Welfare (regulating affordable housing programs)
- Title 47: Telecommunications (regulating antennas and other infrastructure)
- Title 49: Transportation (regulating airports and train stations)

	Pub. L.		USC, Alternate A		USC, Alternate B		CFR	
Layer	Heading	Format	Heading	Format	Heading	Format	Heading	Format
Broadest	**Session***	**1, 2...**	**Title***	**1, 2 ...**	**Title***	**1, 2 ...**	**Title***	**1, 2...**
	Law*	**1, 2...**			Part**	I, II...	Subtitle**	A, B ...
	Title	I, II...	Chapter	1, 2...	Chapter	1, 2...	Chapter	I, II...
	Subtitle**	A, B...	Subchapter**	I, II...			Subchapter**	A, B...
			Division**	A, B...				
	Part**	I, II...	Part**	I, II...			**Part***	**1, 2 ...**
			Subpart**	A, B...				
	Section	100, 200...	**Section***	**1000, 2000...**	**Section***	**1, 2...**	Section	1000, 2000...
Narrowest	Subsection	(a), (b)...	Subsection	(a), (b)...	Subsection	(a), (b)...	Subsection	(a), (b)...

***One of the paired headings used in shorthand designations, as in Pub. L. 101-336 or 42 USC 12101**
**Optional headings used only when needed due to complexity of the title's content

Organizational format of federal statutes and regulations.

Navigating Laws

Thankfully, there are several cross-reference lists in each of the laws.

- Both the USC and the CFR are indexed by subject matter. These indexes are very well done and easy to use, considering the complexity of the task. They are surprisingly useful, with references under "Individuals with Disabilities" that list the ADA, tax credits, and research programs, even though each is in a different part of the codes.
- The USC includes an index of "Acts Cited by Popular Name," where one can simply look up "Clean Air Act" and find out its Public Law and USC designations.
- Both the USC and CFR contain "Tables of Authority" where one can look up CFR parts by the USC sections that authorized their development.
- The USCS (United States Code Service) also includes a "reverse" Table of Authority, where one can look up the authorizing USC legislation if one knows the CFR designation.

An Example

Pub. L.: The Americans with Disabilities Act (ADA) of 1990 was passed by the United States Congress on July 26, 1990. The first citation, Pub. L. 101-336 refers to the entire ADA law as passed, complete with its numerous sections. Its Pub. L. designation indicates that it was the 336th law passed in 1990, the second year of Congress's 101st session.

U.S. Code: Since the law has multiple areas of concern, its various provisions are assigned to four different parts of the United States Code:

- Title 28 for provisions related to the attorney general's duty to prosecute its violations.
- Title 42 for provisions related to buildings.
- Title 47 for provisions related to telecommunications.
- Title 49 for provisions related to transportation.

CFR: The Pub. L. and USC statutes noted above, as with all statutes, do not define the actions that constitute compliance. For specific regulations, one can turn to the CFR, where the ADA Accessibility Guidelines (ADAAG), developed by the Department of Justice, are found in 28 CFR 36. Just to confuse things, ADAAG is also inserted at 36 CFR 1191 and 49 CFR 37, each time in fulfillment of a different statutory directive. That's right, the complete text of ADAAG appears in the CFR in three different places (at least). But to return to that first citation, 28 CFR 36 means that in Title 28 of the Code of Federal Regulations, specific provisions of the ADAAG are found in Part 36.

Local Laws: Ordinances

Ordinances are laws that are enacted by local governments. While their origins are within the societies to which they apply, their intent is, like statutes, to control activities and behaviors. Ordinances deal with issues as local and controversial as dog walking, hours of operation for businesses on Sunday, and sales of handguns. They also establish which building codes will be used and deal with the rights of property owners to build on the land they own.

Regulatory Supplements Developed by Public Process (Standards)

Standards include two kinds of rules:

- Those drafted by government agencies in response to statutes. This type of standard may start as a guideline (see below) and be adopted as a standard at some future time.
- Those drafted by experts in the standard's topic, who work for private, often nonprofit, organizations. Many of these establish industry-wide quality levels for materials and processes. This type is usually written by or at the behest of trade or professional associations, and often becomes accepted as a standard without thorough public comment and review periods when they have been broadly recognized as industry standards for many years.

In either case, standards may describe installation techniques, inspection procedures, or testing methods. Those who write codes can require minimum levels of quality by referencing standards.

Regulatory Supplements Lacking Public Process (Guidelines)

The type of public review required for adoption of standards can be a problem for subjective issues. They are often matters of individual taste and can stress the idea of public consensus to the limit. Guidelines can be very useful in dealing with relatively subjective issues. Guidelines are usually design advisories intended to help project teams fulfill the objectives of regulations and standards. They do this by offering suggested solutions for the mandated requirements, particularly when the mandated requirements are very brief or cryptic.

Guidelines attempt to clarify by example the criteria that design review boards use when they consider new projects for approval. They say, in effect, "the board is more likely to approve new proposals that are like this than like that." Guidelines are often, but not always, published along with the regulation or standard that they supplement. Project teams are well advised to ask if guidelines exist for the codes they are required to use.

As an example, the Secretary of the Interior's Standards for Rehabilitation of historic properties contain only ten statements and barely fill a single page. In 1977 a fifty-page set of guidelines was published "to help [the project team] apply the ... Standards ... during the project planning stage, by providing general design and technical recommendations."

Although mandated by law, guidelines are not enforced with the same clarity as other types of rules. Determining compliance and noncompliance is more difficult with guidelines because they deal with subjective issues. Findings of noncompliance are generally determined by review boards based on deliberation of the issues rather than by reviewers based on a comparison of a project with the requirements.

Sometimes these distinctions can shed some light on the status of a rule. To use the example listed earlier, the Americans with Disabilities Act (ADA), a statute with regulations, is extended by corresponding statutes and guidelines. These include the illustrated and quite lengthy ADA Accessibility Guidelines (ADAAG), which identify certain design solutions that a committee thought were acceptable responses to those statutes and regulations. Legislation was proposed in June 1994 to upgrade the guidelines to standards, revising the name from ADAAG to ADAAS. Since then the guidelines have been examined through extensive public hearings, and during several review and comment periods. Their designation as standards, if it happens, is not expected before the summer of 1998.

Rule Compilations: Codes

Codes of any kind are collections of statutes, regulations, and standards that a particular population finds neccessary and acceptable. In the public arena, just as the federal government has the USC as a repository for its statutes, states have state codes, and counties have county codes. In addition, each may adopt regulations, standards, or guidelines grouped to form codes. The CFR is a code of individual regulations, while building codes are compilations of individual standards.

Model Codes

Although some larger communities (mostly big cities) are empowered to develop their own building codes from scratch through enabling legislation passed at the state level, the codes enforced in most jurisdictions are based on "model" codes. Model codes are sets of suggested rules developed by panels of experts who spend time studying the activities and behaviors that produce acceptable buildings.

Adoption by Reference

Some rules enforced by model codes are not actually found within the model code, but are "adopted by reference." Through this process, model codes require compliance with other sets of rules by simply saying so. As an example, a model code might include such language as "All plumbing work shall comply with the provisions of the Acme Plumbing Code." Standards are commonly adopted by reference. In this way, the code serves as a menu, coordinating many standards.

Making Codes Local

Unfortunately, the experts who write model codes can't know much about each of the individual populations considering adopting them. Therefore, model codes don't become law until adopted by the legislative bodies representing particular populations.

Code adoption does not have to be an all-or-nothing proposition. Three mechanisms allow customization of the model codes.

- When model code publishers intend enforcement of certain provisions to be optional, they frequently put them in appendixes rather than in the main text of the model code. This makes it easier to separate them; adopting legislatures can adopt the basic provisions without adopting some or all of the appendixes. Topics that may be consigned to appendixes include regulation of development in floodplains, construction to withstand hurricanes, or provisions for gray water recycling systems. Where local geography, scale of development, or industry practice makes these provisions irrelevant, they need not be adopted.

- Adopting legislatures can attach other published sets of rules, rules that aren't found in the model codes, through the adoption-by-reference process noted above. This might be done to include still-relevent local laws that predate the model code.
- Adopting legislatures frequently write some rules from scratch. Such home-grown rules are published as supplements to the model codes, making them more responsive to local conditions and needs.

BASIC APPROACHES

Most rules establish minimum levels of quality by combining prescriptive language, which describes specific minimal solutions, with performance language, that describes minimal levels of achievement. Given this combination, it is easier to get approval for projects that strictly comply with prescriptive rules. However, project teams are most likely to get design approvals for innovative design concepts, whose compliance with prescriptive rules is sometimes indirect, when they effectively demonstrate fulfillment of performance rules.

Prescriptive Rules

Prescriptive rules describe the specific design solutions that code officials will accept. Also referred to as specification codes, they are comparatively easy to write and can be as brief or elaborate as their writers feel is appropriate. They are also easy to enforce, since it is easy to determine whether a particular design complies. However, they make the search for design solutions more difficult, since the possibilities are strictly limited. By limiting the number of design options to those listed in the rules, they limit the degree of optimization possible, and thus limit the efficiency of the designs. Considering the demands and slim margins typical of the contemporary real estate market, this is problematic. For this reason, the current trend is to phase out such rules in favor of performance rules.

Performance Rules

Performance rules list required outcomes, leaving design solutions to the project teams. This makes them particularly helpful for projects with accelerated design schedules, such as design-build, because performance rules can establish a degree of control over project quality before specific design solutions have been explored or finalized. They are more difficult to write because they are used to assess the degree to which particular designs fulfill given sets of design goals. They also tend to be much more lengthy than prescriptive rules. Further, they often incorporate other documents by reference, such as standards, to describe tests used to confirm levels of performance.

An Example

As an example of this distinction, BOCA96 Chapter 15, "Roofs and Roof Structures", contains separate sections for performance requirements (1505.0) and prescriptive requirements (1507.0), both of which must be satisfied. In the category of "performance" are such requirements as:

[1505.2: Wind resistance] All roofs and roof coverings shall be secured in place to the building or structure to withstand the wind loads of Section 1611.0. . . and shall be tested in accordance with FM 4450 . . .

Note that nothing in this requirement tells the project team how to achieve this degree of wind resistance. Section 1611.0 merely describes how to calculate load in pounds per square foot based on the expected wind speeds at the site. FM 4450, a standard published by Factory Mutual, simply describes a test procedure for determining if required performance has been achieved.

In the category of "prescriptive" are such requirements as:

[1507.2.3: Asphalt shingles . . . shall be secured to the roof with not less than 4 fasteners per strip shingle or not less than two fasteners per individual shingle. Shingle headlap shall not be less than 2 inches."

This language is very direct; a design either fulfills it or doesn't. The assumption that four fasteners per strip shingle will provide adequate wind resistance is based on previous use of that design solution. However, since both of the requirements quoted above must be met, each is, individually, nothing more than a minimum requirement that must be exceeded if the other requirement turns out to be more restrictive.

Future Directions

In the constant evolution of building codes, prescriptive rules are starting to yield to performance ones. The

IBC will also likely take this evolution toward a significant focus on performance. Additionally, as of this writing, the Design-Build Institute of America (DBIA) is working with both the Construction Specifications Institute (CSI) and the Army Corps of Engineers in separate efforts to write performance-based model specifications.

The Life Safety Code has already identified its intentions in this direction. "With each new edition, the Code continues its evolution from a specification code into what is intended to be a performance-oriented document." [1994 *Life Safety Code Handbook*, p. 9]

The flip side of growth. © Roger K. Lewis.

BASIC TYPES

Planning Rules

Planning rules are primarily concerned with the planning and design of communities and cities rather than individual buildings. Of course, they include requirements for individual buildings, but that is out of concern for the impact of individual buildings on the wider community, not out of concern for the individual buildings themselves.

Planning and Community

Planning rules provide a means by which communities can restrain property owners from using their properties in ways that cause other community members to incur:

- Financial loss through erosion of property values. A common example is erecting a factory in a residential area.
- Erosion of lifestyle through behaviors that extend beyond their properties. A common example is erecting a public building with insufficient parking, so traffic overwhelms available roads, and visitors appropriate use available off-site parking spaces.
- Loss of opportunity through loss of resources. Examples include erecting tall buildings that block views, and initiating logging operations or wetlands destruction benefiting the property owner while compromising a necessary part of a community's ecosystem.

Public interest can be hard to pin down, and figuring out who is on which side of an issue is often hard to track. Those complaining the loudest about the impact of new projects on their properties are sometimes proposing their own insensitive projects soon after. Planning rules are sometimes required to fulfill contradictory goals, such as:

- preserving the value of real estate investments for property owners when different sectors of the public have differing ideas as to what kinds of development strengthen and retard values;
- prohibiting future developments from adversely affecting existing properties (views, privacy) without reducing the future development rights of those existing owners;
- prohibiting projects from degrading the environment without burdening existing owners with environmental rules;
- holding development within levels that can be supported by available infrastructures without inconveniencing the public or inhibiting the economic growth fueled by new development.

The Power to Enforce Planning Rules

Ordinances have been attempting to address community aspirations since the 1890s, when Boston and San Francisco led the nation with the first laws regulating the placement of obtrusive facilities such as slaughterhouses and heavy industry. New York City took the next step, adopting a complete zoning ordinance in 1915.

Planning rules work because, as a society, we have vested our governments with police powers. The support for this has not always been strong. In fact, the legal precedent for this was established by a single vote margin when the Supreme Court heard *Village of Euclid (Illinois) v. Ambler Realty* in 1926. This police power resides primarily at the state level but is delegated, topic by topic, to individual communities through a state's constitution and enabling legislation. Topics are described with a high degree of specificity, such as the power to set building heights or the power to decide whether home offices will be allowed.

States delegate power in one of two very different ways: by listing those topics whose regulation they delegate, or by listing those topics whose regulation they withhold. In states that delegate rather than withhold, local governments that want to enact new laws for a listed topic must petition the state for new enabling legislation. In states that withhold, it is only necessary to make sure the subject of proposed local regulation is not on the list.

In exercising this power, a community must weigh the rights of private landowners against the needs of the community. Partly for this reason, although all cities have comprehensive plans, not all cities have zoning ordinances. Houston, Texas, is probably the best-known example. Its comprehensive plan addresses overall development, use, and density, but not on a site-by-site basis since there is no zoning ordinance.

Like politics (as the late House Speaker Tip O'Neill might have said), all planning rules are local. An attempt was made in the 1920s to create a national model zoning ordinance, but it failed to achieve sufficient support. In its stead, the American Planning Association has published guidelines for creating or revising a zoning ordinance[8] that focus only on format and process, and contain very little in the way of suggested provisions.

Keeping Current

Planning rules must balance long-range stability with agility, not only because they are inherently local, but because they must respond to the changing needs of a community and the effects of current and past rules as such effects become apparent. The impact of proposed revisions on existing development must be considered. As definitions of community have broadened, along with the impact zone of large developments, planning efforts have attempted to stay apace.

Types of Rules

Rules for planning fall into three categories:

- Governmental rules for "land taming": Land does not, by nature, lend itself to private ownership. "Raw" land must be "domesticated" before it is sold or buildings are added. This is the work of developers. Using legal tools rather than physical ones, their surveyors carve the land into discrete parcels. Each is tagged with a legal description and registered in the government records, and then, ususally, road crews bring public roads and utilities to each parcel. Only then are they ready to be sold to retail purchasers. Such rules include subdivision regulations and public facilities manuals.
- Governmental rules regulating community growth by establishing the kinds of projects built, the services they provide, and their aesthetic and functional relationships to other buildings. Such rules include comprehensive plans and zoning ordinances.
- Nongovernmental rules regulating growth and activity by establishing the kinds of projects built, the design of new projects, the way existing buildings are used by their occupants, all of the above, or just one or two of the above. These include development guidelines imposed on builders working in the area, and covenants, conditions, and deed restrictions imposed on individual property owners. They are generally imposed by developers, homeowner associations, condominium committees, and the like.

Governmental Regulation of "Land Taming"

Subdivision Regulations

Although there are no national model subdivision regulations, local versions have similar concerns even if they deal with them in different ways. Subdivision regulations:

- Establish procedures for defining land (measuring it, recording it) so that it may be bought and sold.
- Establish general design guidelines for streets. Such guidelines include street widths, relationships of streets meeting at intersections, lengths of blocks, minimum widths of individual lots, and even naming of streets. It is left to public facilities manuals, as noted below, to establish detail design guidelines.
- Establish procedures for setting up easements for access to properties by fire department and other emergency vehicles and for access to utilities.
- Set aside unique natural, scenic, and cultural features for the enjoyment of all.

Subdivision regulations don't necessarily foster good community design. ©Photo by K. Wyllie.

Exceptions to subdivision regulations, like most other rules, are permitted for hardship cases. Hardship can be argued in many ways and for many reasons, including:

- Physical characteristics of the property make a demand of strict compliance unreasonable. Perhaps the property is only six inches shy of the width required to allow construction of a garage. Perhaps an irregularly shaped "leftover" property makes a literal application of setbacks unfair.
- Naiveté. When purchasers of properties can argue that they were misled as to the properties' development potential, it's possible that reasonable adjustments would be made.

In any case, the requested adjustment must be seen as a reasonable concession.

Public Facilities Manuals

Public facilities manuals are standard construction documents for site work and civil engineering. They include detail drawings and specifications, which are used, verbatim, for any roadwork done within the jurisdiction. They are generally developed by a jurisdiction's Department of Public Works and ensure that the detail design of all local roads will coordinate.

There are facilities manuals and voluntary guidelines that pertain to high-speed roads and federal highway projects, but as they have virtually no impact on the planning and design of buildings, they are beyond the scope of this book. Such documents are published by the American Association of State Highway and Transportation Officials (AASHTO) and the U.S. Department of Transportation.

Public facilities manuals establish:

- Storm management: requirements for handling surface runoff and retention, sizes of storm conduits, design of culverts and catch basins
- Roads: allowable grades, curb radii at intersections, minimum distances from intersections to curb cuts, and the design of curbs and roadbeds
- Accessories: standards for street lighting, retaining walls, guardrails, signage supports, planters, etc.
- Fire-fighting infrastructure: locations of hydrants, requirements for fire department access

Copies can generally be purchased at city halls and county offices. They may also be available for temporary use at public libraries.

Governmental Regulation of Growth

Comprehensive Plans

Planning decisions are far-reaching and can have important implications for property value and the health of a community. Fortunes can be made and lost on such planning decisions as the path of a new highway or commuter rail line or the designation of a certain area for business district expansion. Decisions must appropriately distribute wealth and set the right direction for future growth. Comprehensive plans are intended to maximize economic fairness and chances for long-term community quality.

Comprehensive plans are mandated by all states, so all jurisdictions have and periodically update them. They may be called by different generic names, such as GLUP (Generalized Land Use Plan), or names referencing the jurisdiction, such as "A Plan for the Future of Evansville County." One aspect of the comprehensive plan, the master transportation plan, may be published independently. Comprehensive plans are both written and graphic, containing maps of existing and proposed land use. Above all else, they set regional, long-term visions for decision making. By following the guidelines they establish, even when seeking modifications, project teams avoid claims of "arbitrary and capricious" or "spot zoning." Such claims are the zoning equivalent of insider trading and nepotism.

Zoning Ordinances

The intent of comprehensive plans is made concrete through the provisions of zoning ordinances. Zoning ordinances set parameters for uses and exterior design of individual buildings within the comprehensive plan. When requesting exceptions to a zoning ordinance, explanations must be couched in terms of their effect on the comprehensive plan.

Although there are no national model zoning rules, local versions have similar concerns even if they deal with them in different ways. These similar concerns can be categorized as:

- base districts,
- overlay districts,
- rules common to all districts, and
- administration.

(See pp. 94–96 for more on districts.)

Ordinances set the relationship between a particular property and the larger community, and the disposition of construction within a specific property. The rules they contain establish the following:

- use requirements
- bulk: density, height, and setback requirements
- parking requirements
- loading requirements

As with other types of rules, chapters on administration deal with such issues as the permit, appeals, and certification processes; enforcement and penalties; and procedures for public hearings.

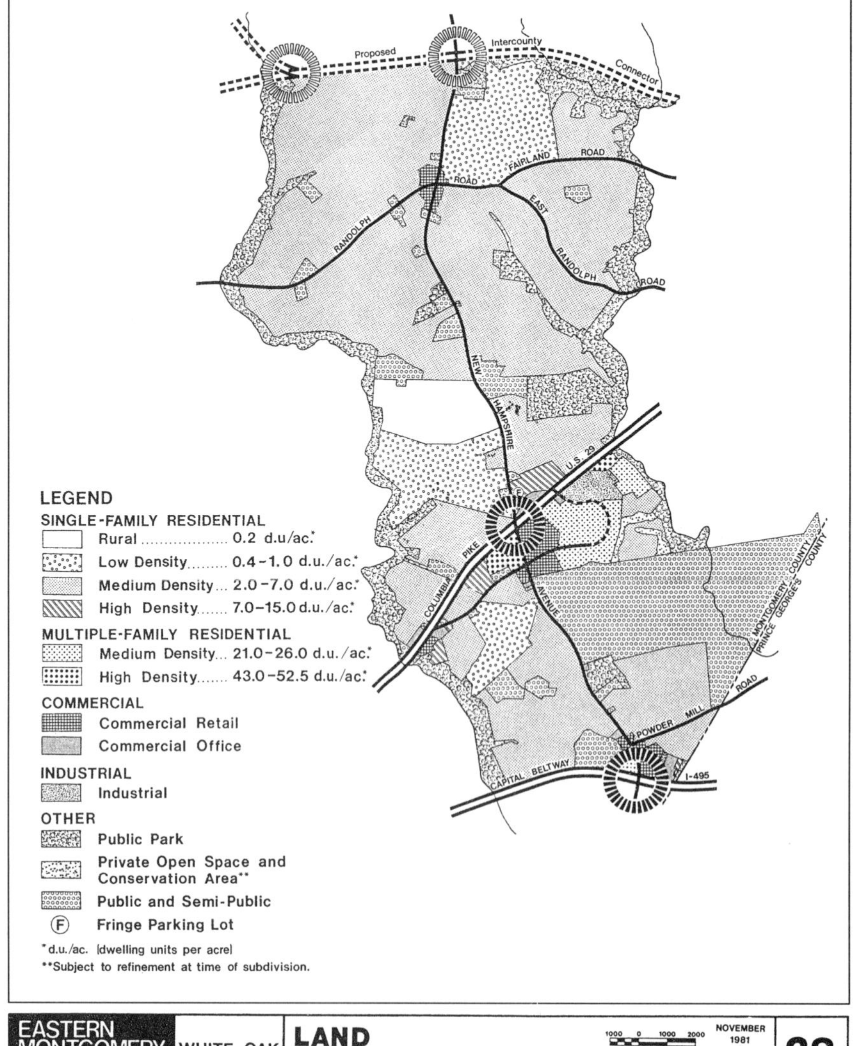

Comprehensive plan map of eastern Montgomery County, Maryland.

Nongovernmental Regulation of Growth and Activity

Development Guidelines

In residential communities, developers and builders play very different roles. Developers acquire properties, subdivide them, do rough grading (sculpting the ground as needed for roads and individual properties), and install such "site improvements" as streets, curbs, sewers and utilities, and street lighting. Builders are firms that then purchase these improved lots, put houses on them, and sell them when completed.

Developer guidelines are the developer's way of exercising some control over the work of the builders to whom it sells land. They establish rules that describe a "look" for a community and apply only to new buildings constructed on vacant sites. One well-known example is the guideline written for Seaside, Florida, by Duany Plater-Zyberk, Architects.

CC&Rs: Covenants, Conditions, & Restrictions

Once a community has been built and its original developer and builders have gone, developer guidelines are of no use. A different set of rules, commonly referred to as "covenants, conditions, and restrictions," are developed to govern alterations and additions to buildings that were originally constructed under the developer guidelines. The rules are empowered not through traditional governments but through the consent of property owners within the community. CC&Rs are usually first established by the developers while there are very few owners. The developers establish themselves as the board of directors of the "homeowners association." As time goes by, positions on the board of directors rotate from the original developers to property owners. Agreement to the rules is negotiated with those early owners. Subsequent purchasers must also agree, or find property elsewhere.

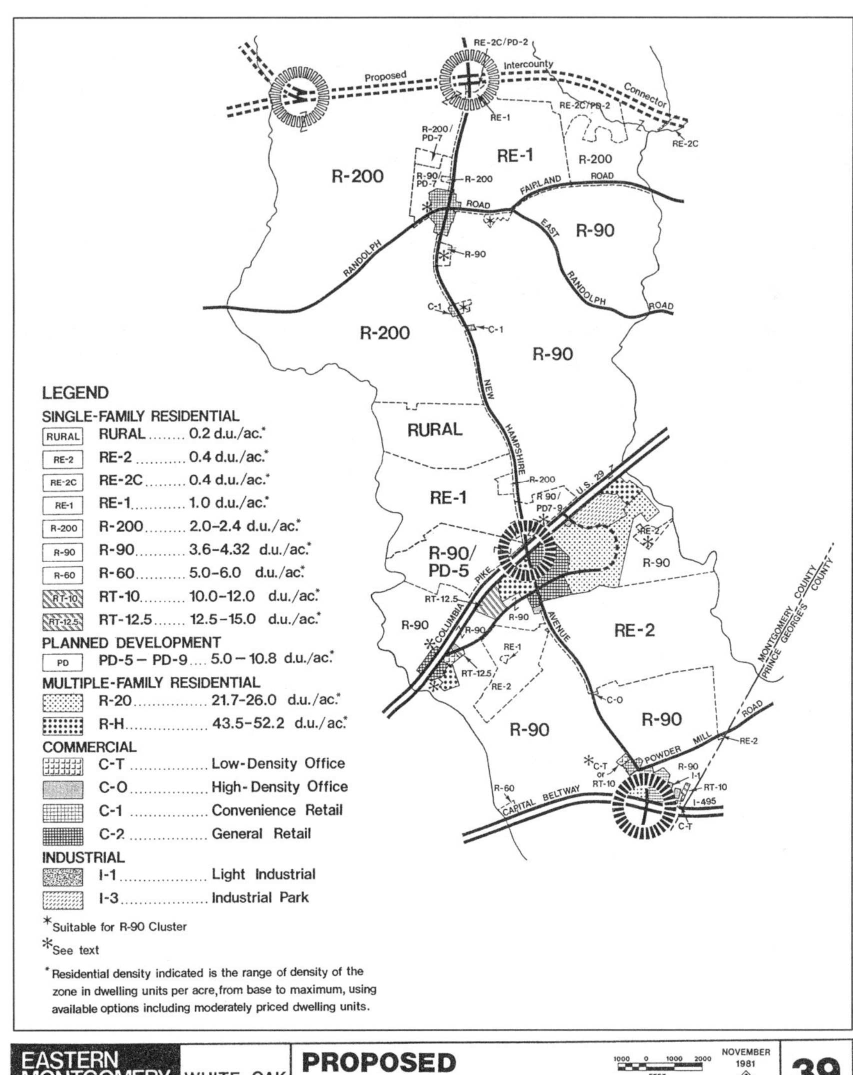

Proposed zoning map of eastern Montgomery County, Maryland.

CASE STUDY

Reston, Virginia, was developed as a new town starting in 1963, with its first "cluster" homeowner associations forming in 1966.[9]

It has about sixty thousand residents in about twenty-two thousand dwelling units. Roughly 55 percent are in condominiums and clusters, with the remainder split equally among single-family residences and rental apartments. Clusters contain anywhere from 13 to 230 units, averaging 60 to 100. Owners of single-family residences have special representation status. Representing the others are 130 homeowner associations at the cluster level and 30 condominium associations.

It has covenants and restrictions at many levels. The town is, of course, bound by the governmental rules of Fairfax County, the Commonwealth of Virginia, and federal statutes and regulations. To those, it adds not only a set of Reston development guidelines and covenants, but also sets of covenants specific to each cluster within Reston.

The rules quoted below clearly establish an attitude about site and building design. Some address general concerns while others are quite specific; some are open to many design approaches, while others would accept a very limited number of solutions.

From the Reston Development Guidelines

- The site plan ... should consider views and unique site features.
- Site plans and house designs should minimize grading, so as to preserve the natural vegetation wherever possible.
- Rear property lines of attached residential units should be staggered to prevent unbroken lines of fencing.
- All sides of a building should be of like or compatible materials and consistently detailed.
- Large expanses of blank walls should be provided with architectural elements to add visual interest.
- Chimneys must be constructed to grade and not appear to be suspended in mid-air.
- Brick facades should be made to "wrap" the corners of the building; brick or stucco quoins must "wrap" the width of the quoin detail.

The focus on exterior design common to development guidelines and CC&Rs is reflected in the following quote: Floor plans will not be reviewed by [the] DRB but may be helpful in considering the exterior design of the building.

From the Reston Covenants

One of the advantages to owning a Reston home is the protective covenants included in the Reston Deed of Dedication. When you purchased your property, you agreed to comply with the property covenants and so help to maintain the design standard that was established for Reston properties.

What are the general design principles?

- Harmony with overall community design
- Effect on neighboring properties
- Workmanship and materials
- Timing of completion

Fairfax County, through its building code, administers safety standards [not] Reston Associates.

The following individual design guidelines are available for single-family detached properties [Author's note: This is just a sampling of the 54 listings]:

- Antennas
- Awnings
- Art Works
- Basketball Backboards
- Decks (Ground Level, Elevated)
- Exterior Colors (Paints/Stains)
- Landscaping
- Lights (Decorative, Post Lamps, Security)

Such rules may restrict exterior materials, colors, and even allowable landscaping or permitted behavior (such as the length of time residents may leave their trash at the curb on collection day).

BUILDING RULES

Context

The second half of the nineteenth century saw a massive migration from rural areas to cities, as a mainly agrarian society gave way to one that was largely industrial. In larger cities, high-density development was an invitation to the spread of fire and disease. In addition, the rising dominance of business, and the financial sector that supported it, encouraged a shift of liability from property owners to their lenders and insurers. Thus there was rising demand for rules that would reduce the potential for loss of life and damage to property. Although minimum standards for design couldn't directly prevent loss, they could mandate the development of buidings that were less susceptible to catastrophe. Building codes were an inevitability.

The catalyst came with the fires that destroyed significant portions of major cities in the United States and Europe. Chicago showed the way when it took matters in hand and wrote the first American building code in 1875, in the aftermath of its devastating 1871 fire. This code remained in place for over a century and was soon joined by New York's code and those of many other cities.

Still, the strength of Chicago's code was not typical. Efforts by other cities demonstrated the problems inherent in trying to write building codes locally, especially for cities with fewer resources to apply to the task. Most municipal governments did not have the resources for developing the specialized knowledge required to write effective codes. Establishing minimum quality standards for design and construction required research, including extended observation of building failures and catastophes. After Baltimore's code failed to prevent that city's destruction by fire in 1904, several private and public organizations, led by BOCA (see below), arose to provide such expertise in the form of "model" codes. For most of the twentieth century, local jurisdictions have adopted and then modified these model codes to meet their needs.

Full-Scope Codes

Full-scope model codes are not published as single volumes but rather are issued as a series of trade-related editions. This gives regulators the opportunity to adopt only the codes that they feel apply to their jurisdictions. The term *building code* properly refers to only one of the codes in these series.

For most of the last hundred years, three publishers of model American codes have worked in parallel. Although they are now in the process of unifying into a single code, their parent organizations have left a lasting imprint. The twentieth-century American and Canadian organizations and their codes have been:

- **BOCA—National Building Code,** published by BOCA, the Building Officials & Code Administrators International, Inc., founded in 1915 and headquartered in Country Club Hills, Illinois. For some reason, and unlike the other codes, BOCA's flagship code is generally referred to conversationally as "the BOCA code" rather than as "the NBC."

- **UBC—Uniform Building Code,** published by ICBO, the International Conference of Building Officials, founded in 1922 and headquartered in Whittier, California.
- **SBC—Standard Building Code** (formerly Southern Standard Building Code), published by SBCCI, the Southern Building Code Congress International, founded in 1940 and headquartered in Birmingham, Alabama.
- **NBCC—National Building Code of Canada,** published by NRCC, the National Research Council of Canada, headquartered in Ottowa, Ontario.

This idea of a single American code is a response to the growing internationalism reflected in contemporary business. Such landmark legislation as GATT (the General Agreement on Tariffs and Trade) and NAFTA (the North American Free Trade Agreement), along with the explosive growth in telecommunications and the increasing ease of global travel, has sounded the death knell for regional model codes as well as local amendments. These trends have resulted in the formation of two new American organizations, each of which is a joint effort of BOCA, ICBO, and SBCCI. Their two unified model codes are:

- **CABO—One and Two Family Dwelling Code,** published by CABO, the Council of American Building Officials, founded in 1972.

  This is an abbreviated code written specifically for one- and two-family residences. As such, it focuses on materials and quality of workmanship, says almost nothing about exiting, and completely ignores occupancy and construction classification.
- **IBC—International Building Code,** an upcoming product of ICC, the International Code Council, founded in 1996.

  This will be the new unabridged code for commercial, institutional, and industrial projects.

Among the disciplines for which codes are written are the following:

Building Codes

These address architectural and structural work, and are generally thought of as basic codes, to which other codes serve as companions. Key issues addressed include:

- Use group (categorizing the project's program)
- Total height and area of a building
- Crisis management, through passive fire-resistance measures, active fire-fighting measures (usually referred to as "fire protection"), and provision of exits for occupants to use during a fire
- Interior conditions regarding size and provision of light, air, and heat
- Standards for acceptable materials (such as masonry and steel) and equipment (such as elevators)
- Applicability to and standards for existing buildings
- Site work, demolition, and work extending beyond property lines
- Structural design, including foundations and superstructures, appropriate to the forces generated by normal use and by wind, snow, and earthquakes
- Other issues, including design of infrastructure. These issues are addressed by companion codes, so there are incorporated in basic codes through reference. For example, IBC Chapter 27 on Electrical Systems refers to the companion IBC Electrical Code.
- Code administration: Permit and appeals processes, submittal requirements, fees, etc.

Mechanical Codes

Mechanical codes regulate the design of heating, ventilating (including exhaust), and air-conditioning systems. Key issues addressed include (underlined issues have significant impact on a project's schematic design):

- Systems for distributing air, including such components as ducts, controls, and fire dampers.
- Installation of mechanical systems, equipment, boilers and water heaters, water and gas piping, mechanical refrigeration
- Combustion air quantity and quality, which must be brought from the exterior enclosure to the mechanical equipment
- Ventilation air quantity and quality
- Chimneys and vents, incinerators
- Kitchen exhausts, fireplaces
- Energy conservation, solar heating and cooling systems

Plumbing Codes

Plumbing codes regulate supply, waste, and storm piping, and related fixtures. Key issues addressed include (underlined issues have significant impact on a project's schematic design):

- Quantities of sinks, toilets, fountains, and other fixtures
- Connecting to public utilities and sewers
- Materials, joints and connections, pipe hangers and supporters
- Waste and sanitary drainage piping. Cleaning and controlling flow with traps, cleanouts, interceptors, separators, and valves.
- Storm drainage
- Water supply and distribution
- Venting and stacks
- Health care plumbing, where applicable

Fire Prevention Codes

Fire prevention codes regulate post-occupancy issues related to fire safety. As such, they are primarily concerned with ongoing building management rather than initial building design. When they do address design, they are generally addressing retrofitting of existing buildings. The codes published by the model code publishers are in addition to fire prevention codes contained in state law. Key issues addressed include (underlined issues have significant impact on a project's schematic design):

- Number and spacing of sprinkler heads
- Fire resistance of materials, compartmentation of fire
- Fire-suppression systems
- Exiting, including signage and lighting
- Emergency planning for special uses: incapacitated or restrained occupants, projection rooms
- Maintaining special facilities: lumber yards, tent structures, bowling alleys, airports, etc.
- Classification and handling of hazardous materials, flammable and explosive materials, corrosives, irritants, etc.

Energy Codes

These codes address energy conservation and thermal efficiency. Key issues addressed include (underlined issues have significant impact on a project's schematic design):

- Building envelope requirements
- HVAC systems
- Plumbing systems
- Electrical systems
- Thermal properties of typical building materials
- Solar heating and cooling systems

Property Maintenance Codes

These codes regulate maintenance and operations for existing properties. They recognize that safety is not assured simply because a building is initially designed and built properly. While other codes are addressed primarily to design and construction teams, these companion codes are addressed primarily to investment teams through their facilities managers.

Key issues addressed include (underlined issues have significant impact on a project's schematic design):

- Light and ventilation, plumbing facilities and fixtures
- Occupancy limitations
- Mechanical and electrical systems, including fire control of mechanical systems
- Elevators and escalators
- Fire safety issues including exiting, fire resistance, fire protection, and elevator use

Even More Specialized Building Codes

SBC publishes (as of 1994) as separate codes an "Amusement Device Code," a "Swimming Pool Code," an "Unsafe Building Abatement Code," a "Gas Code," and a "Housing Code." Within SBC, there are appendixes for a "Hurricane Code" and a "Fallout Shelter Code."

BOCA publishes (as of 1993) a "Private Sewage Disposal Code" and a "Rehabilitation Code."

Limited-Topic Codes

Organizations other than publishers of full-scope building code produce codes for limited-scope topics. Chief among them is NFPA—The National Fire Protection Association, founded in 1896 and headquartered in Quincy, Massachusetts.

NFPA publishes a wide-ranging variety of specialized codes and standards, not all obviously related to fire protection. These include:

The National Electrical Code: NFPA 70

NFPA 70 is a full-scope electrical code for commercial work. In addition, NFPA publishes Standard 70A, a variant for one- and two-family dwellings that corresponds with the CABO code. Key issues addressed in these codes include (underlined issues have significant impact on a project's schematic design):

- Wiring strategies such as circuiting and electrical service, and protection such as grounding and surge suppression
- Wiring methods and techniques
- Wiring materials. including cable and conduit types
- General-use equipment such as lighting, appliances, generators, motors, and space heaters
- Special occupancy requirements such as for health care, theaters, manufactured buildings, and marinas
- Special equipment such as elevators and escalators, photovoltaic systems, computer equipment, and office furnishings

The Life Safety Code: NFPA 101

This code supplements and extends the egress issues regulated by basic codes. Their main interest is in making sure that injuries to people are minimized. Key issues addressed include (underlined issues have significant impact on a project's schematic design):

- Exit path design, including arrangement and size of exits
- Communication systems, including illumination of means of egress

The Life Safety Code states its intentions in negative language: to exclude any issues, such as those handled by building codes or fire prevention codes, not directly related to occupant safety during fires [LSC 1-3.6, 1-3.7]. This does little to clarify the unique role of the Life Safety Code, since building codes also address life safety. What does the Life Safety Code offer that isn't covered by the model codes? Its value is supported by the many jurisdictions that have adopted it in addition to one of the model codes.

The Life Safety Code is distinguished by its specificity—it goes into greater detail than building codes. Model building codes establish a general set of rules for safety, customized by exceptions and clarifications addressing the needs of individual occupancies. By contrast, the Life Safety Code includes an entire and distinct set of rules for each use group. Actually, it includes two sets of rules for each use group—one for existing properties and one for new projects. Finally, it includes provisions for some specialized life safety needs not found in building codes. Examples of these include egress from unusual structures such as observation towers, and exiting from boiler rooms.

Other Significant NFPA Codes

- The Fire Prevention Code: NFPA 1, with rules for the ongoing operations and maintenance of existing buildings
- Sprinkler Systems: NFPA 13
- National Fire Alarm Code: NFPA 72

Building Codes Published by Governments

Sanitation Codes

These rules are generally written on a statewide basis and regulate food preparation and medical projects. They are rarely listed with other codes pertaining to building, and if you don't specifically ask if there are any applicable health and hygiene rules, you probably won't find out until it's too late.

Key issues addressed include:

- Design of mechanical systems to inhibit microbial growth (e.g., prohibition of duct liners)
- Restrictions on materials (e.g., requiring Mylar facings on acoustic ceiling panels, prohibiting porous or absorbent flooring materials)

Codes for Government Projects

The federal government publishes public facilities manuals, statutes and regulations, but not its own building code. (See pp. 57–58.) State governments have a similar agenda, using statutes as a way to adopt model codes for use on projects within the state.

Common Names for Codes

Since code titles rarely roll off the tongue, common shorthand names are used conversationally. In some cases, common names refer to individual sections of larger codes. Examples include:

- The "high rise code" is actually a subset of the rules listed in the building code under "Special Use and Occupancy."

- The term "fire code" is commonly used in reference to the model codes' fire prevention editions, concerned as they are with the operations and maintenance of existing buildings.
- The "elevator code" refers to ASME A17.1: Safety Code for Elevators and Escalators (not to be confused with ANSI 117.1, the accessibility guidelines).

Progress Toward a Single Code

Over a fifteen-year period ending in 1993, the ICC developed a standardized format and chapter headings for the BOCA, UBC, and SBC basic codes. Editions of those codes published since 1993 incorporate this standard format.

With the single format a reality, the ICC is developing a single version of the language as well. Although not yet completed, it is scheduled to be published as the IBC (International Building Code) in April 2000. After that, the three existing model codes will no longer be updated. It may be many years, if ever, before all legislatures adopt the IBC, and it is expected that twentieth-century editions of the three model codes will continue to be published for jurisdictions that have not adopted the IBC.

A working draft of the IBC was issued in May 1997. After a public comment and revision period, a first draft was issued in October 1997. A second draft is scheduled for July 1998, coinciding with the publication of this book. Another period of comment and revision will precede the final publication.

Several IBC companion codes have been completed. These include:

- The International Mechanical Code (IMC), first released in 1996
- The International Plumbing Code (IPC), first released in 1995
- The International Private Sewage Disposal Code (IPSDC), first released in 1995

An International Fire Code, sponsored and administered jointly by the ICC and the NFPA, is expected in 2000.

The integration of the Life Safety Code into the ICC series of publications is not yet resolved. Over the years, the model codes have incorporated increasing amounts of the provisions that were once unique to NFPA 101. The NFPA's lead in state-of-the-art rules for prison design, atrium design, and other difficult life safety areas is no longer as substantial as in years past. The process of integration is complicated by the following factors[10]:

- NFPA 101 is organized by occupancy (residential, mercantile, etc.) rather than function (egress, fire resistance, etc.).
- In practice, NFPA 101 is not modified by local amendments as the model codes are.
- The Social Security Law of 1967 requires health care facilities to comply with NFPA 101 to be eligible for Medicare and Medicaid funding. This affects about 30,000 facilities nationwide.
- NFPA 101 is developed by broad construction-industry consensus. While the model codes solicit input from a broad constituency, code officials make the decisions.

The NFPA is leaving open the door to unification by extracting the means-of-egress provisions from each of its occupancy-oriented chapters. This will allow the model codes to more easily reference it should they decide to do so. NFPA 101 will most likely remain a force in building regulation for the foreseeable future.

Civil Rights Rules

Context

Some rules pertaining to civil rights affect the planning and construction of building projects. These rules prevent discrimination based on disabilities or on association with certain groups of people. Although supplemented by state and local laws, they are based in federal law, in contrast to the locally generated planning rules and nationally developed, locally adopted building rules. Many civil rights provisions have a strong impact on historic preservation.

Accessibility

Context

There are several overlapping laws protecting the rights of the disabled to access buildings. They include:

- The Americans with Disabilities Act of 1990 [42 USC 12101, et. seq. (meaning "and the sections that follow")], commonly called the ADA. This is not the only law, but it does affect the majority of situations and has garnered the most publicity.

- Title V of the Rehabilitation Act of 1973, a precursor to the ADA, which still controls accessibility rights for the federal government's executive branch and state and local governments.
- The Fair Housing Act of 1988, prohibiting disability-based discrimination in owned or rented housing, including hotels. See below.

The ADA is a civil rights statute, not a building code. Since it is federal legislation, it takes precedence over corresponding laws passed by individual states. It prohibits discrimination based on an individual's level of physical ability, with suits heard in federal court.

In a stirring example of the power of governmental advocacy, the preamble to the ADA states, in part [42 USC 12101]:

It is the purpose of this chapter

1. *to provide a clear and comprehensive mandate for the elimination of discrimination against individuals with disabilities;*
2. *to provide clear, strong, consistent, enforceable standards addressing discrimination against individuals with disabilities;*
3. *to ensure that the federal government plays a central role in enforcing the standards established in this chapter on behalf of individuals with disabilities; and*
4. *to invoke the sweep of congressional authority, including the power to enforce the fourteenth amendment and to regulate commerce, in order to address the major areas of discrimination faced day-to-day by people with disabilities.*

In Subchapter III, dealing with private facilities, the General Rule prohibiting discrimination says [42 USC 12182(a)]:

No individual shall be discriminated against on the basis of disability in the full and equal enjoyment of... any place of public accommodation by any person who owns... or operates [it]."

Disability is defined as [42 USC 12102(2)]:

A physical or mental impairment that substantially limits one or more of the major life activities of the individual... or being regarded as having such an impairment.

The law also clarifies those users it does not define as disabled. The statute specifically excludes compulsive gamblers, homosexuals, users of illegal drugs, and those diagnosed with certain psychological disorders [12210, 12211].

More specific rules define discrimination in building construction to include:

For new and modified construction [42 USC 12183(a)]:

a failure to design and construct facilities... that are readily accessible to and usable by individuals with disabilities, except where... structurally impracticable...

For existing construction [42 USC 12182(b)(1)(A)(iv)]:

a failure to remove architectural barriers... in existing facilities... where such removal is readily achievable... [Where not] achievable, [discrimination includes] a failure to make such... facilities... available through alternative methods if... readily achievable.

If there is reasonable suspicion of discrimination, an attorney general may file suit in federal district court. If the facilities are found to be discriminatory, the court may require that the facilities be made accessible, it may award monetary damages to the disabled person, and it may assess fines of up to $100,000 for repeat offenses [42 USC 12188(b)(2)].

Knowing all this about the ADA doesn't help much when it comes to making project decisions. The ADA doesn't elaborate on the kinds of designs a court would consider acceptable. That is the role of regulations and guidelines, as noted below.

Where to Find the Requirements

STATUTES

In the Public Law, the ADA has five parts. Each part was assigned a location within the U.S. Code and the Code of Federal Regulations. Further, a department of the federal government was designated to administer each. This information is summarized in the table below.

REGULATIONS AND GUIDELINES

Background

Design teams need help in figuring out how to fulfill the ADA. A set of ADA companion regulations would provide such guidance. One of the most significant problems in the building design professions today is the lack of such a set of regulations. A brief description may help bring the conflict to light.

Standards and guidelines for accessible design have been around longer than many realize. The initial research was conducted at the University of Illinois

Topic	Public Law	U.S. Code	Code of Fed. Regs.	Administered by
Employment	101-336 Title I	Title 42 subchapter I (starting at 42 USC 12116)	29 CFR 1630	EEOC
Public Services—State & Local Gov't, Public Transportation	101-336 Title II	Title 42 subchapter II (42 USC12131 and 12141)	28 CFR 35, 49 CFR 37 and 38	DOJ, DOT
Public Accommodations and Services by Private Entities	101-336 Title III	Title 42 subchapter III (starting at 42 USC 12186)	28 CFR 36	DOJ
Telecommunications	101-336 Title IV	47 USC 225 (it's brief)	47 CFR 64 subpart F	FCC
Miscellaneous Provisions [mostly applicability]	101-336 Title V	Title 42 subchapter IV (starting at 42 USC 12201)	28 CFR 36, 49 CFR 37 (ADAAG)	ATBCB (DOJ-DOT)

The ADA series of federal statutes and regulations.

under a grant from the Easter Seal Research Foundation and published by the American National Standards Institute in 1961 as ANSI A117.1. Later the President's Committee on Employment of the Handicapped and the U.S. Department of Housing and Urban Development got involved, expanding the research and developing additional standards.

As noted above, the right to access, as legislated through the ADA, is enforced by the federal justice system. Faced with a lawsuit, designers who thought they had designed for accessibility would most likely argue that they followed all the rules required by state and/or local laws. Many such rules exist, including accessibility provisions of the model building codes and the Life Safety Code, as noted below. Unfortunately, such laws are not recognized in federal courts, so a defense based on them would be inadmissible. The only state or local laws that are recognized are those that have been submitted to and certified by the federal Civil Rights Division of the Department of Justice [42 USC 12188(b)(1)(A)(ii)]. As of July 1996, the only state to obtain certification was Washington. It took it three years to do so.

This problem could be alleviated if the federal government provided regulations for accessibility. If followed, they could help defend against accessibility-based negligence claims. Of course, state and local governments could then repeal their own accessibility codes, unless they chose to challenge the constitutionality of the federal ones in court.

The process of developing a federally certified accessibility regulation has been under way for several years now. Four federal regulations are currently in place, two for private and nonfederal public projects and two for federal ones.

- **ANSI/CABO A117.1:** In some ways the granddaddy of accessibility codes. As noted above, this was the first widely recognized national standard. Much of its popularity was due to the extensive graphics used to show requirements rather than just describe them.
- **ADAAG:** The "ADA Accessibility Guidelines" respond to a directive in ADA Title V. They are developed by the U.S. Architectural and Transportation Barriers Compliance Board (ATBCB or simply "Access Board"). The ATBCB is a joint undertaking of the Departments of Justice and Transportation. Correspondingly, there are two parts to the ADAAG.

1. The ADA Accessibility Guidelines for Buildings and Facilities [inserted three times, at 28 CFR 36, 36 CFR 1191, and 49 CFR 37]: Their purpose is "to implement Title III... and require places of public accommodation and commercial facilities to be designed, constructed, and altered in compliance with the accessibility standards" they establish. They clarify and expand the protections afforded by Title V and Section 504 of the Rehabilitation Act of 1973 [29 USC 791 and 794].
2. The ADA Accessibility Guidelines for Transportation Vehicles [inserted twice, at 36 CFR 1192 and 49 CFR 38], regulating the design of buses, trains, etc. to accommodate the disabled.

- **UFAS:** The Uniform Federal Accessibility Standards, also published by the ATBCB [24 CFR 40]. Where ADAAG establishes guidelines for private work and the work of state and local governments, UFAS establishes guidelines for facilities that are "designed, constructed, altered, or leased by or on behalf of the federal government." It follows on, and expands, the protections afforded by the Architectural Barriers Act of 1968.

- **FHAG:** The Fair Housing Accessibility Guidelines, developed by the Department of Housing and Urban Development (HUD) for federally sponsored housing initiatives. HUD also references ANSI A117.1.

In addition, the following nonfederal codes are available to help define design features that may or may not fulfill the Department of Justice's idea of ADA compliance.

- **Accessibility provisions of the model building codes and the Life Safety Code:** All model codes and NFPA 101 include chapters on accessibility. In the IBC, accessibility is addressed in Chapter 11. By adopting model codes, states mandate enforcement of the codes' accessibility requirements. Unless the adoption language or local code amendments specifically exclude those requirements, project teams in such states will have to follow model code provisions in addition to ADAAG and other accessibility codes in effect.
- **State-sponsored accessibility codes:** Many states had written their own accessibility codes prior to the development of ADAAG, some prior to the accessibility provision of the model codes. Most of these will probably be abandoned with the greater use of model accessibility codes.

Progress Toward a Single Guideline

As with building codes, there is an effort under way to unite the state and private regulations with the federal ones. The ATBCB'S ADAAG Review Advisory Committee was formed in September 1994 to develop and submit a set of composite regulations. If adopted by the ATBCB, it has potential for solving the conflict between design and enforcement.

The IBC is attempting to eliminate the redundancy of federal and local rules by making its accessibility provisions sufficiently thorough to replace separate accessibility standards and regulations. In its IBC Working Draft, the ICC stated that "the primary objective of the proposed accessibility provisions in the IBC is to meet or exceed . . . ADAAG and FHAG documents."

In September 1996, the ADAAG Review Federal Advisory Committee published its recommendations. It is being reviewed by the ATBCB, and a final document is expected in early 1998. The unified accessibility regulation, when completed and executed, will provide a welcome relief to project teams. In the meantime, sorting through all the applicable accessibility rules to confirm compliance can be quite a task.

Getting Information from the Source

The ATBCB is available to answer questions. It has a toll-free hot line, (800) USA-ABLE, and can be reached through the Internet at www.info@access-board.gov. The Department of Justice also has an answer line, (800) 514-0301, and can be reached through the Internet at www.usdoj.gov/crt/ada/adahom1.htm.

Fair Housing

Context

Prohibiting discrimination in housing has complex implications for project teams. While the idea of fair housing seems to be a "no-brainer," the implications of that idea are not quite so obvious. The Fair Housing Act of 1968 [90 Pub. L. 284 Title VIII, 42 USC 3601 et seq.] prohibits landlords, real estate brokers, and mortgage lenders from disqualifying prospective residents based on their association with or classification into certain groups. At this point in the evolution of the American experiment in democracy, most people would agree with this idea. Some conflicts are suggested by a closer look at the associations or classifications that no longer justify denial of housing. These are:

- Race, color, and national origin. Where people, regardless of ethnicity, can afford the expenses asked of other residents, the act requires that they be allowed to purchase or rent. This suggests that a housing complex developed for immigrants from or descendants of a particular country would have to accept residents unaffiliated with or even expressing disdain for that country.
- Religion. Without exceptions, this might suggest that church groups building housing would be required to accept residents of any faith or no faith.
- Sex. Without exceptions, this might require that when housing is offered, it accept men and women residents equally. This might be hard on monasteries and convents, on college dormitories, and on housing complexes intended for single mothers or battered women.
- Handicap [42 USC 3602(h)]. Curiously, the ADA refers to the "disabled" while the Fair Housing Act

refers to the "handicapped." The difference appears to be semantic only. Regarding intent, the ADA ensures that disabled occupants be able to use built spaces, while the Fair Housing Act makes sure the handicapped are permitted to own or rent them. Handicapped is defined broadly, including those with physical or mental disabilities, those with chronic alcoholism or mental illness, and those with AIDS and AIDS-related complex.

- Family status [42 USC 3602(k)]. For many years, some properties saw children as noisy and disruptive, and single people as less stable than married couples. Under the Fair Housing Act, posting "No children allowed" is illegal, as is denying housing to pregnant women or unmarried people.

The Fair Housing Act includes exceptions for some of these conflicts. One exemption allows discrimination (perhaps "selective offering" is a more appropriate term, if less precise) for members-only housing built by private organizations, as noted below. This allows for monasteries and convents, enclaves for Irish-American residents, and the like.

Requirements

The Fair Housing Act of 1968 [Pub. L. 90-284 Title VIII, 42 USC 3601 et seq.] is a statute. Most of its provisions deal with the way properties are marketed, advertised, and transacted. Some deal with accommodation of the handicapped and are supported by the Fair Housing Accessibility Guidelines (FHAG) as noted above.

Requirements for accommodation of the handicapped apply, with exceptions, of course (see below), to all single-family residences and to ground-floor units of multifamily residences with four or more units and no elevator. For multifamily residences with elevators, each floor must have accessible units [42 USC 3604(7)].

Beyond accommodations required by FHAG, landlords are required to let handicapped tenants make additional modifications to their own dwelling units and to common areas to meet their needs when such tenants bear the expense [42 USC 3602(3)(A)]. When they move out, landlords may require tenants to restore properties to their premodified condition.

Optimization

WAIVING THE WHOLE ACT

Properties that qualify under the exemption for members-only housing can waive the entire law. This is true also for a few limited types of projects including owner-occupied housing containing fewer than four units, and single-family residences being sold without real estate brokers.

The following quote from the code establishes the rules for members-only housing:

[42 USC 3607(a)] Nothing in this subchapter shall prohibit a religious... organization... from limiting... dwell-ings which it owns or operates for other than a commercial purpose to persons of the same religion, or from giving preference to such persons, unless membership in such religion is restricted on account of race, color, or national origin. Nor shall anything in this subchapter prohibit a private club not in fact open to the public... from [similarly] limiting... occupancy... to its members or from giving [them] preference.

With respect to number or relationship of occupants,

[42 USC 3607(b)(1)] Nothing... limits... reasonable local, State, or Federal restrictions regarding the maximum number of occupants permitted to occupy a dwelling. Nor does any provision... regarding familial status apply with respect to housing for older persons.

PROHIBITING SPECIFIC INDIVIDUALS

The act allows discrimination against individuals who are a direct threat to the health or safety of others [42 USC 3604(9)], or currently using illegal drugs [42 USC 3602(h)].

ADVOCACY RULES

Some rules might be thought of simply as safeguarding broadly supported agendas. These pertain to historic preservation, environmental quality, occupational safety, and consumer protection. These agendas are currently addressed by rules.

Historic Preservation

Context

There are several codes, standards, and guidelines applicable to architectural projects that are in the purview of local historic preservation offices, the National Park Service of the Department of the Interior, the Internal Revenue Service, and individual state and local governments. All of them attempt to

CASE STUDY

This case is a great illustration of the complexities that can be involved with regulation[11]. A property ownership group (Deer Hill Arms apartments) wanted its property to comply with the Fair Housing Act. The local government (Danbury, Connecticut) sued to force the property to be exempt, and won.

Here's what happened: Deer Hill Arms had facilities specifically designed for the elderly. Over 80 percent of the units had at least one resident over fifty-five years old, and the complex's in-place policies demonstrated an intent to provide housing for the elderly. Despite this, the investors claimed that the property should not be categorized as elderly housing and hoped to be able to rent to younger tenants. They said that there were no "significant facilities which would qualify this [property] for the exemption." Evidently, the judge disagreed.

Among the relevant provisions of the Fair Housing Act is the following [42 USC 3607(b)(2)(C)]: "As used in this section, 'housing for older persons' means housing . . . intended and operated for occupancy by persons 55 years of age or older . . . "

Although enforcing the Fair Housing Act would not have prohibited renting to the elderly, preferential consideration of elderly prospects would have been deemed discriminatory. Younger prospective tenants could have sued based on the Fair Housing Act. Assuming that Deer Hill Arms is in a desirable neighborhood, fewer units would have been available for elderly residents.

deal with the preservation of our built cultural heritage, some by establishing techniques for the management of historic properties, others by providing financial incentives for preservation.

Rules for historic preservation are almost exclusively concerned with the exterior of buildings. Inside, unless prohibited by state or local rules, buildings are sometimes gutted, and brand-new buildings are installed within the old walls and roofs. Some state and local rules, recognizing the tendency for developers to build additional floor slabs in old high-ceilinged spaces, and in an effort to preserve the view into a historic building, require that zones of nominal width be maintained against exterior walls. Within such zones, new intermediate floor slabs are prohibited, and new partitions are not allowed to intersect existing windows.

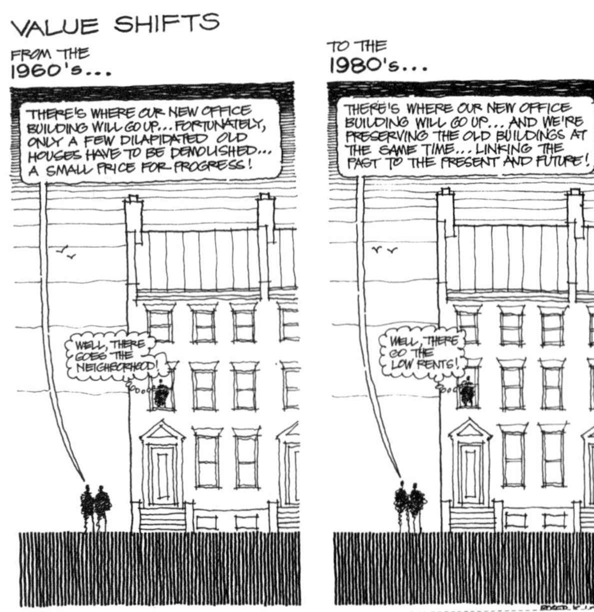

"Value Shifts." © Roger K. Lewis.

Statutes

The Antiquities Act of 1906 was the first national preservation law, predating even the National Park Service. With passage of the Historic Sites Act of 1935, the National Park Service was given authority to designate national historic landmarks, survey historic sites, and acquire historic properties. This initial interest in preservation was pushed to the side in the decade following World War II, when federal construction projects in "urban renewal," flood control, and highways proceeded with little regard for historic value. Partly in reaction to this, it was felt that new statutes were needed.

The National Historic Preservation Act (NHPA) of 1966, as amended [89 Pub. L. 665, 16 USC 470], provides federal matching funds to states, public agencies, nonprofits, and even individuals engaged in preservation. The agenda it advocates is brought to private-sector efforts with the Federal Historic Preservation Tax Incentive, part of the Tax Reform Act of 1976, as amended. This statute provides tax incentives and disincentives for work done with historic properties.[12]

Regulations

The National Park Service has developed a set of standards and guidelines to serve the twofold purpose of determining whether projects qualify for special federal tax incentives and of providing technical advice regarding preservation methodology. The primary regulation related to work on historic properties is currently "The Secretary of the Interior's Standards for the Treatment of Historic Properties," 36 CFR 68, developed in 1992. The primary regulation related to tax implications is the "Standards for Rehabilitation," 36 CFR 67.

The Technical Preservation Service (TPS) of the National Park Service publishes illustrated guidelines for project teams trying to implement the regulations.

In most cases, state and local historic preservation offices have expanded the standards and guidelines, in some cases extensively, through local amendments.

Environmental Protection

While energy codes address environmental controls within a building, environmental protection rules safeguard the environment beyond a project's property lines. They influence far more projects than just those located on delicate sites (such as wetlands and dunes) and environmentally influential sites (such as wetlands and flood control districts). Environmental protection is concerned with maintaining the quality of natural resources, the raw materials on which health and commerce depend. Some of the resources protected, such as water and air, seem of distant concern to an individual building. Still, the cumulative effect of thousands of climate-control systems and site interventions such as paving and regrading is substantial. Rules are concerned with:

1. Long-term factors such as building and site design. Such rules may address:
   - Mechanical equipment: efficiency ratings, energy usage, and released by-products such as carbon monoxide, lead, sulfur dioxide (the source of acid rain) and CFC's (chlorofluorocarbons).
   - Wastes: kinds of materials generated as intentional waste, and any pre-release processes used.
   - Building emissions: by-products of building operations other than heating/cooling and intentional disposal. Such by-products include volatile cleaning solvents and offgassed materials.
   - Water consumption, with the most familiar example being that of low-flush toilets.
   - General energy usage: harnessing daylight where possible, using motion detectors and timer controls, and otherwise reducing overall energy consumption.
   - Storm water runoff: handling it by regulating erosion controls, ground permeability, and storm water retention.
2. Temporary measures such as those utilized during construction.
   - Demolition: allowable release of particulates (dust), handling of hazardous materials such as asbestos and PCBs (polychlorinated biphenyls).
   - Storm water runoff during construction.
3. Existing conditions.
   - Remediation of hazardous or toxic materials. These include many thousands of materials, including asbestos, heavy metals such as cadmium, PCBs, dioxins, and other exotic-sounding materials and chemicals reported in the news media.

Although many state and local governments have adopted environmental rules, it has not dampened the drive of the federal government in this arena. It recognizes that pollution, consumption of resources, and other forms of environmental damage do not recognize state borders. As we learn more about the complexities of ecology and the environment, rules will most likely evolve. Current rules are extensive, with no central clearinghouse to coordinate their enforcement. Even local code officials may not be aware of all rules that apply. As a start, check with the following major pieces of environmental legislation and the organizations that support them:

- Primary statute: NEPA, the National Environmental Policy Act of 1969 [42 USC 4321]. Information is available from the internet at www.epa.gov.
- CERCLA, the Comprehensive Environmental Response, Compensation, and Liability Act of 1980 [42 USC 9600], and its "Superfund" for financing the major cleanups it mandates. It is generally pronounced as a word ("Circla").
- Hazardous waste: RCLA, the Resource Conservation and Recovery Act of 1976 [42 USC 6900], generally pronounced as a word ("Rickla").

- The Clean Air Act [42 USC 7400].
- The Noise Control Act of 1972 [42 USC 4900].
- The Solid Waste Disposal Act [42 USC 3250].
- The Toxic Substance Control Act [15 USC 2600], generally pronounced as a word ("Tosca").

Until recently, most rules set specific maximum quantities of toxins as the targets to be achieved for all sites, regardless of any particular site's position in the environmental chain and its subsequent potential to cause damage if polluted. The EPA, and some local governments are starting to develop rules based on Risk Based Corrective Action (RBCA), whereby the extent of required cleanup is based on the potential environmental effect of toxins at a specific site.[13]

Environmental protection is an evolving field, subject to pressures applied by interest groups on both sides, sometimes triggered by headline-grabbing disasters. Predicting future regulatory directions is a wild guess at best.

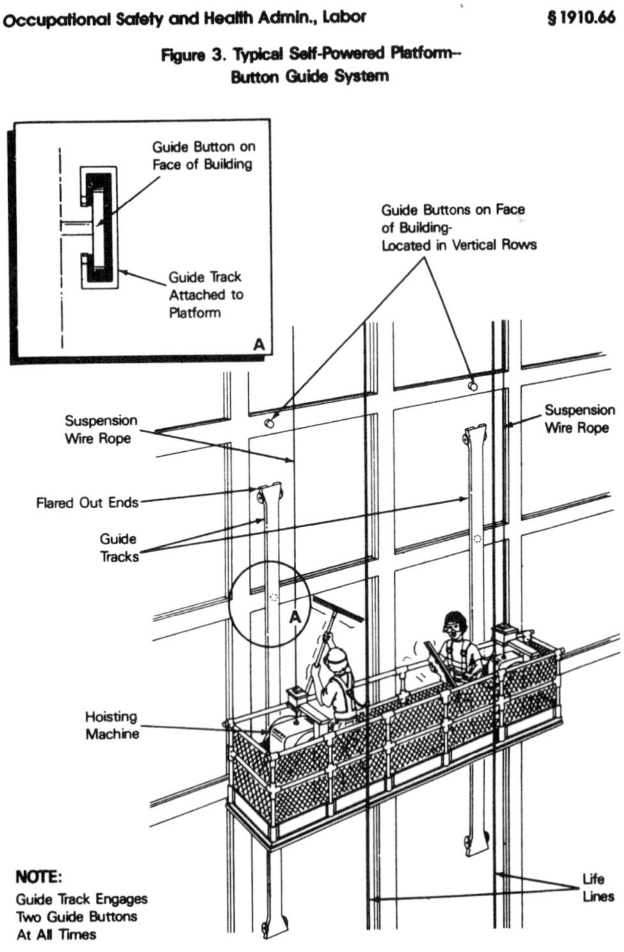

29 CFR 1910.66, App. B, Figure 3. Typical Self-Powered Platform—Button Guide System. The same appendix also includes sketches for a Continuous External or Indented Mullion Guide System and an Intermittent Tie-In System.

OSHA (Occupational Safety and Health Act)

OSHA is a federal regulation that sets standards for working conditions and is intended to protect employees while on the job. The provisions contained in OSHA are a mix of performance standards and design guidelines. Some are quite specific, including minimum dimensions for egress corridor widths. While many such provisions parallel the requirements found in building codes, OSHA adopts the Life Safety Code (NFPA 101) by reference.

Requirements

Federal OSHA standards were developed and are enforced by the Occupational Safety and Health Administration, a division of the Department of Labor. Officially known as the Williams-Steiger Occupational Safety and Health Act of 1970, OSHA applies "with respect to employment performed in a workplace" throughout the United States and its territories [91 Pub. L. 596, 29 USC 651 et seq.]. Provisions for workers in completed buildings are contained in 29 CFR 1910. Provisions for workers on construction sites are contained in 29 CFR 1926.

State standards are preferred when approved by the Assistant Secretary of Labor for Occupational Safety and Health, and are encouraged by the federal government. The federal law sets minimum standards, but approved state versions govern in states that have them.

Building Use

Many OSHA rules deal with the way buildings are used or equipped rather than how they are initially designed. As such, they are of concern primarily to employers and their facilities management staffs. It is the employers' duty to provide proper safety devices on manufacturing equipment and to control employee exposure to hazardous materials or dangerous levels of sound and radiation.

Building Design

Permanent workplace features regulated by OSHA that are the project team's responsibility include:

- Means of egress [29 CFR 1910.37].
- Toilet facilities [29 CFR 1910.141].

- The design of fire detectors, standpipes, and automatic fire suppression systems [29 CFR 1910.158 to 1910.164].
- Provisions for window washing equipment. Last-minute redesigns of several high-rise commercial structures were the result of design team ignorance of the impact of window washing safety device design on curtain wall design [29 CFR 1910.66(e)].
- Many less global issues including the design of industrial stairs and fixed ladders, of refuse collection and disposal, of storage rooms, minimum lighting levels, and maximum sound levels.

Construction Sites

One of OSHA's most influential arenas is the construction site, where adherence to its provisions has saved many lives. Many of the measures it mandates are removed as a building is completed and becomes safe for normal occupancy. For this reason, responsibility for fulfilling its rules usually remains with the general contractor rather than the architect or engineer.

OSHA sends inspectors to construction sites periodically to confirm compliance. Typical provisions concern temporary railings at edges of floor slabs, configuration of rubbish chutes, and temporary elevators.

CPSC (The U.S. Consumer Product Safety Commission)

This organization regulates consumer products and materials that people are likely to use or contact. Its empowering legislation, the Consumer Product Safety Act of 1972, requires it to "protect the public against unreasonable risks of injuries and deaths associated with consumer products." Among the consumer products it monitors are some that are commonly found fixed to buildings. They include:

- Tempered safety glass in and adjacent to doors and in skylights. These rules are generally prescriptive rather than performance-oriented. All model building codes include CPSC glass requirements [16 CFR 1201] by reference.
- Products that are targeted by fire and burn prevention efforts, including many decorating fabrics [16 CFR 1630 and others] and cellulose insulation [16 CFR 1209 and 1404].
- Hazardous substances and products affecting Indoor Air Quality (IAQ, in contemporary lingo) [16 CFR 1500 and others].

Material and Workmanship Rules

Many organizations write rules intended to establish acceptable levels of quality for materials and workmanship. They write such rules not because anyone empowered them to do so, but because they have expertise in the subject matter. In many cases, the availability of industry-produced standards has eliminated the need for governmentally imposed regulations. Model building codes refer to them as "reference standards" and list applicable ones in their appendixes.

Standards Written By Research And Testing Organizations

- ASTM, the American Society for Testing and Materials, is best known for its extensive series of standards for testing procedures and workmanship. This is a tremendous resource for any member of the project team interested in knowing what "in a workmanlike manner" really means, or what the "industry standard" is for quality of workmanship for any particular part of the construction.
- NFPA, the National Fire Protection Association, has standards for many elements of construction, not all related to fire and life safety.

7 **Application of Gypsum Board**

7.1 *General:*

7.1.1 *Method of Cutting and Installation* — Cut the gypsum board by scoring and breaking or by sawing, working from the face side. When cutting by scoring, cut the face paper with a sharp knife or other suitable tool. Break the gypsum board by snapping the gypsum board in the reverse direction, or cut the back paper with a knife or suitable tool.

7.1.3 When gypsum board is to be applied to both ceiling and walls, apply the gypsum board first to the ceiling and then to the walls.

7.1.6 Drive the screws to provide screwhead penetration just below the gypsum board surface without breaking the surface paper of the gypsum board or stripping the framing member around the screw shank.

7.1.9 The external corners shall be protected with a metal corner bead or other suitable type of corner protection that shall be attached to supporting construction with fasteners or a crimping tool nominally 6 in. (152 mm) on centers.

A few select excerpts from ASTM C 840-94, "Standard Specification for Application and Finishing of Gypsum Board." This 13-page document offers highly detailed definitions of several "industry standard" ways to use gypsum board.

- UL, Underwriters Laboratories, is well known for its two-volume *Fire Resistance Directory* and standards for doors and dampers. See Chapters 6, 8, and 10 for more on this.
- FM, Factory Mutual, is the stand-alone engineering and research division of an insurance company. Its mission is to develop standards for its insureds, but it markets to a broader audience. Most of these standards are for fire protection products such as suppression and alarm systems.
- ANSI, the American National Standards Institute, publishes standards for accessibility and for ceramic tile work.

Standards Developed by Professional Associations

- ASHRAE, the American Society of Heating, Refrigerating, and Air Conditioning Engineers, Inc. publishes one of the most useful standards for the planning of new projects, Standard 90, *Energy Conservation in New Building Design.*
- ASME, the American Society of Civil Engineers, publishes the industry standard for minimum design loads, including considerations for seismic design, as ASME 7. Standards for masonry buildings are published in ASCE 5 and 6.
- ASME, the American Society of Mechanical Engineers, publishes standards for conveyances and piping. A key document is the elevator and escalator standards, ASME A17.1.

Standards Developed by Trade Associations

There are thousands of these standards. Since the trade associations represent many manufacturers and a diverse range of product offerings, their standards tend to be reasonably unbiased and useful. These associations recognize the value of regulating themselves.

Examples of standards written by the trade associations include:

- Millwork (cabinetry) standards by AWI, the Architectural Woodwork Institute
- Clay masonry standards by BIA, the Brick Institute of America
- Plywood standards by APA, the American Plywood Association
- Glazing standards by GAA, the Glass Association of America (formerly FGMA, the Flat Glass Manufacturers Association)
- Ductwork standards by SMACNA, the Sheet Metal and Air Conditioning Contractor's National Association

Some descriptions of materials and assemblies, although called standards, might more accurately be thought of as recommendations. Such standards may go beyond current industry practice to a higher level of quality, one to which the authoring organization aspires despite the realities of current industry practice. Sometimes such recommendations are adopted by code officials who either do not recognize them as somewhat experimental or have not thoroughly considered the implications of using such progressive standards. It behooves design teams to evaluate standards against the strengths and weaknesses of existing industry practice before adopting them.

	Custom	Premium	Economy
Tops, Exposed Ends, and Bottoms			
Stop Dado, glued under pressure, and either nailed, stapled, or screwed (fasteners will not be visible on exposed parts)		√	
Dowelled, glued under pressure (approx. 4 dowels per foot of joint)		√	
Spline or biscuit, glued under pressure (approx. 3 per foot of joint)	√		
Thru Dado, glued under pressure			√
Cabinet Backs—Wall Hung			
Full bound, captured in grooves on cabinet sides, top, and bottom		√	
Side bound, captured in grooves on cabinet sides, glued and pinned	√		
Mill Option			√
Note: When not in violation of design, surfaces of intersecting body members may be set back not to exceed 1/8" provided setback is consistent.			

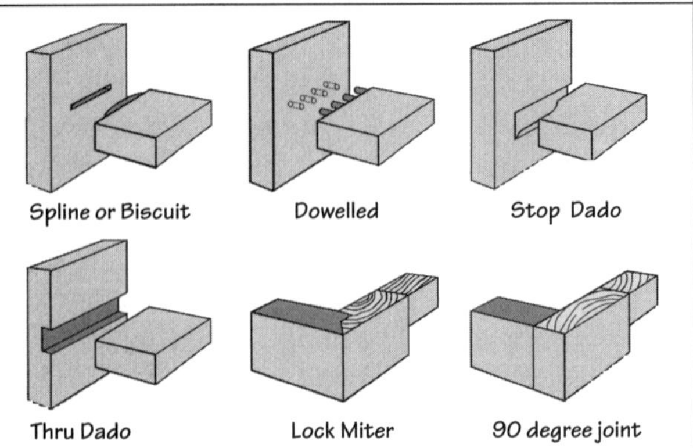

An excerpt from the AWI Standard for cabinetry—Specification 400A-S-9, "Joinery of Case Body Members."

SEEN IN CONTEXT

A Comparison Matrix

	Planning	Building	Accessibility	Historic Preservation
Purpose	People's environments extend beyond their own four walls. Planning rules allow for decisions on a larger scale than that of the individual building.	To provide safe places to live and work.	As a society, we have decided that people should not be excluded from participating in common activities due to physical status.	As a culture, we have decided that buildings constitute a record of our cultural heritage, which we have decided is worth maintaining.
Primary Goal	Value: Stability in property values and neighborhood ambience, a rough correspondence between available infrastructure and demand levels.	Safety: The minimum standards of construction quality necessary for safe occupancy and emergency exiting.	Equal access: By all people, whether momentarily, temporarily or permanently at the edges of the range of potential physical ability.	Preservation of our history: The reasonable but caring maintenance of our architectural heritage, including evolution where appropriate.
Method	Establishes design requirements and a review and appeals process whereby existing residents and occupants can have a say in new projects that might change the nature of their neighborhood.	Establishes design requirements enforced through plans review and construction inspection, for buildings.	The ADA (Americans with Disabilities Act) is a civil rights law that allows individuals to bring lawsuit to force adaptation as needed to assure access. Regulations have been developed to provide guidance.	Establishes guidelines for designating a built resource (building or district) as historic, qualifying it for tax incentives. Establishes design requirements for designated resources, enforced through plans review and construction inspection.
Regulated Categories	Contextual (external) considerations including permitted use, building size, parking and loading requirements. May include incentives for providing public amenities.	Building (shell & interior) considerations including fire resistance, egress, health (ventilation & light, sanitation), structural parameters (gravity, wind & earthquake), quality of materials.	Site work and exterior facilities, new buildings, additions, alterations, and historic preservation.	Stylistically notable or historically notable. Preservation (repair), rehabilitation (adaptation), restoration (authentication), and reconstruction (complete rebuilding).
Affected Design Issues	Height, area, and shape of building; position on site.	Interior layout, design of fenestration, choice of finishes, type of structure.	Circulation (access to spaces), use (proximity and adaptability of components), communication (signage and alarms).	Architectural style, choice of materials and assemblies, use of color.
The Authors	Local governments: Urban planners and legislators who know local issues, with input from current residents/occupants.	National organizations: Private companies that employ experts in building failure (fires, sick building syndrome, etc.), and/or local code officials, and the public at large.	The federal government and state governments, and contracted nonprofit organizations.	The Department of the Interior's (DOI's) National Park Service, state governments, and local historic preservation offices.
Procedures for Becoming Law	After initial drafting, they are further developed through public hearings.	Adopted by elected representatives of local residents/occupants.	Since the ADA is a federal statute, no further adoption is needed. Conversely, there is little consensus on application of available design standards (ADAAG, UFAS, FHAG, ANSI, state versions, model code versions).	State programs for projects receiving federal assistance are certified by the DOI. Local governments may independently adopt guidelines for privately funded projects. No guidelines apply unless properties in question have been designated as historic.
Addressing Local Concerns	Inherently respond to local conditions since they are written by local people.	Local residents/occupants elect representatives who either write local amendments to the model codes themselves or appoint others to do so on their behalf.	Exceptions are made for hardship, either due to the cost of compliance or because of compromises to authenticity required of historic properties.	Preservation is not federally mandated, other than for projects receiving federal funds. State programs are more local by nature. Even then, hardship is considered for specific sites.
Enforcement	Local governmental officials.	Local, and sometimes state, governmental officials.	State governments and the U.S. Justice Department.	The U.S. Department of the Interior and State Historic Preservation Offices.
Exceptions	See the section entitled "Optimization" in this chapter			
Benefits to Project (other than legality)	Preservation of neighborhood ambience.	Reduced property insurance premiums.	Increased social equality; appeal to a greater market segment.	Tax incentives and grants; sometimes increased property value reflecting a project's historic pedigree.

Rules matrix.

The History of Construction Regulation

Many books on regulation start with condensed histories of construction rules in Judeo-Christian cultures as a lead-in to the current system in the United States. I apologize for the bias of this list toward building codes.

The History of Rules

THE WEST

- Babylonian law's Code of Hammurabi in 2000 B.C.
- Israelite law's Deuteronomy 22:8, 8th–6th century B.C.[14]
- Roman law's Twelve Tables of 450 B.C.
- British law's "Henry Fitz-Elwyne Assize of Buildings" of 1189 A.D., Act of Parliament of 1676, and Metropolitan Building Act of 1844

UNITED STATES

- Chicago's building code of 1875, the first in the United States
- Boston and San Francisco's use statutes of the 1890s, foreshadowing zoning ordinances
- Chicago's Commercial Club proposal of 1895, the first municipal plan in the United States, followed by the 1905 McMillan Plan for the District of Columbia
- The United States's relatively short-lived first attempt at a "National Building Code" in 1905
- Passage of the Antiquities Act of 1906, the first national preservation law in the United States
- New York City's Zoning Ordinance of 1915, the first in the United States
- The Historic Sites Act of 1935 establishing historic landmarks
- The National Historic Preservation Act (NHPA) of 1966
- The Architectural Barriers Act of 1968
- The National Environmental Policy Act of 1969
- The Williams-Steiger Occupational Safety and Health Act (OSHA) of 1970
- The Americans with Disabilities Act of 1990
- The United States's second attempt at a national building code, the "International Building Code" in 2000.

The Founding Dates of Significant U.S. Organizations

- Building sciences organizations: UL and ASHRAE in 1894, ASTM and NFPA in 1896
- Code-writing organizations: NFPA in 1896, BOCA in 1915, ICBO (originally PCBOC) in 1922, SBCCI in 1940, CABO in 1972, and ICC in 1994
- Governmental organizations active in construction regulation: the National Park Service in 1919, the Department of Housing and Urban Development in 1965, the Environmental Protection Agency in 1970

APPLICABILITY

DO RULES APPLY TO YOUR PROJECT?

Must every project conform to rules? The answer depends on who's asking and why—which project and which rules are being considered. When the actions people take match societal needs, no rules are needed. Rules are needed when there is a difference between societal goals and the actions of individuals within the society. To persuade the individuals to alter their plans and conform to rules, there must be some form of enforcement. So, as every willful child knows, the essential question is: Which rules are going to be enforced, and what is the penalty for noncompliance? Codes apply when written or adopted by whatever bodies of government have been constitutionally empowered to do so, as further addressed below. But first, a brief exploration of enforcement may clarify things.

Enforcement

Any person or group can write rules, but written rules are meaningless unless enforced. In order to enforce rules, one needs assessment and enforcement bodies empowered by law. Such bodies are empowered to check that applicable rules are being met, and someone must be available to rectify situations involving broken rules. In general, rules affecting the built environment are enforced through the permit and inspection process. If a project team doesn't comply, no permit is issued and construction will be prohibited. Once a permit is issued, if construction varies from what was approved, inspectors can reject it and require reconstruction. In a more extreme case, where construction proceeds without a permit, the governments that wrote the applicable rules can take stronger action (see p. 56), since enforcement procedures are always written into the rules. A project team that wants to know the consequences of noncompliance, can find them on the rules themselves, generally in an introductory chapter on administration of the rules. But even within the rules, there is almost always flexibility. Project teams should consider all the options allowed within the rules and take advantage of the conditions under which exemptions or other relief are granted.

The concept of applicability and enforcement of rules through permmits works well for new project or changes to existing ones. The question is more complex for existing projects when no improvements are being contemplated by the investment team.

Exemptions for Work on Existing Buildings

Context

Common maintenance, as well as more extensive alterations, repairs, additions, and changes of occupancy, are all influenced by codes. Should noncompliant buildings be maintained or demolished? Could alterations or additions increase hazard levels, increase combustibility or add potential fuel, add occupants, increase exit distances or block or eliminate exits? Answers are not simple. Still, some basic work is required for any project being modified through addition, alteration, or change of occupancy, regardless of the extent of the modifications.

New work related to existing buildings falls into one or more of the following categories:

- Repairs: Maintenance work that involves "reconstruction or renewal" without materially affecting structural, mechanical, plumbing, or electrical work
- Additions: Work that increases floor area or height
- Alterations: Work that doesn't increase area or height but is more than a repair
- Changes of Occupancy: Work that, while not necessarily involving any construction, forces a reexamination of an existing building's design to accomodate a new use.
- Relocated Structures: Work involved in moving a building to a new site

Repairs do not change a building's design, and therefore need not meet the standard set by codes. So long as existing construction is merely restored to a previous condition, plans need not be reviewed. Conversely, additions and alterations to existing buildings are required to meet the requirements for new construction, while existing portions that remain may or may not be exempt. BOCA does not require compliance with current codes [BOCA96 3404.2], while the IBC leaves that decision to the local code official [IBC 3402.1].

Requiring that they be brought into compliance with current laws would probably make many projects unfeasible. Project teams have the option of following an alternative set of rules, discussed below, if they so choose [BOCA 96 3408]. This is intended to address this very real dilemma. Some jurisdictions use a threshold for full compliance that is expressed as a percentage of the current value of the property. For example, if the work being proposed is valued at more than 50 percent of the appraised value of the property, full compliance would be required.

Complicating matters further, many projects have no clear line delineating affected areas from those unaffected by new work. While architectural work, such as walls, doors, and finishes, may be fairly distinct, the same cannot be said of infrastructure. Connecting utilities often requires reaching into an existing building.

This ambiguity is not quite so bad with rehabilitation projects carried out under the standards established by the Secretary of the Interior. These standards require new work to be clearly distinguishable from existing work so as not to dilute historic authenticity. But that leads to another topic, work done to projects of historic value, addressed below.

Requirements

General

Exits: Codes permit existing stairs, fire escapes, and other archaic means of egress to remain, even if not compliant with current rules. Since fire escapes are prohibited in new construction under any circumstances, once removed they may not be replaced. However, if stairs are removed, their replacements may match the old stairs exactly if compliant stairs cannot be made to work in the existing space [BOCA 3402].

Lead-based paint: Due to its "stealth" toxicity and long-range debilitating effects, codes require lead-based paint found on residences and day-care facilities, whether indoors or outdoors, whether on buildings, garages, or fences, to be removed or covered [BOCA 93 3402.8].

Project Categories

Repairs

As noted, repairs are maintenance projects involving "reconstruction or renewal." Work cannot be classified as a repair if it involves any of the following:

- "Cutting away" (whatever that means) all or part of any partition, bearing or nonbearing
- Cutting or removing any structural members
- Changing or removing any egress components
- Changes to the building that require changing any egress components
- Changes to, including simple replacement of, any mechanical, plumbing, or electrical work

In most jurisdictions, work that qualifies as a repair does not required notification of or filing applications with the building department [IBC 105.2.3, 201.4].

Additions

Additions are projects that increase height or floor area. Codes generally treat additions as new projects, requiring them to meet the mandates of current codes, but not requiring the same of the existing building to which they are added. The massing of the original portion plus the addition is expected to comply with the height and area limitations for new projects [BOCA 93 3403].

Structurally, codes require additions to carry their own loads, permitting only a nominal increase (5%) in the forces imposed on the existing structure [BOCA 93 1617.2].

Alterations

Alterations are projects that cannot be classified as either repairs or additions. In other words, they involve new work but don't expand the building. Only areas affected by new work are required to meet rules for new construction.

With respect to egress, BOCA 93 [1001.2, empowered by 3404.2.2] says only that alternative but equivalent means of egress can be used as approved by local code officials when "strict compliance . . . is not practical."

Structurally, codes allow alterations to continue under the live load assumptions that were in effect when the building was first built. If the altered project fulfills the same or an equivalent occupancy, the structure that has worked for so many years should continue to be sufficient [BOCA 93 1617.4-5].

Changes of Occupancy

Sometimes a project team wants to change the occupancy for which a building is used, with or without new construction. Even this triggers regulatory processes. In general, projects that involve changes of occupancy must demonstrate that they fulfill "the intent" of the new occupancy's code requirements and present no more hazard than the original occupancy [BOCA 3405.1].

Structurally, the rules for alterations apply, unless current codes require heavier loads of the new occupancy. In this case the structure must be upgraded to meet current requirements [BOCA 93 1617.5].

Relocated Structures

Relocated structures are expected to fulfill either the requirements for new construction or the alternate rules set forth in code sections addressing existing buildings [BOCA 93 3407.1].

Alternate Rules for Work on Existing Buildings

BOCA 96 [3408] offers an alternate set of code requirements intended to offer incentives for maintenance and repair without discouraging more complete renovations. This option is not allowed for riskier industrial and hazardous occupancies. Project teams interested in pursuing this option have to fulfill two requirements. They must:

1. Conduct and submit the results of a structural analysis.
2. Complete an evaluation form [BOCA96 Table 3408.7] of fire safety, means of egress, and general safety using scoring formulas provided by the code. Projects with positive final score values are considered to be in compliance with the code. The evaluation considers:
   - Building height and area
   - Compartmentation and openings, including management of mixed occupancies
   - HVAC systems
   - Fire and smoke detection, signaling, and suppression

- Exit capacity and distance
- Elevator control

Work on Projects of Historic Value

What about new life for historically significant buildings? At what point do we decide to alter historic properties to meet contemporary standards of safety and value, when doing so might compromise the links to our cultural heritage that they represent [BOCA 93 3406]? Often, cultural roots are valued as long as they don't inhibit contemporary wishes. A building without central air-conditioning would be a hard sell, and one without indoor plumbing would be illegal.

What about our conflicting interest in making opportunities available to all members of our society when it requires widening doorways and adding ramps and elevators that are likely to obscure the history lessons that only original buildings can offer [BOCA 93 1110].

Historic preservation is sometimes more of a national priority than a local one. The Secretary of the Interior does not have inspectors or police to send from town to town to ensure that historic preservation guidelines are being met. Instead, the federal government encourages compliance through tax incentives. When a project's economic viability is marginal, these incentives can reduce financial risk and make the project more attractive to investors or buyers. (See pp. 46–47 for more about preservation tax statutes.)

Still, if tax incentives were the only inducement to compliance, many project teams might decide that the tax incentives didn't outweigh the benefits of demolition or noncompliant development. While taxes may be the primary means of enforcement at the federal level, local governments have other options. Historic preservation is mandated in many local ordinances with penalties ranging from denial of a building permit to required reimbursement for the costs of critical repairs or the other construction undertaken by the local government. On the other hand, it is worth noting that the model building codes generally empower local officials to waive building code requirements for identified historic buildings that they judge to be safe and historically valuable [BOCA 93 3406.1].

Exemptions for Unimproved Existing Buildings

Context

Projects that predate rules generally do not need to be rebuilt to comply. Codes contain "grandfather clauses" that define which older projects are waived from current requirements. However, as codes are constantly revised to reflect new knowledge regarding building performance and new cultural attitudes regarding risk-taking, policies must be developed for existing buildings that were built to different standards, ones that usually do not effectively address contemporary issues. With respect to existing buildings, there are two basic conceptual attitudes codes could take:

- Encourage proper maintenance and repair of older buildings, knowing that such work keeps noncompliant and potentially unsafe projects on the market. This presupposes that most older buildings, if properly maintained and repaired, will be relatively safe. You can almost hear the refrain: "They've lasted this long, haven't they?"
- Allow maintenance, but prohibit any efforts to repair projects whose designs don't meet contemporary standards. When buildings need repairs, their owners will be forced to bring them into compliance with current codes.

As with most dichotomies, the solution is probably somewhere in the middle. The key to a reasonable stand is finding a way to mandate a level of disrepair that would be acceptable to all parties involved. Users might want compliance to be required for most projects, so that the majority of buildings would quickly be upgraded, except for the fact that their costs of occupancy (rent or loan payments) would be forced upward accordingly.

Meanwhile, investment teams would want most projects to be exempt, so they could make the decision to repair rather than have it forced on them through rules. Obviously, some would postpone repairs until long after they could reasonably be considered due. Nice as it might be to increase our stocks of fully compliant buildings, raising the standards may not accomplish it, and users would pay a price in projects fallen to dangerous levels of disrepair.

Applying Codes

BUILDING CODES

When new regulatory provisions are proved to be vital to health, safety, and welfare, and modification to comply involves a generally acceptable level of hardship, should they be applied to all buildings? Often the answer is yes. Installation of smoke detectors is an excellent example of this situation.

Zoning Ordinances

Existing buildings that violate zoning ordinances by encroaching on current setbacks are generally allowed to remain. Developers who want to tear their buildings down and rebuild on the site would normally be required to conform to current provisions. Even if they decide to rebuild on the same foundations, they are generally prohibited from doing so. A developer who decides to replace an entire building one section at a time, however, will often be allowed to continue with the nonconforming design since the existing building is never removed.

Accessibility Rules

Most buildings predate the ADA. However, if this civil rights legislation allowed access only to buildings built since its passage, the disabled would face major obstacles for many years to come. The better buildings of all ages, those that last or are maintained for years, would be forever beyond reach. The ADA requires compliance in all applicable buildings, new or existing, "where readily achievable." Enforcement is waived only for hardship, so the ADA can't be said to have grandfather clauses in any standard sense.

Historic Preservation Guidelines

For unimproved existing buildings, historic preservation rules apply only to properties that have been, or are located within areas that have been designated as historic. Of course, where entire street facades or neighborhoods become registered as historic places, there may be individual owners who don't agree with the election to historic registry who will nonetheless be required to comply.

Requirements

Even when no new work is being considered, investment teams are required to keep their properties maintained and in proper repair, and to periodically inspect and test them for compliance. Special codes address these needs specifically, such as BOCA's National Property Maintenance Code and National Fire Prevention Code.

Structurally, when repairs show existing structures to be deficient, codes require them to be replaced to conform to current levels [BOCA93 1617.3].

Residential (One- and Two- Family) Exemptions

Many regulatory provisions are intended for large-scale and public projects. While most residential (or other very small scale) projects are not entirely waived from compliance, owners are usually allowed to comply with the more lenient provisions of CABO's *One And Two Family Dwelling Code.* Even in full-scope codes, exceptions to almost every section substantially reduce the burden of compliance.

Willful Disregard

Of course, rules are sometimes simply ignored. Yes, this is illegal, but it is also often self-defeating. Investors may build without a permit, being careful that construction workers don't arrive until after inspectors have stopped cruising the streets for the day, and making sure that Dumpsters are hauled away before daybreak. As a result:

- Such projects usually take much longer to build and can incur substantial revenue losses, the unrealized economic gains lost due to delayed completion.
- One claim filed against the investors due to a design deficiency that codes might have prohibited, or that an inspector might have caught before the claim, can very quickly exceed any financial gains temporarily reaped by designing to a lower standard.
- Violations can be exposed during presale inspections. When revealed at such crucial times, they can severely weaken investors' negotiating positions.

Arlington County, Virginia
Department of Community Planning, Housing, and Development

LEGAL NOTICE

WHEREAS, violations of

Article ________, Section ________ of the Zoning Ordinance
Article ________, Section ________ of the Building Code
Article ________, Section ________ of the ________ Code

have been found on these premises, IT IS HEREBY ORDERED in accordance with the above Ordinance/Code that all persons cease, desist from, and

STOP WORK

at once pertaining to construction, alterations, or repairs on these premises known as ____________________.

Any person or persons acting contrary to this order or removing or mutilating this notice are liable to arrest unless such action is authorized by this Department.

Date __________ Building Official __________

A Stop Notice ordering a halt to construction.

- Design professionals can face disciplinary action, with forfeiture of memberships in professional societies such as the American Institute of Architects, or forfeiture of licensing in extreme cases. This can be true even if all they did was look away when other members of the project team proceeded with code violations.

Poverty is not an acceptable alibi—owners who cannot afford to build projects probably shouldn't build them. Proper capitalization is critical to achieving minimal standards of quality.

WHICH RULES APPLY?

Context

WHICH RULES ARE APPLICABLE?

In some ways this is the easiest issue, and in some ways the most work. To find out, read or ask. If you read, start with the state and county ordinances. They can be found in most public library systems. These rules cover all legal issues of local concern, from restriction on the sale of alcohol on Sundays to requirements that dog owners clean up after their pets.

Mixed in among all the other issues are rules for construction, sometimes presented in full, but more commonly presented in the form of citations refernceing model codes or standards. This is true for federal projects as well. The United States Code contains rules for public buildings in general [40 USC 250] and for construction work specifically [40 USC 600]. Federal projects are required [40 USC 619] to comply "with one of the... model building codes... [in its] latest edition" but not any particular one. In practice, however, federal projects use the UBC for general construction and the Life Safety Code where it is relevant. Further, their design teams are required to "consider" (not the same as "comply with") local zoning laws, and construction teams are required to submit to inspection by local code officials.

Some requirements are found in every state and county ordinance, such as the model code adoptions noted above. Other provisions might be less expected, such as language requiring some projects to be approved by a design review board or requiring sidewalk hot dog stands to meet the higher standard of the state hygiene code. In addition to adoption of major codes and procedures such as these, most jurisdictions write amendments to the model codes to adapt them to local needs. After the unified IBC Code is published, and the question of which model code to use is laid to rest, local amendments will continue to have their say, possibly with renewed purpose and vigor.

Although this research can take some time, it's the surest way to make sure you've found all the applicable rules. Keep in mind that the worst time to find out about the existence of a relevant rule is when the construction permit application is rejected. Where warranted, project teams should include attorneys familiar with state and local requirements.

Those who look through the ordinances and still have questions can call the local government in the jurisdiction of the project. Since these governments are the ones that make the decisions, or implement and enforce them, or both, they will be the ones to know. Those who find reading to be too laborious or who are afraid that they may miss key provisions by overlooking the critical pages shouldn't be surprised to find a cold reception from code officials. These administrators quickly run out of patience with callers who haven't first looked for the answers. Some jurisdictions have started charging callers for the time they spend answering questions that are answered in the rules.

Many jurisdictions use several overlapping rules. In these cases, project teams have more work to do. If BOCA has been adopted for general architectural work, and NFPA 101, the Life Safety Code, has been adopted for egress issues, someone on the team will need to compare the two, line by line, and make sure that the project's egress design meets the requirements of BOCA's Chapter 10 on egress, as well as the provisions of NFPA 101 Chapter 5. No, there aren't any shortcuts to this. A compilation set has not yet been written, and probably never will be. This is partly because every jurisdiction has its own ideas about which edition of which combination of rules should be in effect, and partly due to the added complexity imposed by local amendments. The document closest to a compilation set is the matrix in the Appendix of this book.

WHICH EDITIONS OF THE APPLICABLE RULES?

Knowing which model code has been adopted isn't enough; it is also critical to know which edition. Just because a new version has been published doesn't mean that local legislators have had time to review it, discuss it, and hold a vote to change their ordinances to require its use. Even if they have, there's no guarantee that they decided that its "new and improved" provisions were in the interest of their constituents, considering local conditions.

Often, the difference between older editions and more recent ones is that the more recent ones are more sophisticated. Early versions of new codes and of new provisions of older codes tend to be blunter and simpler, which makes them less wordy and easier to understand. It also makes them much less amenable to optimization. The additional time recent editions have had in development and in use lets their drafters be more precise in the way they wield control, consequently granting the project team more design flexibility.

Regulating New Work in Existing Buildings: General Concerns

Why You Should Care

The American Institute of Architects' Vision 2000 study, a survey that asked architects to forecast the state of professional practice in the twenty-first century, projected that 93 percent of projects in the next century would be in existing buildings. Note that this survey was taken of people in a profession that, since its founding, has overwhelmingly focused on new projects over renovations. Additionally, 20 percent of downtown areas are currently within historic districts. It behooves anyone with a role in the building industry to take renovation, preservation, and expansion work seriously.

Assessing Existing Structures

As an important part of establishing feasibility, existing buildings should be evaluated for the following issues:

- Historic value: If the building is recorded on state or national registers of historic places, it is subject to compliance with preservation rules.
- Room for expansion: When an addition is being considered, can the building be expanded without negatively affecting adjacent properties or urban amenities? Is the existing building already too big for its site, overshadowing adjacent buildings or the street? Does it already strain the capacity of existing utility mains? Is there room on the site to add parking for the patrons generated by the addition?

  When a different use is being considered, would it be sympathetic, or even safe, to the kinds of uses found in the existing neighborhood? Would the new usage require more parking than the old one?

  These issues are addressed by zoning ordinances.

- Structural type and integrity: Is the structural system conducive to contemporary needs? Is the column spacing marketable for office space planning? Is it feasible to core (cut holes through) structural elements to install new infrastructure? Can elements be removed to allow for insertion of marketable atrium space? Will it support the weight of new precast concrete curtain walls in place of the original glass and metal panel system?

  Is the project sound? Will it physically support the loads imposed by contemporary uses? Have the effects of termites, rust, or dry rot reduced the carrying capacity significantly? If applicable, how well would the structure deal with severe lateral loads relative to what is now known of building behavior during earthquakes?

  These factors are all addressed by building codes.

- Construction type: materials, fire resistance, potential for adaptation. Aside from structure, is the building adaptable? Is it made with materials that have since been proved to be hazardous, inadequate to the task, or otherwise problematic? How well are internal cavities fire-stopped to prevent the spread of smoke and sparks? What degree of fire resistance can existing assemblies be relied upon to provide, if any? Can new materials be added to those existing without leaving objectionable transitions? Is there room to accommodate the risers, distributions, and mechanical rooms or closets needed by new infrastructure? Are structural components provided with adequate fire protection?

  Most of these factors, and many others, are addressed by building codes.

- Architectural design: type, shape, and arrangement of spaces and equipment. Is the building adaptable to contemporary uses? Is there room for the amenities that most tenants expect? Does the building inhibit use by those with special needs? Are spaces accessible to the disabled? Can switches and outlets be reached? Do fire stairs include, or can they be expanded to include, areas of refuge?

  Most of these factors, and many others, are addressed by building and accessibility rules.

- Impact of rules: grandfather clauses, extent of scope. How much can the building be improved without triggering compliance requirements? How much can be done to comply with basic requirements without triggering variance procedures? Each rule must be reviewed for its attitude regarding preexisting conditions. All rules address this, usually in an introductory or concluding section.
- Life cycle cost impact of renovation: Is it worth reusing? Is it energy efficient, or can it easily be made so? How much usable life can be expected of the structure, the systems, the conveyances? What needs to be done to bring maintenance costs into balance with renovation costs?

  The answers to these questions also depend on building, energy, and accessibility rules.

Regulating New Work in Existing Buildings: Projects of Historic Value

Context

Statutes and regulations pertaining to historic preservation speak to the differing interests of owners and users, influence a number of project types, and are applied through the efforts of many enforcement agencies and advocacy groups.

Interest in Preservation

From the Public's Perspective

Historic preservation is a concern with properties in either of two categories: stylistic value and social/political value. Buildings with stylistic value incorporate significant architectural innovations, worth preserving for their aesthetic significance. Examples include the Johnson Wax headquarters building in Racine, Wisconsin, by Frank Lloyd Wright, as well as beautifully designed but anonymous projects like the Victorian houses in Cape May, New Jersey.

Buildings with social/political value are those where important human dramas took place. Even modest projects have been recognized as important in this regard, such as a cluster of very humble former slave quarters in rural southern Maryland, the last of their kind.

Many projects embody both stylistic and social/political aspects, since many of the important events in our past have occurred in grand settings. One example is a Federalist house known as the Octagon, designed by William Thornton, architect of the U.S. Capitol, where James and Dolly Madison once lived and the Treaty of Ghent was signed.

From the Owners' Perspective

General

Owners must agree before buildings are added to the National Register, except for properties located in historic districts where a large majority of the other owners agree. When properties are designated as historic landmarks, owners' rights of ownership and use are preserved; none are transferred to the Department of the Interior. Further, there is no associated obligation to provide access to the public. Many landmark properties remain closed to the public following their designation.

Investment Value

Increases in the value of eligible properties are very significant.[15] Although public infatuation with historicism goes in and out of style (pardon the pun), the rise in values is a response to the stability eligibility brings. Investors know that historic preservation guidelines will force their neighbors to develop responsibly.

In a capitalist economy, where supply and demand are two sides of the same balance, limited supply tends to increase demand, and therefore value. There will always be a limited supply of well-maintained historic properties.

Tax Incentives

Tax codes change every year, and predicting upcoming revisions is a gamble. Recent tax codes have provided incentives for the first few years after construction to owners who improved historic properties [36 CFR 67]. One important limit to this incentive, in terms of transferable value through eventual resale, is that it only benefits owners within this limited time frame.

Project Types

Background

The Department of the Interior and its Certified Local Government (CLG) programs recognize a diverse range of properties, including:

- Individual buildings
- Entire street facades
- Entire neighborhoods or districts

To fall under the purview of preservation guidelines, projects need only be significant. A major function of local historic preservation societies is surveying existing properties to determine eligibility. Eligible properties are protected, whether listed in registers or not, and penalties can be assessed against owners for damaging historic features. To qualify for tax incentives, however, properties must be placed on the National Register of Historic Places [36 CFR 60]. Registration is the end result of a process whereby State Historic Preservation Offices, sometimes with help from citizen groups, nominate properties to the register, submitting historical and architectural documentation.

There have been occasional instances of groups later trying to have the registration revoked after finding that the very guidelines they fought to impose had, in the process, thwarted some of their project goals. Revocation is not generally possible, except where buildings have changed to the point that they are no longer eligible. Since registered properties are monitored, and deliberate destruction is subject to penalty, revocation is not a realistic option.

Project Categories

Preservation

Focuses on maintenance and repair, and assumes no expansion or change in use. This category requires the least change from the project's current configuration. The French Quarter of New Orleans, whose current physical composition and condition match those seen in old photographs of the district, is an example of a preserved district.

Rehabilitation

Allows alteration and change in use if consistent with historic character. Most architectural projects involving recognized historic properties fall into this category. The Old Post Office in Washington, D.C., with its corporate offices and ground-floor food court, is an example of a rehabilitated building.

Restoration

Returns building to one period in time, removing work from other periods. This tends to keep buildings from adapting to changing societal, business, or market needs but does preserve a sense of what life was like during the preserved period. Harper's Ferry, West Virginia, is a completely restored town whose primary industry is tourism.

Reconstruction

Museum work—recreates previously demolished portions of a property or entire properties. A corn crib long ago destroyed that was recently rebuilt at Mount Vernon, George Washington's family home, from old photographs and the original sketches drawn by the president himself, is an example of a reconstructed project.

Organizations

Many organizations, both public and private, are involved in the preservation effort. Designers must be aware of these groups and might look to them for feedback on projects involving historic sites and districts.

International

The World Heritage Convention maintains the World Heritage List, recommends preservation guidelines to its member nations, lobbies for international projects, provides technical assistance, and runs educational programs. Private efforts include the International Council on Monuments and Sites (ICOMOS) with its United States Committee, which offers educational programs including scholar/expert exchange programs, and the Association for Preservation Technology, International (APT). Organizations with a special focus include Documentation and Conservation of Monuments of the Modern Movement (DoCoMoMo), specializing in European examples of twentieth-century modernism.

National

Public groups include the Advisory Council on Historic Preservation, which advises the President and Congress. The National Park Service of the Department of the Interior administers national historic landmarks, documents them through the HABS/HAER[16] program, and handles the matching grants program. Public buildings and special projects are handled by the General Services Administration (GSA) and the Department of Housing and Urban Development (HUD). The latter operates the UDAG program (Urban Development Action Grants), whose provisions include rules for historic preservation [36 CFR 801].

The primary private group is the National Trust for Historic Preservation. In addition to the properties it owns and operates, it is known for efforts in public advising and education such as its technical service, its seminars and publications, and its national demonstration projects.

STATE

The various State Historic Preservation Offices (SHPOs, often pronounced as a word, either "shpohs" or "ship-pos") handle all public matters regarding preservation. Privately, there are state historical societies, state lobbying organizations, and historic preservation organizations.

LOCAL

Landmark District Commissions and Historic District Commissions extend the work of the State Historic Preservation Offices and the Department of the Interior to the local level. Privately, Historic Societies and Historic Preservation Organizations advocate for the public in policy debates, own and operate properties, and are involved in education and outreach efforts.

WHICH SECTIONS APPLY?

Take the time to scan the Table of Contents in each applicable set of rules. It's probably the fastest way to discover which articles apply. An architect who sees an article on atriums in the section "Special Spaces," who is intending to include one in his or her project, had better make a note to read that article. Once this winnowing process is completed, there may not be so many applicable sections. This might also be a good way to identify overlapping concerns among different rules, as noted above.

Readers should also make use of the matrices in the Appendix of this book. They can help identify specific sections of specific rules that are relevant to specific design issues.

USING RULES

ANALYZING RULES

There are a huge number of issues to be resolved in planning and designing a building. Many of them are influenced or restricted by rules. How does one begin a code analysis?

Model building codes all contain their own recommendations for conducting code analyses. Along with the preface to BOCA93 is "A Guide to Use of the BOCA National Building Code." Similarly, the appendix to NFPA 101 begins with "a suggested procedure for determining the Code requirements for a building or structure." While neither of these follows the organization or content typical of zoning ordinances, accessibility guidelines, or historic preservation standards, the thinking is illustrative.

Without first discussing the issues that you will need to analyze, it wouldn't help to offer suggestions for doing an analysis. Therefore, you will find such suggestions in Chapter 11: "Method."

DEFINITIONS

Definitions used in rules are specific to each set of rules. The definitions found in standard English dictionaries can be misleading for the context used in rules. Words with rule-specific definitions are defined in the rules that use them, generally in a separate chapter for definitions near the beginning of the book. In addition, when such words are used within the text, their special meanings are usually indicated by printing the words in italics.

A few key terms are used throughout this book, so to follow the industry example, meanings will be established here.

Occupancy and Use Group

The occupancy of a building or a portion of a building is the primary programmatic purpose for which it is intended. Use group is a classification system for occupancies. For example, apartment and monastery occupancies are classified as Type R-2 Residential uses, and car showrooms and department store occupancies are classified as Type M Mercantile uses. This is determined not by any rules but by the project team based on its intentions for the building. Model codes contain descriptions of the various use groups, with examples of occupancies corresponding to each. Project teams can read through them until they find a description and examples matching their project.

Once this determination is made, however, rules respond by asking for design features that correspond with the chosen use group, such as maximum building height, maximum floor area, and exiting requirements.

Many projects contain several use groups. When this is the case, the project team must note each and apply the rules for each use group to the part of the project housing it. (See also pp. 192–193 and 180–185.)

Construction Classification

This is a broad rating for the combined fire resistance of an entire building, based on the fire resistance of the assemblies within it. The following concepts are key:

- Construction classifications are expressed primarily in relative labels, such as Type IIB, even though they may connote base time requirements. To find out what the label means in terms of required ratings, one must consult a building code. This is not true of NFPA 101, whose three-digit classifications, such as 332, directly express the number of hours of fire resistance required of exterior walls, interior multistory columns, and single-story columns.
- In general, construction classifications are not material-specific, so long as the materials present can't contribute fuel to a fire. In other words, any incombustible material can be used in any classification. Therefore, a Type 2A building might be made of steel, concrete, or masonry, so long as its components meet the required fire resistance ratings. The one exception is heavy timber construction, which occupies its own class in every code. Further, when wood is used nonstructurally as nailers, sleepers, or trim, even it may be noncombustible classifications such as BOCA Type 1 and 2.
- Construction classifications are differentiated from one another by their degree of combustibility (the extent to which their components will ignite when exposed to heat or flame) and by their incorporation of protection (the use of encasing materials to shield critical components from a fire). See the matrix below. (For further information, see pp. 134–137 and 139–140).

Egress

Egress is a technical term for exiting under crisis conditions. Regulating building design to achieve efficient egress is a major concern of all the model codes and of the Life Safety Code published by NFPA. Its opposite, ingress, is of no concern to codes. When people are unable to find an entrance, or are able to enter only slowly or a few at a time, or have trouble getting to the rooms they're looking for, they may be frustrated but are not in any danger. Still, since most buildings are entered through the same passages used for egress, egress concerns shape the experience of occupants entering most buildings. The differences between

Proposed IBC Nomenclature (5/97 Draft)

	TYPE I		TYPE II		TYPE III		TYPE IV	TYPE V	
	A	B	A	B	A	B	HT	A	B

UBC Nomenclature

	TYPE I	TYPE II			TYPE III		TYPE IV	TYPE V	
	NON-COMBUSTIBLE				COMBUSTIBLE				
	Fire-resistive	Fire-resistive	1-Hr.	N	1-Hr.	N	Heavy Timber	1-Hr.	N

BOCA Nomenclature

Noncombustible					Noncombustible/Combustible			Combustible	
TYPE 1		TYPE 2			TYPE 3		TYPE 4	TYPE 5	
Protected		Protected		Unprotected	Protected	Unprotected	Heavy Timber	Protected	Unprotected
1A	1B	2A	2B	2C	3A	3B	4	5A	5B

SBC Nomenclature

TYPE I	TYPE II		TYPE IV		TYPE V		TYPE III	TYPE VI	
			1-Hr.	Unprotected	1-Hr.	Unprotected	Heavy Timber	1-Hr.	Unprotected

NFPA Nomenclature

TYPE I		TYPE II			TYPE III		TYPE IV	TYPE V	
443	332	222	111	000	211	200	2HH	111	000

Bearing Elements	The Hourly Ratings They Require									
Exterior multistory walls	4	3 (4 UBC)	2 (4 UBC)	1	0	2 (4 UBC)	2 (4 UBC)	2 (4 UBC)	1	0
Interior multistory columns	4	3	2	1	0	1	0	H	1	0
Interior single-story columns	3	2	2 (1.5 BOCA)	1	0	1	0	H	1	0

Comparison of construction classifications.

designing for egress and for ingress are mostly limited to one-sided issues, such as the visibility of exit signs, locations for panic hardware, and the direction in which doors swing. (See Chapters 7 and 9 for more on egress.)

Fire Resistance

Fire resistance is the ability of materials or assemblies to withstand the effects of fire. The amount of resistance required is a function of use group and construction classification. A fire resistance rating applies to entire assemblies rather than single materials or components in an assembly. Even in the case of a simple concrete column, the rating applies to the combination of concrete, reinforcing steel, etc. The rating is determined by subjecting an assembly to ASTM's E119 test, in which an assembly is actually built and set on fire. Fire resistance is expressed as the number of hours an assembly can be expected to survive when subjected to fire, with survival defined by the following criteria. An assembly can be said to have survived if at the end of the stated time period, all of the following are true:

- Structural components are still supporting their design loads.
- Assemblies are still sufficiently intact to prevent the passage of flame and hot gases, and although not specifically mentioned, smoke.
- The temperature on the sides of the assemblies farthest from the fires is still below 250 degrees Fahrenheit.
- Vertical assemblies (walls and columns) can withstand the blast of fire department hoses without water spray emerging through them.

Fire Protection and Prevention

These terms refer to methods employed to minimize the potential for and effects of fires.

Protection

Fire protection is focused on the preconstruction planning of facilities and intended to ensure that new facilities will be safe. In contrast with prevention, protection is primarily focused on equipment, construction detailing, and other physical issues. Its methods are intended to shield critical construction components from fire, and include:

- Finding fires with detectors
- Communicating the presence of fires to occupants and firefighters with alarms and other signals
- Signaling affected building systems, such as air distribution equipment, in order to activate or deactivate them.
- Containing the spread of fires within buildings and resisting the intrusion of fires outside buildings through careful design of enclosing assemblies
- Minimizing fire damage by shielding and protecting building components from fires
- Extinguishing fires by combined use of fire suppression (e.g., sprinklering) and water distribution systems (e.g., standpipes and hose cabinets)

Prevention

Fire prevention is focused on the operation and emergency preparedness of existing facilities. Its purview includes fire protection but adds:

- Fire drill procedures
- Maintenance procedures for fire suppression, detection, and signal systems
- Building operations related to maintaining occupant load and preventing encroachments in the exit path

References

Additional information, such as addresses and phone numbers of the organizations whose standards are referenced, are listed in the appendices of many rules, and always in the appendices to codes. With codes, referenced publications of each organization are also listed in full.

GETTING APPROVALS

Sifting through rules only gets a project team partway there. Team members must get their ideas approved before, during, and after construction. The multiple steps to this process are described in the opening "Administration" chapters of each regulation.

Authority

A key concept is that approvals are local, regardless of where the set of rules came from. The government employee at the county building or city hall with jurisdiction over the project is the one who decides whether to let the project proceed. Such reviewers may contact the model code publishers for clarification, but they are expected and empowered to use their own judgment with respect to project specifics.

Planning Approval

For New Development

Long before the construction of buildings begins, a piece of land must be prepared. This involves several steps, each with its own approvals process. They include:

1. Zoning approval: This review confirms planned use and density of a proposed development.
2. Subdivision approval: Many developments start with a large piece of land, subdividing it into individual lots for resale. Plans are made for shared amenities, such as roads, and for bringing utilities within reach of each lot. These plans are reviewed in a two-stage process of preliminary and final site approval.

   In some cases, a large piece of land is amassed from several smaller ones. This is called assemblage and involves the same basic process as subdivision.

3. Recordation: Once final site approval is given, plans are recorded in government files, and individual lots can be sold.

"Obtaining Building Permits." © Roger K. Lewis.

For New Construction

When applying for a construction permit for a building project, the first level of approvals is invariably from zoning review. In some areas this may involve obtaining approval from several separate review processes for such issues as driveways and curb cuts, permitted use and massing, and signage. Other related approvals may be required for environmental impact.

Building Permits

The next step is getting a building permit. Within any set of rules, the minimum required level of documentation is listed, including even the number of copies that must be submitted. Many jurisdictions issue separate demolition permits.

Permits in all larger cities and many smaller ones involve rotations, whereby each set of drawings is approved separately by several different plans reviewers, each with his or her own list of issues to approve. Stops are made for zoning review; architectural review; and structural, mechanical, plumbing, electrical, and fire safety reviews. The exact number and order of stops along the way vary among different building departments, but zoning is always first, and architectural is almost always second.

For most projects, plans must be dropped off, to be picked up several weeks later. If the set has been approved by all reviewers, a building permit can be obtained and construction can start. If one or more reviewers withholds approval pending the submittal of additional information or the making of design changes, the drawings must be revised and resubmitted.

Some jurisdictions allow smaller projects to be "walked through." This is an express or "expedited" process whereby an applicant is allowed to carry drawings from station to station, waiting while they are reviewed and possibly answering reviewers' questions or agreeing on the spot to changes they require. This process often takes anywhere from four hours to a full day, so it makes sense for applicants to be in line when the office opens.

According to standard construction industry contracts, builders are responsible for obtaining building permits, but architects commonly bring the drawings to the permit office themselves.

The construction market in many larger cities supports a small number of "permit expediters." These are special consultants who obtain permits for other mem-

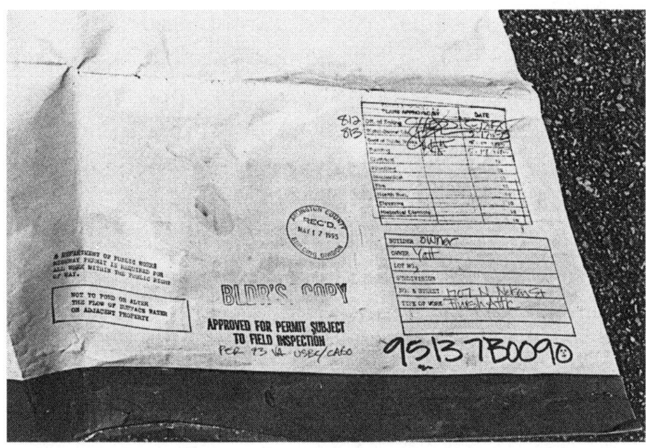

The back side of a record set of construction documents, showing the approvals and caveats associated with building permits for even simple residential projects. Photo by the author.

bers of the project team, saving them a considerable amount of time waiting in line. Some expediters are former employees of the building departments; others are simply individuals with training as architects or construction managers.

On issuance of the building permit, the applicant is given one of the submitted sets of drawings to bring back to the construction site. This "record set" is sealed by each of the reviewers and must be kept available at the site until final inspection, for the use of visiting inspectors.

Building Inspections

During construction at various stages, the work must be inspected. Generally this happens at several points along the way while work to be concealed is still exposed. Inspectors may consult the record set of documents but are empowered to order changes to the work as they see necessary, even if plans reviewers found no fault with the drawings.

A typical schedule includes the following [17]:

1. Site and Foundations
   - Underground utilities and grading
   - Foundation setbacks and forms
2. Under-floor work
   - Foundation wall
   - Joists
   - Drain, waste, and vent lines and water supply
   - Ducts and mechanical equipment
3. Rough-ins
   - Mechanical, plumbing, and electrical rough-ins
   - Windows and stairs
4. Diaphragms: roofs, shear walls, and insulation
5. Wall covers: wall finish and lath
6. Final inspection
   - Building
   - Mechanical, plumbing, and electrical

Such inspections are not intended to catch everything, and government inspectors generally cannot be held liable if a defect remains unchecked.

Certificates of Occupancy

After construction is complete, a certificate of occupancy ("C of O" to the initiated) must be received from the government inspector before users are allowed to move in. This certificate is based on different considerations than the architects' Certificate of Substantial Completion and may be issued before or after that document.

Arlington County, Virginia
Department of Community Planning, Housing & Development

Date:

CERTIFICATE OF OCCUPANCY

Certificate Number:

Permission is hereby granted to (name of owner):____________

to use (suite/apt #):_____of the building located at (address):________

____

to be used for the following purposes:

Use Group and floor area:____________

Seating capacity: __________ Number of children:__________

Zoning district:________ Use Group:__________

Case No. 1 __________ Case No. 2 __________

Type of Construction: _________ Occupant load:__________

This certificate does not take the place of any license required by law. Any change in the use, ownership, or occupancy of this building or land shall require a new certificate.

The building or the proposed use of the building or land complies with all provisions of law and of all county ordinances and regulations.

By ____________________ By ____________________

Building Official Zoning Administrator

This certificate shall be conspicuously posted at all times.

A typical Certificate of Occupancy.

OPTIMIZATION

OPTIMIZATION STRATEGIES

What options exist when regulatory provisions don't support a project team's design approach? This question contains an inherent assumption: that rules should necessarily support design work. Teams might benefit from asking whether conflicting regulatory provisions should support the affected aspect of their design approach. There are always two options for handling conflicts between rules and design approach:

- Find ways to overcome the rules, or
- View conflicts as opportunities for catching design program inadequacies.

Since rules generally provide only minimum acceptable levels of performance, failure to comply might signal problems in some aspect of the project's function. Teams should consider whether the issues addressed by conflicting rules should be added to a project's design goals. Only when issues are found to be inapplicable should the team look for ways to optimize rules.

Code officials are empowered to offer administrative waivers from regulatory requirements for situations allowed by rules when appropriate documentation has been submitted. Photo by the author.

That being said, rules can and should be optimized to enable design excellence. Further, one could reasonably say that rules are designed to be optimized, that the societies that craft them want optimizations to be used. As evidence, one need only look to the number and range of exceptions, incentives, and appeals procedures contained in construction industry rules.

See the figure on pp. 69–71 for a comparative summary of the techniques described below.

Explain the Design

Demonstration

Demonstrate that the design meets the intent of the rules.

- An effective way to do this is to show fulfillment of performance criteria when prescriptive criteria can't be met. Since performance rules inherently accept a much broader range of solutions, your work will mostly require design cleverness, followed by strong arguments and diagrams or other materials capable of explaining or demonstrating the anticipated performance of your solution.
- Keep in mind that the quantities designed often either exceed the minimum quantities required or are well within maximums. Such quantities can refer to widths of egress path, numbers of exits, numbers of toilet fixtures, areas per compartment, or other factors. When designs are changed, elements mandated by rules will not necessarily need to be changed if there is a margin between the amount required and the amount designed.

For renovation work, this is particularly critical, since the quantity provided in the existing structure may very well exceed requirements. For example, don't assume that if occupant load for a new design exceeds the existing occupant load in a building, exit access corridors will need to be widened. Don't assume that all existing stairs are required for egress. It makes sense to do complete code analyses of existing buildings to confirm their minimum and maximum quantities.

Selection

Some projects fall under the purview of several different sets of rules. Generally, the strictest (most exclusive) rule for any particular issue is held to govern, but not always. In such cases, project teams may be able to select the particular set of rules to follow.

Change The Design

A team can always change its design approach to comply with the letter of prescriptive rules. Unfortunately, when there is good reason for a design approach, modification simply for compliance may not serve the project's goals, even if it is expedient.

Incentive

By taking advantage of listed exceptions, a team may achieve design ends while scoring political points. Many, many regulatory requirements are immediately followed by listings of permitted (and often encouraged) exceptions. It may take a bit of cross-referencing and page flipping, but design options may be broadened significantly. In addition, this approach often produces better buildings, since many of the allowed exceptions involve quality issues that governments would like to make mandatory, but can't for numerous legal, political, or financial reasons, such as inclusion of sprinklering systems.

A key form of this method is "incentive zoning," whereby allowances are reduced, and incentives are given that, if used, can increase them. The incentives are things that a community wants but doesn't feel can be mandated. (See the case study on Washington Mutual Tower on p. 154.)

The advantages of incentive zoning were summarized by Judith Getzels and Martin Jaffe[18]:

> *If controlled, a bonus system offers flexibility and an opportunity for a city to achieve desirable public benefits. With a bonus system, the developer can choose from a selection of design features. A comprehensive bonus schedule spelled out in the ordinance offers a high level of predictability. As-of-right bonus systems can usually be administered with a minimum of expenditure of time and with few demands on staff experience. Able developers, well-drafted bonus provisions, and a strong real estate market can make the application of incentive zoning a useful tool for a municipality.*

"Incentives." © Roger K. Lewis.

Concealment

Hiding solutions that conflict with other project goals is particularly appropriate to some of the issues raised by building codes and standards and historic preservation guidelines. For example, how does one meet regulatory requirements for a two-hour rated column when the design intends an exposed steel structure, since steel loses its effective strength when exposed to heat? This approach to optimization raises a question for architectural modernists, who believe strongly in "letting materials express themselves." They wonder if it is ethical to "deceive" users of buildings by concealing functional elements behind cosmetic ones.

Mies van der Rohe isn't reported to have lost any sleep after designing exterior wall structural members that were sprayed with fireproofing and then clad in metal panels. The uninitiated assume that the metal they see is the structural column when it is, in fact, only a thin enclosing skin. Praxiteles went ahead and designed the Parthenon on the Acropolis in Athens in stone but had the workers carve protruding boxes (triglyphs) over each column as if they were actually the ends of wood joists.

Clearly, project teams that face this issue have to resolve it for themselves, one way or the other. It is likely that different members of the project team will take very different positions on this emotional issue. In the end, it may turn on the ethics of maximizing value compared with the ethics of a purist aesthetic philosophy.

Concession

When nothing works, teams can give up on contentious aspects of their designs and follow basic requirements. This method has the advantage of being the most expedient. It doesn't take too long to figure out what must be done when one simply reads the basic requirements and fulfills them, without pursuing exceptions. It may, however, severely limit design options. Make this an option of last resort.

Change the Rules

Substitution

Substitutions allow projects to substitute different criteria in place of the rules that would normally apply. They accomplish the spirit of the rules using alternative design solutions. When the relevant rules are prescriptive, you'll have to try to infer the performance requirements the code's writers were trying to achieve by reading between the lines. Getting substitutions approved may require life safety diagrams, computer simulations, precedent examples, etc. A common form of substitution is the zoning variance.

Mitigation

Through mitigation, a project team offers to design compensatory measures into projects to offset limited areas of noncompliance.

Customization

In some cases, it is appropriate for project teams to develop special sets of rules for specific projects. Once accepted by the appropriate governing bodies, the new sets of rules replace the sets mandated by government, but only for the projects for which they were drafted.

Exchange

This is one of the most controversial optimization techniques used today. Through it, a project team obtains relief by using another project's excess regulatory allotments. For example, if rules allow an existing project more area than it occupies or expects to ever need, that project can transfer its excess rights to a new project. The old project generally receives monetary compensation from the new project for permanently giving up its development rights. (See pp. 89–90 and 154 for concrete examples.)

Reclassification

Arguments can sometimes be made for applying a different set of rules than would conventionally have jurisdiction over a project.

Bypass the Rules

Administrative Relief

When the request is minor and poses no threat to public health, safety, or welfare, regulatory officials can often give on-the-spot approvals. This requires no hearing, but it helps when the design team has done its homework and can argue the justifications convincingly.

Judicial Relief

When all else fails, there is always the option of filing a lawsuit. Judges can override the actions of government officers. As with most litigation, teams opting for this form of relief should be prepared to divest themselves of large amounts of time and money.

One advantage of judicial relief is that it can provide time-sensitive crisis management. Judges can issue temporary injunctions forestalling such irrecoverable actions as the demolition of historic properties.

Amendment

Revising model codes takes too long to be of any help to a project currently being planned. Still, for major issues, getting involved in drafting model codes is more effective than harboring resentment. Model codes are written and revised with the help of practicing engineers and architects. Design professionals with suggestions to make are usually invited to submit their ideas, testify at hearings, or join code-writing committees. Waiting until code revisions are enacted is not an effective way to improve them.

The state, county, and municipal committees that prepare zoning ordinances and local amendments often include citizen activists or allow a period of citizen comment before adoption. In most jurisdictions, concerned citizens have opportunities to be heard and make a difference comparable to those listed above for design professionals.

Fraudulent Permits: "Bait and Swap"

This is ILLEGAL, and therefore not an option. Still, the number of project teams that attempt it is sometimes amazing.

In some cases, they apply for building permits with drawings that show code-compliant designs that they have no intention of executing. After inspectors approve the construction and leave, the owners or their contractors change the construction without getting approvals. Perhaps the space below a stair is shown enclosed to prevent its use as a storeroom, even though the owner intends to cut a door into it after final inspection.

In a more mild form, but one that is no less problematic, spaces are deceptively labeled on permit drawings. In a typical example, a bedroom through which occupants may need to pass to reach a required fire exit is labeled as a family room instead, since codes prohibit the use of a bedroom as any part of an egress path. In other cases, a smaller scope of work is claimed than the work to be contracted.

Legally, these actions constitute fraud—the willing deception of the government, and through it, the public. If regulators find out, such behavior puts project owners at risk of fines, criminal prosecution, and in extreme cases, forfeiture of the noncompliant properties. When they don't find out, worse might happen. Someone might be injured or killed because the building was built to unacceptably low standards, as could happen with any of the cases noted above. What starts out as a way to get something past the inspectors can end with tragedy and irrecoverable loss.

Approach	Method	Explanation	Project team attitude	Process	Issued as...	Conditions for granting	Examples related to: Zoning Regulation	Building Codes	Accessibility Guidelines	Preservation Guidelines
Explain the Design	**Demonstration**	When it isn't what was required, show functional equivalency.	*It does what you wanted.*	**Administrative** (no public hearing required)	Approval	Automatic if consistent with regulations	Computer simulation showing that height won't block views.	Demonstrating adequacy of egress by timed exiting.	Build a mock-up or do computer simulation to show clearances.	Unnecessary Guidelines are not prescriptive.
	Selection	When several sets of standards apply, the project team may be able to play one standard off another.	*I know it's a hotel, but it's also a historic landmark.*	**Quasijudicial** (requires hearing, review board approval)	Approval	Automatic if consistent with regulations	Overlay districts, which amend the rules otherwise applicable in a district.	Juggling between the many competing regulations (NFPA, ADAAG, OSHA, etc.).	Since certified programs are rare, there is not yet consensus on which rules should be enforced.	These rules generally relate to the regulations included for zoning historic overlay districts, so some flexibility is possible.
Change the Design	**Incentive**	Every regulation has listed exceptions that grant more liberal permissions in exchange for amenities desired by the regulators.	*You scratch my back and I'll scratch yours.*	**Administrative** (no public hearing required)	Approval	Automatic when listed incentives are observed	Incentive zoning: Public amenities for FAR increases.	Increased area for sprinklering or increased street frontage.	The threat of being charged with discrimination in court obviates the need for incentive regulations.	Very much tied up in tax law.
	Concealment	When achieved compliance is covered to fulfill design.	*I'll put what you want under what I want.*	**None**	Not needed	Not needed	Put required parking stalls behind screen wall.	Wrap fireproofed structural steel with steel cladding.	Amenities can't be hidden, but can be designed to coordinate with other design elements.	Use operable panels to cover historic features that must be left exposed.
	Concession	Put a stop to imagination and fulfill basic requirements.	*OK, let's just get it over with.*	**None**	Not needed	Not needed	Provide no public amenities, build basic allowance.	Comply with numbers listed in matrices, without pursuing footnoted exceptions.	Use off-the-shelf accessibility solutions.	Guidelines are not prescriptive.

Optimization, Part 1.

Approach	Method	Explanation	Project team attitude	Process	Issued as...	Conditions for granting	Examples related to: Zoning Regulation	Building Codes	Accessibility Guidelines	Preservation Guidelines
Change the Rules	**Substitution**	Used to achieve equivalency when enforcement of existing rules won't have the intended effect.	*No fair! (with a good explanation)*	**Quasijudicial** (requires hearing, review board approval)	Approval	Where the project is so unusual that compliance with stated rules is meaningless or won't achieve intended effect, unfairly burdening the project	"Variances," such as for irregularly shaped sites where stated rules don't work.	Egress design based on timed-exiting calculations under prescriptive codes.	Variances for hardship.	
	Mitigation	When all aspects of compliance are achieved except minor, well-defined ones, dealing with the effects of those aspects may be sufficient.	*What can I do to make it up to you?*	**Quasijudicial** (requires hearing, review board approval)	Approval	If objections are mitigated by compensatory actions	Allowing infrastructurally intensive uses when arrangements can be made to provide needed infrastructure, such as parking during peak hours.	Acceptance of additional lateral bracing in lieu of required collar ties.	Acceptance of alternate locations or arrangements for obstructed participants.	Case by case negotiations.
	Customization	When the project team offers to comply with a set of rules they write, when it achieves the goals of the required regulation.	*Tell you what we'll do...*	**Quasijudicial** (requires hearing, review board approval)	Acceptance of alternate rules	Consistency with intent of comprehensive plan	Planned Unit Developments (PUDS).	Replacement of portions of the mandated code to address environmental sustainability.	Since ADA is not a code, and ADAAG is only a guideline, project teams can develop their own rules under the ADA agenda.	The older the building and the sparser the available documentation, the more difficult it is to apply standard rules.
	Exchange	Allows for projects to buy unused regulatory allotments from adjacent properties that don't plan on using them.	*Who can we get it from?*	**Quasijudicial** (requires hearing, review board approval)	Permanent acquisition of rights	Proximity of conveying property to project; release of conveyor's rights must be permanent	Transferable Development Rights (TDRS): Use another property's unused FAR allotment.	Not allowed, since safety in one project can't be downgraded for the sake of another.	Providing the equal of required facilities in a nearby project.	Not every property can be saved, so compromises are sometimes reached to save one building at the loss of another.
	Reclassification	Permanent classification of a project under a different part of the regulations than would normally be enforced when the new classification is more appropriate.	*We're not going to go behind your back—but you need to change the rules for us.*	**Legislative** or **quasijudicial**, depending on state law	Approval	Consistency with program or use	Rezoning.	Determining the construction classification of an existing building.	N/A: Accessibility rules do not have class alternates.	N/A: Preservation rules do not have class alternates.

Optimization, Part 2.

							Examples related to:			
Approach	**Method**	**Explanation**	**Project team attitude**	**Process**	**Issued as...**	**Conditions for granting**	**Zoning Regulation**	**Building Codes**	**Accessibility Guidelines**	**Preservation Guidelines**
Bypass the Rules	**Adminis-trative Relief**	Petitioning a zoning or building plans reviewer or commissioner for relief	*Just approve it!*	**Adminis-trative** (no public hearing required)	Waivers by administrative order	Minimal impact and/or consistency with intent of regulation	Parking waivers.	Acceptance of preexisting conditions.	Waivers for hardship.	
	Judicial Relief	Petitioning a judge for relief.	*Tell them to just approve it!*	**Judicial** (relief through lawsuit)	Injunctions, declaratory judgments	All other remedies exhausted	Used when environmental damage may result or when property rights are in jeopardy.		Common, since the ADA is civil rights law.	Common when destruction or damage is imminent.
	Amend-ment	When project teams suggest amendments to the regulations that are subsequently adopted.	*We can improve the system for everyone*.	**Legislative** (requires hearing and town, city, or county council approval)	New code provisions	Perceived improvement over existing regulations	New ordinance provisions introduced through public hearings	Revisions of the local amendments to the model codes through hearings; revi-sions of the model codes through their committees.	Revisions of the rules through public hearings.	

Optimization, Part 3.

Planning Rules

Appeal to the Local Zoning Officers or Board

Of the appeal strategies explored in the matrix, three are of particular note:

Reclassification: Changing the Site's Zoning District

This is a bold move. To get approval, one of two strategies can be used:

- The proposed reclassification can be shown as equal to or better than the zoning ordinance's way of fulfilling the Comprehensive Plan.
- The Comprehensive Plan can be shown as flawed. This can be done by arguing that the particular aspects causing problems for the project team are one of the following:
  1. Obsolete due to changes in the community.
  2. Based on incorrect assumptions.
  3. So poorly drafted as to reasonably be considered unenforceable.

  Of the three, the first is most likely to yield results.

Reclassification is particularly difficult in established communities, since the planning issues of those communities are at the heart of any Comprehensive Plan. Major changes are more likely in immature or proposed developments.

Land use plans tend to be developed along patterns familiar to our society. Traditionally, business districts have tended to develop first, and therefore find themselves at the centers of communities. Retail uses tend to be established along major transportation routes, surrounded by residential areas. The edges of developments have tended to house loud or noxious uses, uses in need of ready access to inter-city transportation, and uses that don't need the amenities provided in areas with higher property values. This tradition has a certain logic to it, making it difficult to fight.

When the fight is unavoidable, it helps if the project site is adjacent to a district with the desired zoning or when it is similar to the one currently zoned. In addition, complying with the Comprehensive Plan generally makes sense from a marketing perspective, but not always. Some "outpost" projects, like apartments made from old lofts in industrial districts, have been very successful.

Variance: Liberalizing Just One Aspect of the Current Zoning District Rules

For example, rather than try to rezone a property to a different zoning designation (such as rezoning a low-density commercial-manufacturing CM-1 into a higher-density commercial C-4), comply with all the requirements of the current zone except the problematic one. For example, if the problem is designing within lot coverage requirements (see p. 134), focus on getting approval to change just that, perhaps arguing that an increase from 60 percent to 65 percent lot coverage is reasonable based on adjacent sites zoned for 70 percent. Variances are not generally allowed for changes as fundamental as permitted use.

This type of effort is generally granted through administrative waivers. Zoning ordinances explain the conditions under which variances are approved and the process necessary for obtaining one, which often involves meeting the chief zoning officer rather than holding a public hearing.

Customization: PUDs

Most jurisdictions allow investment teams on larger projects to propose their own rules. If accepted, these rules supersede the designations established by the government. Through this process, a site is redistricted as a Planned Unit Development or "PUD" and its special rules are recorded for later projects at the same site in future editions of the zoning ordinance.

The special rules developed by a project team must respect precedent established by existing properties and by legal history. The rules must be presented to and accepted by the zoning board before being implemented.

Minimum property sizes for PUD developments, established by ordinances, range from a half acre to several hundred acres depending on the total development potential for the site. Sites currently sized and zoned for high-volume development more easily justify special rules than small properties or those zoned for low density.

Convince the Client to Acquire a Different Property

Giving up on a property is often a simple question of value. At some point, the benefits gained cease to outweigh the invested time and effort. Of course, this is much easier to do before a property is purchased. Developers generally put properties "under option" before purchasing them. This involves paying the seller a fee to agree to temporarily take the property off the market. That gives the developer sufficient time for a proper analysis before committing to buy.

Building Rules

Building departments may or may not have appeals boards, depending on their size. Even when they do, the complexity of an appeal may be difficult to justify. If an appeal is not an option, for either reason, recourse is available through administrative relief provided by building commissioners and fire marshals. A project team can appeal to them or the plans reviewers, and try to convince them of either the equivalency of the design approach, or the notion that the hardship involved in satisfying the code provisions is disproportionate to the consequent benefit.

- **Demonstrated Equivalency**

  Your design will be as safe as it would be if it followed the letter of the code.
- **Hardship**

  That following the letter of the code would result in the project's being canceled, and that a slightly less than acceptable situation is better than the current totally unacceptable situation.

When there is a board of appeals, it establishes and distributes the procedures it uses. An IBC appendix addressing boards of appeal contains provisions for membership of the board but minimal operational issues. It does, however, require hearings to be open to the public and sets a two-thirds vote of a board as the minimum necessary to modify or reverse a code official's decision. With boards of appeal comes the full range of optimization options noted in the matrix [IBC Appendix 1-3].

Accessibility Rules

Fulfilling accessibility requirements can get complicated for several reasons.

- As noted above, ADA compliance related to construction is enforced by the Department of Justice, not local regulatory officials.
- There is no nationally accepted code or substantial number of certified state codes to support the ADA. Still, local governments will most likely require compliance with their adopted code.
- Accessibility is a moving target. It involves designing so that people who lack the abilities required to

operate traditionally designed buildings are nonetheless able to operate them (open doors, etc.). The question becomes, in part, do we plan for the design parameters of the person, or of their equipment (wheelchair, cane, etc.)? Do we then redesign when new, and hopefully better, equipment becomes available?

With accessibility rules, it is more difficult than with zoning or building rules to know what constitutes fulfillment, and when a variance will be needed.

PRESERVATION RULES

To be able to optimize historic preservation rules, one must understand their politics, including the agendas of historic preservation agencies and active citizen groups, and work within them. This brings us to the politics of regulation, addressed in the very next section.

GETTING LOCAL

Confronting regional and personal idiosyncrasies.

REGIONALISM

Differences in content and interpretation of rules from one place to another yield different built environments. When rules change, some businesses and residents are inevitably forced to relocate in response. Because of this, rules adopted in one area cast a shadow, affecting adjacent areas as well. When one jurisdiction tightens rules, the others in the region get the overflow unless they similarly tighten their rules. Crafting good rules requires a sensitivity to their secondary and tertiary implications. Jurisdictions sometimes attempt to customize their own environments while ignoring larger regional realities.

Regionalism in Planning Rules

Regarding planning, regionalism on small (intra-county) and moderate (interstate) scales is manifest in efforts intended to prohibit certain community realities. These efforts are regional in nature, since the realities they attempt to prohibit do not disappear, they simply move to regions that are either willing to accept them or haven't yet moved to prohibit them. Developers and architects can't fight it because they are too focused on the needs of their current projects. The regional perspective must be maintained by those who draft statutes and regulations.

A common target of regionalism in planning is diversity. Some rules seek to limit economic diversity by prohibiting affordable housing; others attempt to prohibit certain behaviors by limiting uses. The official line against affordable housing is that its presence catalyzes general reductions in property value. The claim against problem behaviors (as might be found among the residents of drug rehabilitation and psychiatric care facilities) is the danger they pose to the rest of the community.

Sometimes these results are accomplished in indirect ways: developing affordable housing becomes much more difficult if the minimum allowed lot size is raised, or if permitted use laws prohibit unrelated people from sharing housing. There are no federal prohibitions of either practice, as noted above regarding civil rights rules [42 USC 3607(b)(1)]. By barring unrelated adults from sharing a residence, one recent statute prevented single-parent co-workers and their children from pooling resources in that way to improve their economic situations.

Ultimately, these strategies have left some communities complaining of shortages in such nominally middle class community staples as teachers and nurses, and the other lower wage workers vital to a functioning community.

While the NIMBY (Not In My Back Yard) attitude is supposed to protect community values, it frequently simply repositions them. The NOPE (Not On Planet Earth) mantra takes it even further. Such federal laws as the Fair Housing Act effectively prevent some regionalism, but not all. Communities must recognize that societal realities must eventually be faced; needed projects either must become unneeded or must be put somewhere. As apparent distances shrink and as more solutions require multiple participants, regions will have to start working together.

Regionalism in Building Rules

Regarding design approach, rules vary as the elements that constitute successful design vary. Three regional model codes developed in the United States in response to the different regulatory conditions posed by hurricanes, earthquakes, and industrialization. Individual jurisdictions have generally used local amendments to take regionalism one step further. For

CASE STUDY

The politics of regional planning are complex. In a recent case, owners wanting to expand an assisted-living facility for the elderly beyond its existing eight residents took the zoning board of Howard County, Maryland, to court after it denied their application. They claimed that the board had violated the Fair Housing Act in doing so.[19]

Zoning authorities cited traffic congestion and inadequate parking as the reasons for denying the request. Both opponents and supporters of the expansion used arguments based on community need to justify their positions. A county councilman opposed to the expansion said, "I am not looking to do away with homes [for the elderly]... but I don't want to change the character of the neigborhoods." The president of the Neighborhood Association had no objection to the existing facility but felt that expansions should be limited. In opposition, the director of the Maryland Office on Aging said she hoped the ruling would not become a stumbling block in providing the additional facilities needed to allow increasing numbers of elderly residents to remain in their home communities. The owners of the assisted-living facility felt the decision violated the vision of the community's original planners, who sought to create a place where people of all races and ages could live together.

Ultimately, the court decided based not on the relative strengths of the arguments being proffered, but on the separation of powers between judicial and executive branches of government. It upheld the zoning board's decision, finding that the expansion of existing facilities did not justify violating zoning ordinances when other similar facilities were available. It said, "Congress clearly did not contemplate abandoning the deference that courts have traditionally shown to such local zoning codes." Here, at least, legal logic transcended appeals based on emotion and affiliation.

example, Boston requires permit applicants to submit material samples so that reviewers have the option of performing their own quality testing. With the merging of BOCA, UBC, and SBC, we are more likely to see the rise of strong local supplements.

POLITICS

The characteristics that make for a successful design approach vary from place to place as regulators and their constituents vary. There are many local governments, each with its own needs, attitudes, rules, and methods of enforcement. Somehow, politics seems to always play some role.

Politics in Planning Rules

Developers know that concessions bring preferential treatment. For example, when they agree to build playgrounds in public parks, it is likely that requested relief from density requirements will be granted. It isn't noted in any rules but it happens, and it's generally too subtle to fight. Land use lawyers know about it, keeping quiet when it helps them and complaining when they must fight it. Few cases ever get out of the back room to be exposed in the media. Its cost is generally viewed as simply the cost of doing business.

Before condemning developers, it must be noted that code officials and the public have agendas in such politics. The cost of governmental services is a function of capital improvement costs and operating costs. The public perceives the private sector, rightly or wrongly, as wealthy, and the public holds the power of approval. Without public acceptance, the private sector's business plans are frozen. When developers can be pressured into making capital improvements, tax dollars go further toward operating costs. Back-room negotiations are one way that many communities get private developers to pay for improvements the public wants but won't pay for. Rising tax rates have never been popular, and this brand of politics might be seen as the cost of the "free" lunch. And, of course, politicians appreciate the accolades they get from their constituents for bringing home the amenities.

One of the drawbacks of closed-door politics is that the quality of the results depends completely on the talents of the negotiators. For this reason, and because of the lack of accountability inherent in the process, many communities have moved toward incentive zoning. (See pp. 153–154.) It also is not universally loved, but does provide an alternative.

Politics in Building Rules

The particulars addressed by building codes are also subject to politics, if in different ways. While many municipal and county building departments provide a balanced review process, criticisms are not infrequently lobbed at them by builders and design professionals. In researching this book, the author encountered many such comments, a few of which are listed below. Some may be reality, some may simply be perceived as reality—a reaction against authority. Justified or not, they indicate a need for greater attention to ethics and for better and more open communication among all parties. Such criticism includes:

- There is no appeal outside the building department, and governments generally cannot be sued. Plan reviewers play the dual roles of prosecutor and judge, eliminating the possibility of a neutral arbiter.
- Code officials claim to have no liability (and generally don't) and are therefore sometimes not sufficiently responsible for the design changes they require.
- With the move toward outsourced review processes, permit applicants sometimes pay double fees—one time to the permit office, and a second time to the deputized reviewer or testing agency.
- Imposed rules can differ from one reviewer to another on the same project. Applicants facing additional demands sometimes simply leave, coming back when a different reviewer is available.
- Reviewers have been known to moonlight, accepting side jobs from those they are sworn to regulate—a clear conflict of interest.
- Reviewers are not immune to the "buddy system," granting permits more easily to their friends (social or business) or those related to their friends.
- Outright bribery of officials, while driven underground as a result of stricter laws and more effective enforcement, still occasionally happens. When an official is disciplined, temporary reductions in staffing levels can be problematic.
- Code officials are more often engineers and former contractors than architects. Some architects feel that this leads to reviews more focused on the functioning of individual components than on systems or the project overall.
- Politics is also involved in the drafting of local amendments. In one example, Chicago's own building code allowed lead pipe to be used for domestic water supply as recently as the mid-1960s. The plumbers unions fought to keep that provision, claiming that lead pipe was of better quality. Their opponents said that the plumbers simply saw lead pipe as an expensive product that could justify higher bids. This was years after the health risks associated with lead poisoning were well documented.

"Naturally, I'm against it." © Roger K. Lewis.

Without statistical information about the frequency of these types of perceived or actual problems, no real conclusions can be drawn. Still, to the extent that these criticisms have any validity, several actions are suggested:

- Applicants could attend code seminars sponsored by local governments. At such seminars, code officials explain their interpretations and review recent oral decisions that are not yet recorded in published amendments.
- Design professionals could bring permit applications in person, rather than using "expediter services" or leaving permit procurement to owners or builders. As

with many other things in life, getting to know reviewers personally generally benefits the process. Among ethical people, contact fosters trust, and trust improves business. Although design professionals who visit or call the permit desk to discuss every design move get a bad reputation, there is no need to wait until applying for a permit to get feedback.

- Don't do things that are guaranteed to make code officials angry. Project team members who flagrantly violate rules because they don't think the rules are right simply manage to shoot themselves in the feet. Code officials don't get hurt by such actions, projects do. Let common sense prevail over frustration.
- Seek to understand and apply the intent of the rules, not just the letter. Be able to explain your logic. Use a reasonable standard of care.

Politics in Historic Preservation Rules

There are at least three kinds of bureaucracies that jurisdictions can establish to handle historic preservation:

- Informal bureaucracies: informal offices of preservation that administer federal preservation policy without set local rules
- Formal bureaucracies: formal offices of preservation that administer federal preservation policy without set local rules
- Codified bureaucracies: formal offices of preservation that administer federal and formal local rules

Most places fall under the first or last category. The middle one is an unusual situation.

An overriding consideration is the difference between federal preservation rules and local ones. Federal rules have the luxury of focusing almost exclusively on preservation issues. Local preservation rules are caught trying to balance preservation issues with local political maneuvering that is frequently intense and generally contentious. This also argues in favor of the informal bureaucracy, which isn't charged with enforcing local preservation policies.

The informal bureaucracy consists of an individual or a small staff that considers each proposal on a case-by-case basis. Compliance in these cases becomes not so much an issue of fulfilling requirements or getting variances as an issue of working with officials periodically as designs are developed. Informal bureaucracies tend to involve a low level of politics. For project teams that understand the spirit of preservation, this is the preferred system.

Codified bureaucracies are established in response to the adoption of formal preservation rules. Since rules derive their power from enforcement, agencies are needed to provide such enforcement. This brings a degree of formality but also an escalation in political maneuvering.

At least in theory, formal rules establish a level playing field for all participants. Codified bureaucracies, even when inefficient or politically driven, are likely to achieve some degree of success in catching the worst proposals. In theory, informal bureaucracies might be capricious and inconsistent without a fixed set of rules; in practice, they are often more flexible, and ultimately may be more fair, in their handling of proposals.

This situation may simply be a function of the scales at which each is likely to occur. In the smaller jurisdictions that are more likely to use informal bureaucracies, the most common projects are those best handled by the customized attention such systems can provide. In the larger jurisdictions whose larger constituencies are more likely to demand fixed rules, codified bureaucracies are the type more likely to avoid conflicts of interest or charges of preferential treatment.

CASE STUDY

Washington, D.C., is a city with heavy concentrations of both historically significant buildings and politics. The complexity of its political structure for preservation projects is a story in itself. [20]

All projects involving historic preservation must be approved by the D.C. HPRB (Historic Preservation Review Board). The members of the HPRB are political appointees whose job is to vote on proposals. They are supported by a staff, headed by a director. The director is an administrative appointee, not an elected official, who nonetheless occupies a very political position. The process works as follows:

The director assigns a preservationist to each project. The preservationist's job is to prepare a report evaluating the project and recommending a position. The report is forwarded to the HPRB for action. Generally, the HPRB accepts the preservationist's recommendations.

Often the preservationist will recommend more study. This is often a stall tactic intended to give the staff more time to cultivate constituent support. The staff prefers to have a consensus among the neighbors before going before the HPRB, as it streamlines the process and avoids conflict. Unfortunately, the delays it causes move control of the project from its developers to governmental administrators and can substantially affect project scheduling. Since maintaining a schedule is critical to financial and market feasibility, the process can substantively affect the proposed project.

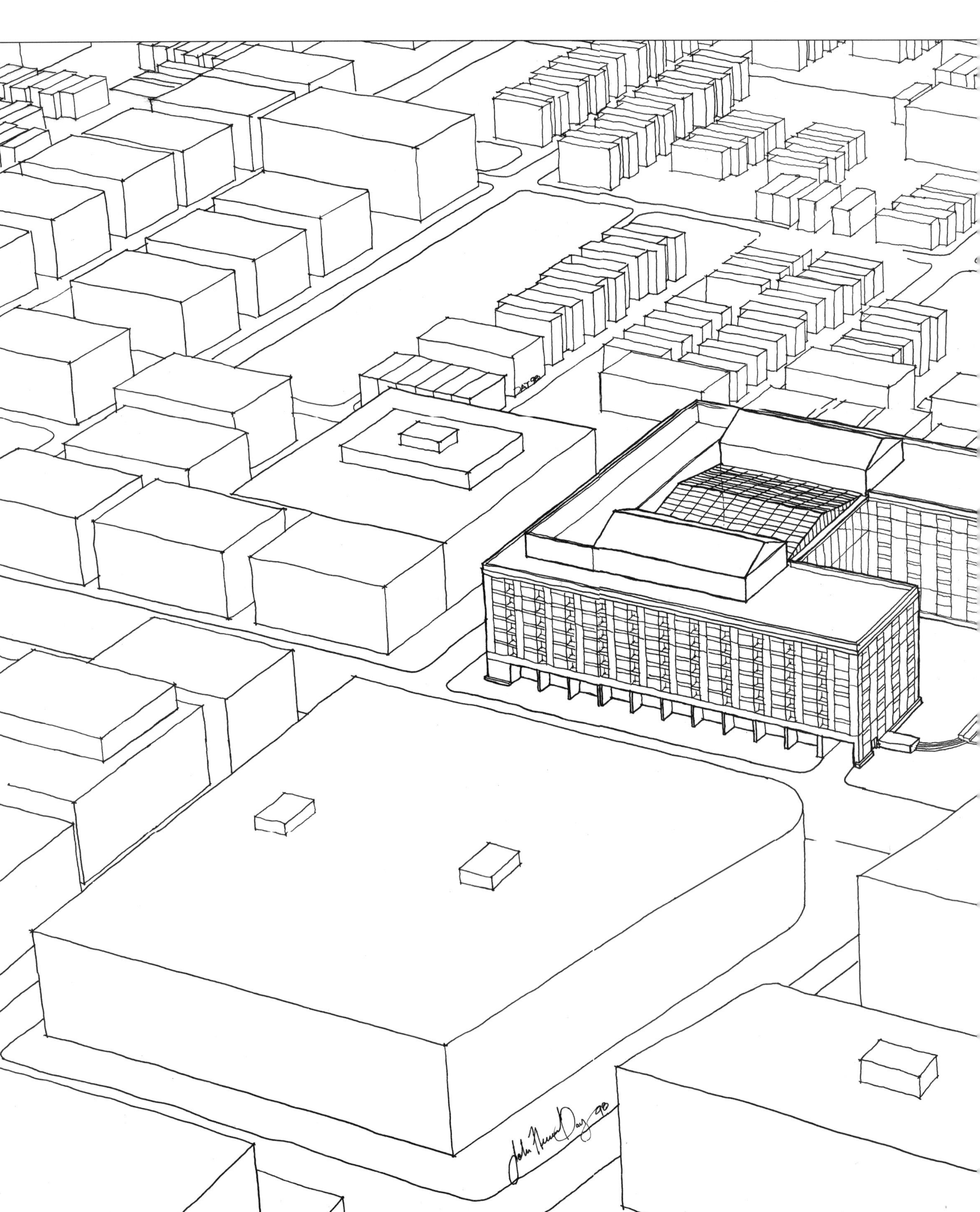

PART 

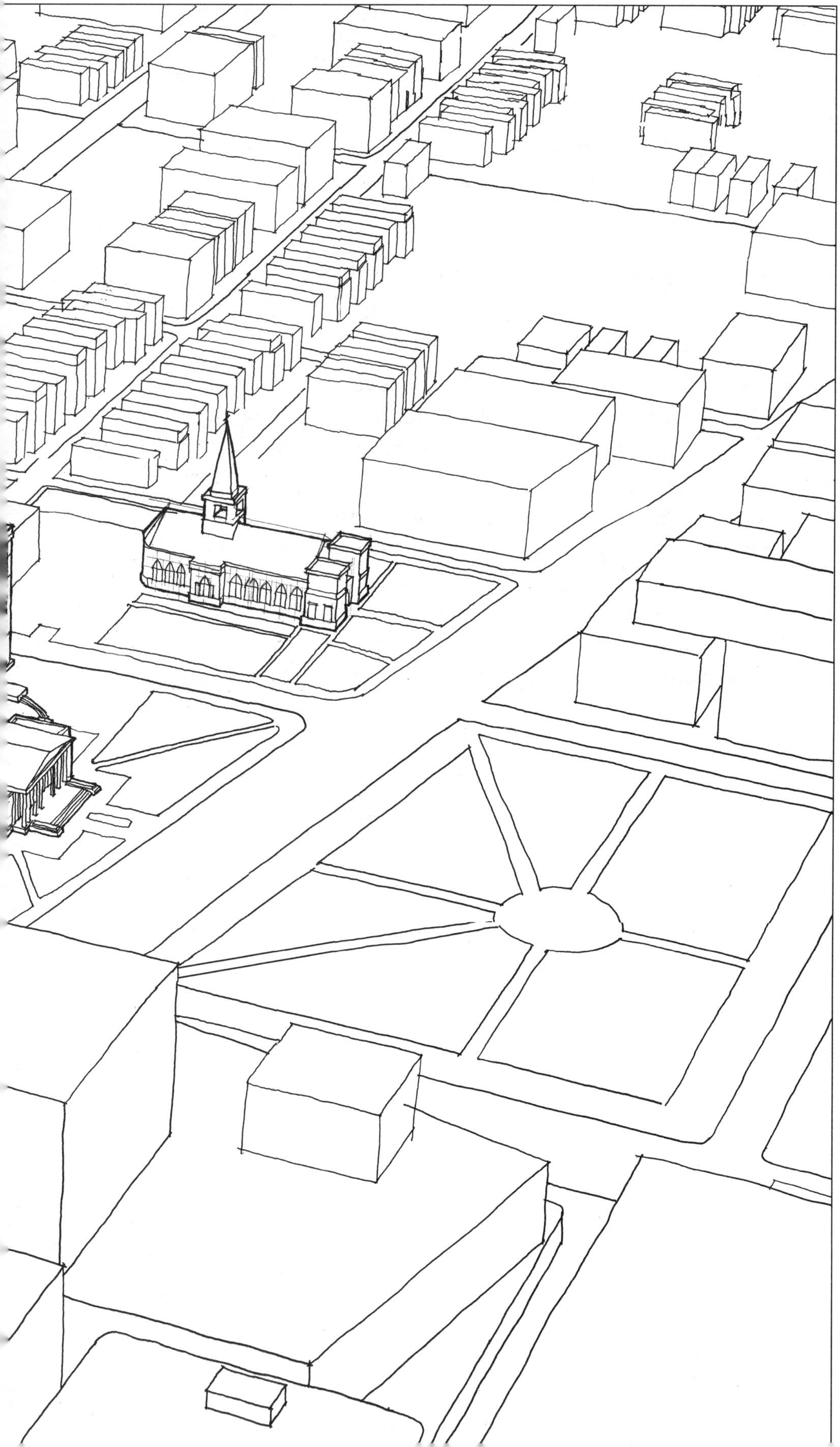

SITE SELECTION AND DESIGN

With Part II, this book begins to work through the design process, issue by issue, starting with site selection and design. This part addresses regulatory aspects of the following design questions:

- Are the proposed sites eligible for development? If not, what kinds of sites should be?
- Would proposed programs be acceptable for their proposed sites?
- How should buildings be positioned on their sites to fulfill marketing needs, program, and aesthetic intent without compromising community identity, inhibiting fire department access, or having adverse environmental consequences?
- How should major site amenities be developed, including roads and parking for cars and service vehicles?

As this book is focused on the design of buildings, issues related primarily to nonbuilding aspects of site design are beyond its scope, even when addressed by rules. This book will therefore not explore regulatory issues related to:

- Signage design
- Parking design not directly related to building design
- Landscape design

3 USE

Determining Fit between Location and Program

INTENDED SITE

Context

Why This Issue Is Important

The rights to private ownership and use of property are fundamental under British- and French-derived legal systems (in particular contrast to Native American systems). It takes a lot to alter those rights. While governments may restrict the way in which properties are used (through "permitted use" provisions), they have rarely questioned the basic right to private use. Conversely, the rights of individual property owners occasionally conflict with the needs of a community. This conflict seems to have surfaced with more frequency in the last few decades and has been contentious from the beginning. Rules attempt to address the rights of various parties to control land use.

There is no point in trying to develop a site that is off-limits to use. Ownership does not guarantee the right to develop a property. One of the first steps in determining project feasibility before initiating a project, and preferably before acquiring a property, is to confirm the basic right to use the site. Property owners should assess the balance between their interests and those of the community in which their proposed site is located.

Where to Find the Requirements

There are usually no prerequisites to a landowner's fundamental right to use owned land. In general, codes and ordinances presume this right. Still, since it could come into question if it was expected to negatively affect a surrounding community, prudent owners put some effort into researching past long-term planning and land-use studies.

A project team might start with a copy of the comprehensive plan for the community in which the proposed site is located. These studies will help identify use problems arising from:

- Conflicts with public needs
- Conflicts with regional conditions such as earthquakes or snow
- Conflicts with site conditions such as difficult topography

PUBLIC NEEDS AFFECTING FEASIBILITY

Some privately held properties find themselves in actual or potential conflict with community values or plans. When this happens, project teams must consider the implications of the development they are considering. Communities may have the right to overrule a landowner's plans when necessary. One hopes that such actions are not taken lightly. When necessary, however, the rights of existing property owners to use land may be reduced or eliminated by governments on behalf of communities either through regulatory takings or by right of eminent domain.

Regulatory Takings

Property owners do not have an unqualified right to harm the communities in which their properties are located, either materially or financially, directly or indirectly. They also do not have the right to violate laws

"My Property." © Roger K. Lewis.

related to construction. Beyond this, property owners generally have the right to use their properties as they see fit and reap the financial rewards and losses of ownership.

By the same token, governments do not have the right to unreasonably limit owners' rights to use their properties. These rules are rooted in the Fifth Amendment to the Constitution, which says, in part, "nor shall private property be taken for public use without just compensation." For many years, this clause was applied only when actual ownership was transferred, when a property was physically taken from its owner (see pp. 84–85). "Regulatory taking," whereby owners lose any ability to benefit from property even while retaining title, was first recognized in a 1992 Supreme Court case, *Pennsylvania Coal Company v. McMahon.*

In a regulatory taking, the government is required, per the Fourteenth Amendment, to compensate owners for mandated loss of use. This legal change reflects an evolution in the thinking of our society. There was a time when we believed that regulations were necessary to maintaining the balance between individual and societal rights. We considered them simply part of the cost of doing business in a large and complex society. Now, apparently, we're no longer willing to accept that idea. Rulings handed down in several significant court rulings in the 1980s show a marked change in orientation.

On the other hand, we are very accepting of uncompensated imposed rules when their effects accrue to the benefit of landowners. This happens, for example, with land-use regulations, since they are intended to preserve or raise property values. One must assume that in many cases they achieve their intentions. If rules truly triggered compensation, landowners would pay the government when their properties increase in value.

Environmental Impact

On-site activities sometimes have adverse off-site effects that can usually be eliminated or held to acceptable levels. Material damage can result if the landowner's actions have an effect beyond the property line. This can be the case when an owner intends to use property for a strip mine or landfill. When off-site effects can't be eliminated or held to acceptable levels, sites may be ruled partially or completely off-limits to development.

Financial damage occurs when owners' activities expose the community to liability. For example, owners that destroy wetlands may force communities to spend more on pollution mitigation or flood control. The liability can also follow on the heels of permissions. Once developers won permission to build high-rise apartment buildings right at the edge of Lake Michigan, residents tried to force the city government to build breakwaters because storm waters were battering lower-level apartments.

Liability can accrue even when landowners' actions are potentially self-damaging, as when they build housing in coastal areas battered by hurricanes, in the floodplains along the Mississippi River, or on mud-slide-prone hillsides in California. Such projects incur double losses, since governments pay to help out victims of repeated disasters and then again if they opt to exercise their right of eminent domain to condemn such properties at taxpayer expense (see the case study below).

Sometimes owner activities that would be perfectly acceptable elsewhere will not be for the property in question. For example, construction of a house on a natural dune might cause permanent damage to a shoreline and, given sufficient political pressure, expose the community to the financial obligation and long-term damage caused by politically expedient short-term solutions: reinforcing the shoreline or building a breakwater. It is important to consider the proposed activity in light of the conditions at, needs of, and impact on the specific property and surrounding area.

Changing the Rules

In another type of taking, a community that imposes new zoning limits on existing properties may find itself having to compensate owners for their loss of value. This has happened in several waterfront cities where new height limits were imposed on properties occupied by low-rise structures to forestall their replacement with view-blocking high-rises. The owners of such properties and the courts have, in some cases, interpreted such actions as takings, and consequently required compensation for their resulting relative loss of property value.

This also occasionally happens when review boards decide that any development would be unacceptable. In one particularly egregious example, a commission stalled approval of a proposed residential project to give time for a neighborhood design review board to be established. Once established, the DRB refused to approve the proposal despite independent reviews praising its sensitive design, low-scale massing, and historicism. The owner argued that the decision constituted a taking, and sued for compensation or reversal.

Strategies

Issues that might trigger a taking based on environmental impact can generally be known ahead of time. Environmental impact studies can flag concerns that could jeopardize landowners' rights to their sites. Much depends on whether the study is done before or after the landowner acquires the site. When new rules are being considered, the best strategies are either getting involved in hearings before they are changed or selling to a new owner whose intended use won't be adversely affected by the new rules.

Condemnations

Governments reserve the right to fix or dispose of dangerous buildings, with or without owners' agreements. For example, the County Code of Arlington County, Virginia, contains the following:

3-14. Unsafe Buildings.

(1) Right of Condemnation. *Buildings and their equipment that fail to comply with the building code*

CASE STUDY

A developer in South Carolina claimed that newly adopted laws, intended to protect a fragile coastal environment, rendered his property investment worthless. He sued the government to recover his costs, claiming that the new laws constituted a taking. The court agreed that the land had been rendered incapable of development, and therefore worthless, by the new rules. The state then bought the property from the developer. After a period of time, it sold the property to another developer, who was planning to develop it. This case brings up several questions:

- If the new legislation was needed to protect the coastal environment, why did the government resell the property to another developer, who would similarly be barred from the development he was planning and might also sue to recover his purchase costs? They knew that he wasn't planning on leaving it as a wilderness area.
- If the new legislation was not needed to preserve the coastal environment, was it due to insufficient study before its original passage? Was it later bypassed due to insufficient study? Was it felt that the environment was not in danger, or simply that its destruction was not of public concern?

This case brings into question many issues related to control of land use, the purpose of legislation, the effectiveness of our legislative processes, the uses of taxpayer funds, governmental accountability, the relative levels of involvement maintained by direct investors compared with community members, and the quality of planning rules. As with most legal cases, this one did not resolve any of those questions beyond the peculiarities of that specific case [*Lucas v. South Carolina Coastal Council,* U.S. Supreme Court case 505 U.S. 1003, in 1992].[21]

in effect at the time of [their] construction. . . which through deterioration, improper maintenance, or for other reasons have become unsafe, unsanitary, or. . . are otherwise dangerous to human life or the public welfare. . . hall be made safe through compliance with the building code in effect when constructed or shall be. . . removed, as the building official shall deem necessary.

(4) . . . . Whenever the building official believes dangerous, unsafe, or unsanitary conditions menace the health and safety of the building's occupants or the public, . . . the official shall. . . order the party responsible to abate, raze, or remove the nuisance, and shall initiate a legal action to compel the responsible party to do so.

It continues, with teeth:

(4)(a)(1) The building official may remove, repair . . . when the owners . . . after reasonable notices . . . and time. . . have failed to remove, repair . . .

(4)(a)(2) [When] the county removes, repairs . . . the cost or expenses thereof shall be chargeable to . . . the owners . . .

And just in case the owners are still uncooperative:

(4)(a)(3) Every charge . . . which remains unpaid shall constitute a lien against such property.

Occasionally, new construction is found to be "dangerous, unsafe, or unsanitary" when a passing inspector happens to notice work that does not meet code provisions. On checking, it is discovered that the work was done without a permit or under a fraudulent (misrepresented) permit. Such violations of permit rules are considered criminal acts.

Takings by Eminent Domain

Background

Governments occasionally decide that they need all or part of a particular piece of land for a project intended to benefit the community as a whole. Sometimes they need to use the land, such as for a new highway or a street widening. Other times they need to reduce liability, such as when they acquire properties in the flight path of an airport expansion. The right of the government to buy such land, even when the owner doesn't want to sell, is called the right of eminent domain. Its exercise must be enabled by citizens, who directly or indirectly authorize their governments to purchase land either through federal and state constitutions or through voter referenda.

Some properties with inherent conflicts will be neither available nor on the market because the communities in which they are located will have already acquired them.

Specific requirements are found, on the federal level, in 40 USC 258.

Optimization

Requirements related to regulatory takings and eminent domain issues arise from local governments' future plans, plans of which the governments are prob-

CASE STUDY

In one particularly extreme case of condemnation following the filing of a fraudulent permit, the owner of a multimillion dollar town house near downtown Chicago got a permit for the repair of an exterior stair in back of his building. He then had his contractor violate the scope of the permit by fully enclosing the stair and rear porches. To make matters worse, he violated the provisions of the code through use of small, fixed windows that blocked the rear half of the house from required access to natural light and ventilation. He also had the contractor add a story to the building containing a rooftop party room in violation of allowable-height and number-of-exit provisions. When the city inspectors saw what had been built, they ordered him to modify it to comply with their codes. When the owner failed to demolish the noncompliant work, the city started proceedings to take the property. Objecting all the way, the owner contacted the American Civil Liberties Union and asked it to sue the city of Chicago on his behalf, claiming that the city's decision was a violation of his civil and Fourteenth Amendment rights. Wisely, the ACLU declined the case.

ably not yet aware. These requirements are not published until it is almost too late for compensatory planning.

The most important strategy for site rights issues is early analysis. The sooner potential problems are identified, the sooner they can be addressed.

Regional Conditions Affecting Feasibility

While "location is everything" in real estate, it can be examined at several scales. For a project team, picking location might involve deciding between the northeast and northwest corners of a single intersection. It might also involve deciding between the mountains of the Pacific Northwest and Florida's Gulf coast.

While it is possible to keep people away from a given piece of land, it isn't possible to place gross restrictions on development in an entire region. Therefore, no properties are restricted or declared off-limits to development by government in response to regional conditions. Further, no use restrictions are imposed in response. That isn't the end of the story, though.

Rules may require increased construction as a result of regional conditions, raising project costs and indirectly denying or reducing feasibility. The following regional factors influence construction requirements, and thereby choice of project region.

For the purposes of this section, "regional conditions" are any environmental factors that are beyond the control of the landowner. By this definition, a region could include two properties or two states. One might think of these extremes as immediate region and extended region.

Earthquake Zones

Background

Every piece of land, no matter where, is classified for earthquake activity. In this way, all model building codes address seismic design, not just the West Coast–born UBC. Besides, even though California is best known for its earthquakes, two other concentrations are located near the South Carolina coast (an area also prone to hurricanes) and along the Mississippi River north of Memphis.

Applicability

All projects are classified by Seismic Design Category. The least restrictive is Category A; the most restrictive is Category F. The category applicable is a function of location, use, and design. IBC 1613.3.1 initiates the determination of Seismic Design Category, a very complex undertaking involving several maps, formulas, and matrices. The basic elements are:

- Maximum ground motion in the region as recorded on maps found in building codes.
- Seismic Use Group (Importance Factor): Classifies the project by its importance for human health and safety. Buildings whose failure would pose little danger are designated as Category I, while critical buildings like hospitals and power stations are designated as Category III [IBC 1613.1.4].
- Site Class, ranked A through F: based on stability of dominant soils. Type A is solid rock; Types B and C are stable soil; Types D, E, and F are risky and require additional testing before design parameters are set [IBC 1613.2.1.1].

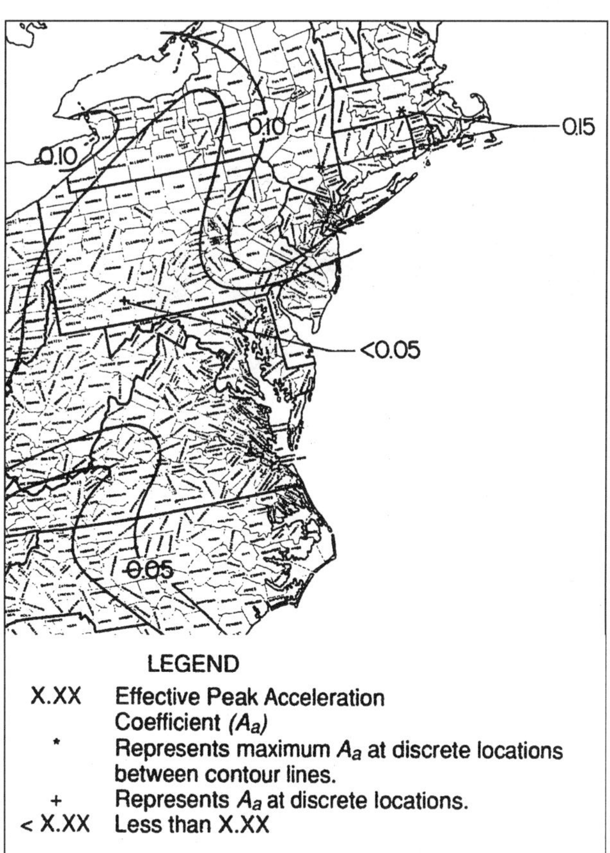

Detail from a ground motion map identifying areas of high risk for seismic activity. © BOCA.

Requirements

Codes impose special design restrictions on projects in regions subject to high ground motion. See Chapter 5 for information on massing for the impact of seismic concerns on building height and shape. Restrictions affect the following issues:

- Architectural and infrastructural detailing [IBC 1613.5]
- Structural form, including building height, materials, and shape [IBC 1613.4]; rules specifically address:
  1. Structural materials and assemblies
  2. Building shape
  3. Analyses that must be done
  4. Required safety factors
  5. Detailing issues
- Miscellaneous construction (not buildings)
- Seismically isolated structures

Snow Belts

Background

Rules for snow are used to establish the loads for which structural systems will need to be designed. Although structural work is relatively inexpensive compared with other building systems, significant increases in design loading can affect project feasibility.

Applicability

Qualifying regions are indicated in Figure 1608.2 and Table 1608.2 in the IBC. In mountainous areas, design snow loads are listed multiple times for different elevations. In the lower forty-eight states, design loads start at zero and go as high as 100 pounds per square foot in central Maine, with other high values through much of New England and the northernmost tier of states in the Midwest and Plains. Alaska has design values ranging from 25 to 300 pounds per square foot.

Requirements

Loads listed are indirectly related to building design. The amount that falls from the sky is deposited unequally over a roof depending on the roof's shape. So while snow load originates with regional factors, it is more a function of building design [IBC 1608] (see pp. 146–150).

Wind Conditions

Background

Some regions have consistently high wind speeds. Despite its reputation, Chicago is not among them. The areas of highest wind conditions in the forty-eight contiguous states are along the southeast coast from Texas to North Carolina (the "hurricane coast"). These

Where snowfalls are unusually heavy for the natural snow-shedding abilities of regional architectural styles, loads can reach dangerous levels. Photo by the author.

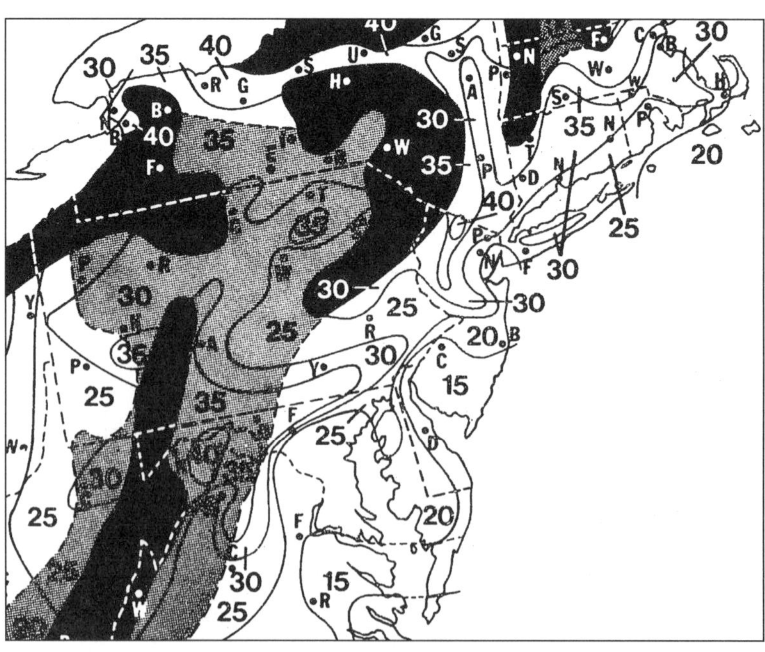

Detail from a snow map indicating loads for structural design. © BOCA.

are matched by winds in all of Alaska except the center of the state, and all U.S. islands including Hawaii, the Aleutians, and Puerto Rico. In addition, there are localized areas of high winds in Central Appalachia, along most of the Great Lakes shoreline, and scattered throughout the Rocky Mountains.

Applicability

Regionally determined wind factors include:

1. Regional wind speed per IBC Figure 1609. The wind speeds shown on this map are fifty-year storms speeds (see explanation at p. 112) for sites in Exposure C, at a height of thirty-three feet above the ground. They are adjusted for other exposures and heights as described below.
2. Exposure: The scale of natural or built windbreaks surrounding the project per IBC 1609.4. This function describes four settings, as follows:
   - Exposure A: Where prevailing winds are limited by at least a half-mile stretch of tall buildings (those averaging seventy feet high) before reaching the project.
   - Exposure B: Where prevailing winds are limited by buildings and/or trees of a residential scale before reaching the project.
   - Exposure C: Where prevailing winds are limited only by grasses and "scattered obstructions" before reaching the project.
   - Exposure D: Where prevailing winds travel across open water for at least a mile before reaching the project.
3. Building height: Tougher requirements are placed on higher portions of buildings per IBC Table 1609.6A. Rules consider increases in both wind pressure and gusting due to height.
4. The importance of the project for human health and safety, per IBC Table 1609.5.1. This follows the same logic as Seismic Use Group. Buildings whose failure would pose little danger are designated as Category I while critical buildings are designated as Category IV.

Of these factors, two (height and importance) are related to program, one to immediate region (exposure), and one to extended region (wind speed), but that last one is far and away the most significant in calculating design load due to wind.

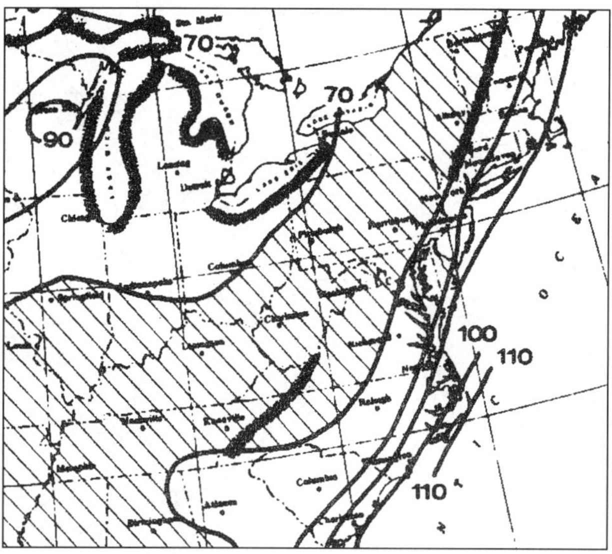

Detail from a wind speed map. © BOCA.

Requirements

Design requirements are found in several places, depending on the project conditions:

- Residential projects: Refer to either of the following:
  1. SSTD 10, "Standard for Hurricane Resistant Residential Construction," published by the Southern Building Code Conference International
  2. *The Wood Frame Construction Manual for One and Two Family Dwellings, High Wind Edition,* published by the American Forest and Paper Association
- Rules for projects with average roof heights of sixty feet or less, classified as Category I or II and in Exposures A or B, are found in the model building codes [IBC 1609.6].
- Rules for taller, more critical, or more exposed buildings than those described above are found in Section 6 of ASCE 7: *Minimum Design Loads for Buildings and Other Structures,* published by the American Society of Civil Engineers.
- Projects in hurricane regions are governed by special rules found in IBC 1609.6.2.2 or SBC Appendix J.

Optimizing Rules Related to Regional Conditions

There are only two, and they are mutually exclusive:

- Choose another region, immediate if sufficient, extended if necessary.

INSPECTOR'S COMING... START USING LONGER NAILS!

"Inspector's Coming. . . ." © Roger K. Lewis.

- Design as required, with the associated costs. If these costs are included in project budgeting from the beginning, their effects on feasibility will not come as a surprise when the project is bid.

Site Conditions Affecting Feasibility

For the purposes of this section, "site conditions" are any natural or man-made environmental factors that are likely to be wholly contained within a property. In some cases, conditions causing conflicts may be even smaller, affecting only part of a property and leaving the rest available for development. (These more limited conditions are addressed at pp. 111–112.)

This section explores conditions both liberating and limiting. Liberating conditions are those that allow an otherwise off-limits site to be developed, and include air rights and Transferable Development Rights (TDRs). Limiting conditions generally can't be avoided through positioning, including landfills and hazardous materials. With some affected properties, these conditions may have made them unfit for development, stripping them of commercial value and resulting in their being taken off the market. For other properties, especially those holding hazardous wastes, "buyer beware" is still good advice.

Air Rights

Background

In some cases, properties that are already developed can provide sites for new projects. The issue turns on the concept of limits to property ownership.

Property ownership was clarified long ago as having a limited reach beyond the surface of our planet. Below ground, some legal systems, particularly where mining is a state business, exclude mineral deposits far below ground from landowner control. Above ground, several American court cases have ruled against landowners wishing to keep airplanes and orbiting satellites from trespassing on their properties during flyovers. Still, rights are respected for a considerable distance both above and below grade.

Within the area given over to owner control, and except as prohibited by ordinance, owners have the right to grant others use of it. Owners can transfer use of horizontal space by leasing all or part of the land surface (or building area when there is an existing building) to someone else. They may also transfer use of vertical space by leasing all or part of the space above a certain height to someone else.

Applicability

For any project to take advantage of air-rights space, doing so must be feasible. Feasibility depends on cooperation from the following factors:

Market Feasibility

Sufficient market demand must exist to justify undertaking an air-rights project, considering the associated difficulty and cost. Sites that have a unique advantage are likely to be successful. In the case of Chicago's McCormick Place Convention Center expansion, built in the airspace over a rail yard, the original building ensured the market for any building that was sufficiently large and close. The rail yard airspace was the only possibility on both accounts.

Availability of Minimal Ground-Level Space

Some land must be available at the surface for access and for structural support.

Access to air-rights projects is impossible without either a helicopter or the right to use a piece of land at the surface. Regardless of other feasibility issues, helicopters are not allowed as means of egress, so transferred surface land must be sufficiently large for a stair or an elevator connecting with the airspace.

Buildings constructed in airspace do not hover. Their structural loads must be transferred to the ground just like those of any surface building. Arrangements must be made for footings and columns to be built below the air-rights space.

Structural Feasibility

Walls or columns acting as stilts tend toward high slenderness ratios, making them more prone to buckling under load. Support structures connecting air-rights buildings to the ground must be designed for this. In some cases, existing buildings are found to have been designed with sufficient excess capacity to support air-rights additions. While unusual, if found to be the case, and if permission to bear on such buildings is granted, this can be a real boon for project teams.

Budgetary Feasibility

The cost of building in air rights is higher than the comparable amount of land-based construction due to the need to build access and structure across the space below the air right.

Zoning Feasibility

Existing development on the site must not be occupying all of the density allowed by zoning ordinances. The combined total development including air rights must not exceed that permitted by ordinance.

Building Code Feasibility

Limits on density imposed by building codes do not generally limit developments in air rights. Since they occupy their own "buildings" (see pp. 180–182) heights and areas allowed by codes start to accrue anew.

Requirements

There are no rules associated exclusively with air-rights developments. Jurisdictions tend to handle such projects on a case-by-case basis. It is critical, therefore, for project teams to contact local code officials very early in the project's development to find out what approvals will be necessary.

Transferable Development Rights (TDRS)

Background

Some jurisdictions permit properties to be developed beyond their generally allowed limits using leftover margins contributed by other properties. This provides a mechanism, called Transferable Development Rights (TDRs), for equitably compensating owners of historic properties and farms for limiting the future development of their properties. (See pp. 132–133 for information on establishing total size of a building.)

Unused Floor Area Ratio (FAR) or Unit Per Acre rights are sold, and the selling property waives the right to future use of that lost density. Sometimes "TDR banks" are used to facilitate the sales by buying the rights from one property when its owners are ready to sell and selling them to another when its owners are ready to buy.

The concept was initially instituted for historic properties, many of which are land rich and cash poor. Owners of prime sites in New York City can get prices in the millions of dollars for agreeing to transfer development rights. Later the idea was extended to farms to con-

"McDonald's Farm." © Roger K. Lewis.

serve a region's agricultural base and limit the advance of suburban development. All agricultural districts allow residential development, if on a modest scale. Even five units per acre can drastically alter a landscape. By selling that right to developers of urban housing, farmers can get cash to supplement their agricultural incomes while simultaneously ensuring that nobody will pressure them into selling for a new development.

A typical historic example might be a nineteenth-century church sitting on a one-acre property carrying an FAR of 12, whose owners are unwilling to, or prohibited from, tearing it down to build a high-rise. A 22,000 SF church would occupy only 0.5 FAR, leaving over 500,000 SF of allowable FAR (43,560 SF x 11.5). Assuming that the urban area is designated as a transfer zone and a receiving zone, and the deal was approved by the zoning board, the church could literally sell its right to those 500,000 SF to a nearby property. In this way, a half-acre property's allowance could grow from about 261,000 SF to about 762,000 SF. A two-acre property would grow from about 1.04 million SF to about 1.54 million SF.

Applicability

Many properties qualify for some aspect of TDR use. Ordinances often identify "transfer zones" and "receiving zones." Transfer zones are frequently agricultural, and receiving zones are generally urban, and may be at some distance from each other.

Requirements

Transfer is required to be permanent and is recorded on titles. Transfer arrangement is best handled by lawyers representing the owners of properties involved in the transfer.

Yes, FAR limits are set to fulfill the intent of comprehensive plans, so projects equal to the property's FAR plus the transferred FAR can overtax available infrastructure. In general, districts seem to deal with any problems this may cause after they've happened. A further limit to transference is the constructibility of a project. It won't make sense to acquire enough additional FAR to build a 150-story building on a quarter-acre site, since it would not be structurally feasible.

Landfills

Background

In essence, landfills are areas where soils have been disturbed. This means that settlement and compaction are less predictable than for undisturbed soils. Although people have been piling garbage, trash, construction debris, and excess soil on land for centuries, in the postwar era, landfills have operated in a more rigorous way, recognizing the value of marketable land. Current waste-management industry techniques include compaction in lifts, ventilation and release of gases generated by decomposition, use of liners and drainage, and monitoring of incoming fill for presence of toxic wastes. Additionally, all recent landfills are monitored by the U.S. Environmental Protection Agency and certified when they become suitable for development.

Not all landfill is the result of waste disposal. Tokyo Airport, Boston's Back Bay, Manhattan's Battery Park City, and much of the Netherlands are examples of landfill created expressly to support new development. In these cases, at least, fills were engineered with their end uses in mind.

Notable failures have necessitated complete removal of filled materials. While such failures are not common, neither is the attempt to build multistory commercial buildings on landfill.

Further discussion of the regulatory implications of building on landfill beyond the scope of this book, but good information can be obtained from the EPA, state departments of environmental protection, and state water authorities.

Applicability

It isn't nice to be surprised by the knowledge that a recently purchased property contains a landfill. Thankfully, that isn't likely to happen. Sellers are responsible for existing conditions, and current practice for the transfer of real estate calls for Phase I Environmental Assessments. (See pp. 285–286.) Commercial landfill locations, which charge fees in exchange for accepting landfill materials, are registered with the EPA. Private landfills serving individual corporations or interests are likely to be smaller but may still be sufficiently large or toxic to render a property unsuitable for development.

Requirements

Although generally not prohibited, building on decommissioned landfill, like renovating existing buildings, can be fraught with surprises and requires appropriate management. Careful geotechnical testing can help identify potential problems. While there are no building rules specific to construction on landfill, general rules for foundation design account for soil capacity, cohesiveness, underground water, and other subsurface issues. Where landfill conditions meet these requirements, buildings can be feasible.

Optimizing Landfill Rules

If possible, locate buildings at edges of landfill, limiting development of filled areas to less demanding facilities, as one would do with floodplains. Where future development is expected to include buildings rather than just landscaping, the landfill design should be adjusted toward that goal. Ultimately, environmental protection rules appear to allow few exceptions.

Hazardous Materials

Background

As in the game Battleship, hazardous materials lurk, often undetected, until investigations detect or symptoms reveal their presence. And as in Old Maid, about all one can hope to do is avoid being on their receiving end. Still, hazardous materials are no game. All regulations, and even standard construction industry contracts, assign responsibility for dealing with hazardous materials to property owners. They are responsible for what they own, not what they did. It doesn't matter if the damage was caused before the present owner bought the property. Further, the problems caused by hazardous waste can't usually be avoided through careful placement of buildings on the sites.

Properties are sometimes acquired or exchanged without due attention paid to hazardous materials. This can happen particularly with public properties if politicians agree on the procurement of a contaminated site for political purposes without adequate assessment of the risks. Such sites can become money magnets, consuming more and more capital resources while being made safe for use.

Still, rules require less remediation than they once did, recognizing that:

- In some places, most of which are urban, hundreds or even thousands of contiguous properties are contaminated. Cleaning one property or another won't make much difference in the grand scheme of things. If, as a society, we're not going to abandon entire regions, we may as well use the properties within them.
- The enormous costs of remediation work would make many, if not most, projects infeasible. If we were to stop developing as a result, that, too, would damage society.

Environmental engineers use truck-mounted coring rigs like this one to extract soil samples when condutcting ASTM Phase II assessments. Photo courtesy of The Robert B. Balter Company.

In response, the term *brown-fields* has come to denote a contaminated property that is rehabilitated without being cured. Hazardous materials must be made nonthreatening, but that doesn't necessarily entail removing materials. On brown-fields, they are typically left on-site but encapsulated and monitored. Encapsulation involves sealing the materials under or within a barrier, such as the paved surface of a parking lot or an underground slurry (concrete) wall.

This does not mean that fixing problems will be cheap, or even that it will cost less than the value of a property.

Applicability

Rules are applied fairly clearly: if hazardous materials are present, the rules apply. As noted for landfills, current practice for the transfer of real estate calls for Phase I Environmental Assessments and makes the seller responsible for the condition of the land. Unfortunately, there are no guarantees. Although federal agencies (primarily the EPA) develop the rules for hazardous materials, and although the same agencies track sites until cleanups are completed, they are not clairvoyant and have no maps identifying undiscovered deposits. People who have owned properties for many years can one day discover hazardous materials on their land as a result of underground migration. If any are discovered before purchasing a property, project teams are strongly encouraged to get the seller to abate them, in accordance with EPA rules, before buying the property.

Some properties are almost guaranteed to contain hazardous materials, the most obvious being former factories and properties more than a few years old with storage tanks, particularly when concealed underground. Factories are frequently contaminated with heavy metals, such as cadmium, chromium, lead, and mercury, whose detrimental effects are not significantly weakened over time. They may also be contaminated by liquid spilled from leaking tanks once used to store solvents, gasoline and diesel, or heating oil. Tanks can also be found on former gas stations and many properties heated with oil. Unlike heavy metals and asbestos, these liquid contaminants eventually break down on their own. They are classified in a two-tier system:

- BNAs: Base Neutral Aromatics, including petroleum products, typically encountered as fuels. They are somewhat evaporative liquids, not as irritating as VOCs.
- VOCs: Volatile Organic Compounds, including benzene, toluene, and xylenes, typically encountered as solvents and cleaning products. They are highly evaporative liquids and therefore quite irritating.

Requirements

The first step is environmental assessment. (See pp. 285–286.) If contamination is found, ASTM Phase III remediation may be required.

Optimizing Hazardous Materials Rules

Assess properties before purchase. It is cheap insurance against virtually unlimited liability.

INTENDED PURPOSE

Context

Why This Issue Is Important

One can't assume that the purpose a project is intended to serve will necessarily be allowed on a given site. If permission is denied, the project is over before it has begun. Therefore, the critical first step in planning or designing any facility is determining that the intended use is permitted.

Use is possibly the most important factor in establishing the character of a neighborhood. Commercial, residential, industrial, and mixed-use districts each have a very distinct ambience, derived substantially from their distinct regulatory provisions.

Some real estate developers add value to the properties they acquire without developing them. They petition to have the zoning district of the sites changed, thereby changing permitted uses. They can then sell the properties at a profit to other project teams, who will build. For example, a property zoned for five houses might be sold as five lots for $50,000 each. If rezoned to allow twenty-five town houses, each might sell for $20,000, a difference in total profit of $250,000, or 100 percent.

Why Regulatory Bodies Care

CONCEPT: THE REASONING BEHIND THE RULES

Zoning ordinances refer to "Permitted Use", building codes refer to "Use Group," and covenants might refer to "Use of Property." All three terms describe the purpose a project is intended to serve. The primary distinction between them is the degree to which the different sets of rules try to control use.

Zoning ordinances and covenants dictate whether any particular use will be acceptable for any given property. Their rules are written with the assumption that certain neighborhoods or districts should contain only certain uses. These uses are listed in zoning ordinances. Whether the uses allowed in any given district are broad and diversified or narrow and focused depends on the zoning boards' long-term visions for their districts.[22] If a project team's proposed use is on the list, the team can proceed with the project. Permitted use has no further direct implications for zoning, since other zoning issues are categorized by zoning district, not use.

Conversely, a building code's rules are written with the assumption that any proposed occupancy can be made safe and durable if safety and durability are priorities. They therefore regulate construction quality to be appropriate for buildings housing the proposed uses. Many building code provisions are based on the occupancy proposed. Thus, project teams determine use group (which must be allowable under the permitted use provisions of zoning codes) and then determine the specific building code provisions that pertain to such a use.

One further factor to be considered in determining permitted use is fair housing law. Certain uses may be viewed as discriminatory to some potential users. (For further elaboration, see pp. 44–45.)

Permitted Use Requirements

Uses are permitted outright, permitted conditionally, or not permitted, as described below. When they are permitted either by right or by condition, projects proceed. When they aren't, projects either relocate, are abandoned, or become acceptable through the processes previously described (pp. 66–73).

Allowed Activities

Activities are smaller than uses. Uses are broad identifying purposes, while activities are individual behaviors that occur on a day-to-day basis. Some rules affect the way occupants can use their projects even more than they affect project teams' design decisions. These rules regulate activities occurring in and around buildings. Examples vary widely but commonly include:

- Prohibitions against outdoor clotheslines, forcing laundry to be dried in appliances or brought to commercial cleaners.
- Requirements that residents put trash out only on the day of pickup and have it concealed again by nightfall.
- Prohibitions against keeping farm animals, even if only as pets. Some recent test cases have involved owners of Vietnamese potbellied pigs.
- Restrictions on the use of residences for home businesses and on residential buildings for churches.
- Restrictions on the use of ground-floor outdoor spaces for sidewalk seating, particularly for restaurants.
- Restrictions on after-hours operations at commercial properties. This particular issue is most likely to be found in a lease agreement but can be just as restrictive on tenants as if it were a statute.

As for requirements, implications, and optimization of allowed activities, there are too many local variations to deal with here. Do some checking, become aware of the rules, and plan for them.

Historic Properties

Historic preservation standards consider use to be as integral to a building's character as its material aspects. Recognizing, however, that historically accurate uses may no longer be marketable, they allow for new uses that are spatially compatible and stylistically sympathetic to the existing building. For projects classed as preservation or rehabilitation, the standards [1996] require the property to "be used for its historic purpose or be placed in a new use that requires minimal change to the defining characteristics of the building and its site and environment." For projects classed as restoration, the definition of new use is one that "reflects the property's restoration period." This refers to the historic period to which it is being restored, not the period at the time of the restoration work.

Where to Find the Requirements

A Precursor: Agreeing on Intended Use

Before rules can be consulted, members of a project team must share a common understanding of a project's intended use. Most construction industry contracts include project descriptions, which should clearly define intended use as a key parameter for determining nature and scope of services, compensation, and schedule. Before signing these contracts, project teams need to discuss and agree on use.

This description can be found in some contracts on the first page, under the catchall phrase "For the following project." The most recent edition of the American Institute of Architects' basic unabridged owner-architect agreement, the B141-1997, allows for much more specificity.[23] In its Article 1.1, titled "Initial Information," it is noted that "the Owner and the Architect have mutually arrived at the following initial information . . ." It allows for use-related decisions to be recorded or cited on the following line:

1.1.2 The physical parameters are: (Identify or describe, if appropriate, proposed use or goals and program.)

On Maps and in Public Records

By reviewing zoning maps or checking tax records, project teams can determine the designation of each plot of land. Armed with this information, they can turn to zoning ordinances.

In Zoning Ordinances

Local governments, through their regulatory bodies, determine permitted use. The process begins with a list of use zones, called zoning districts. As many districts are designated as the local government feels are appropriate.

Districts vary as to how narrowly they are defined, with some allowing few permitted uses, and others being much more inclusive and generalized. Each district carries its own list of allowed uses. In many cases, one district may cross-reference the listings for another district rather than be redundant. For example, an R5 residential zone may allow as permitted uses "all uses permitted for the R4 district, plus those additional uses listed herein." Worse, the description of the R4 district may reference "all uses permitted for the R3 district, plus . . . " It is important to trace all such references to their ends.

Some uses will not be found listed for any districts simply because they are too new. As societies evolve, new types of uses develop and old ones disappear. A century ago, no ordinance would have permitted use designations for drive-through banking or day-care centers, uses we almost take for granted now. In contrast, there may be no cities left with use designations for Turkish baths, once a common feature in cities with large immigrant populations.

New use designations may soon be written for several recently developed use types, including:

- Alternate forms of senior housing such as retirement and assisted living communities, and hospices, to supplement the traditional convalescent homes and "old age" homes.
- Co-housing, a style of shared community that has experienced enormous growth in parts of northern Europe.
- Special forms of outpatient medical treatment centers developed in response to the needs of "managed care" and its health maintenance organizations (HMOs).

In some cases, ordinances reverse themselves. Single Room Occupancy (SRO) hotels were once common, yet all but disappeared in the sixties, seventies, and eighties. They staged a comeback in the nineties as fewer affordable units were developed and as more young people delayed marriage. While they were out of favor, many zoning boards wrote them out of their ordinances and are now faced with readopting them.

PUDs

For properties with Planned Unit Developments, use is determined in the same way as it would be for other districts. Each PUD is indicated separately on zoning maps, under such labels as PUD-1 or PUD-2, just as a

"This Valuable Lot." © Roger K. Lewis.

residential district might be indicated as R-5. Rules are found in zoning ordinances in the same way as are rules for conventional districts.

HISTORIC PROPERTIES
Zoning ordinances usually contain special provisions for historic properties and may include historic overlay zones. In addition, there are rules for use in the Secretary of the Interior's Standards and their associated guidelines. A preliminary assessment of a property should be done to determine its status under the standards. A search in the Landmarks Registers can establish a property's pedigree and help identify documentation necessary to establishing original use.

SOURCES OF PERMISSION

As only zoning ordinances and CC&Rs attempt to control use, it is to them that project teams must look for permission. Permission is rooted in the concept of zoning district. Districts can be established as either base or overlay districts. Each district has its own rules for development.

Base Districts

Base districts have a continuous border within which every property is seen as having certain common attributes. A base district is a community type, identifiable by unique characteristics that are seen as worth retaining. Any given city or county may have many identifiable areas of each district type. Zoning maps are prepared containing lines that carve the city or county into divisions by their correspondence to these district types.

Examples of base districts may include "neighborhood shopping district" or "commercial-light manufacturing." Further, some communities establish a "mixed use district," and in many cities, the owners of a particular parcel of land may petition for a Planned Unit Development, or PUD. One way or another, each piece of land in a city or county is labeled as one district or another, regardless of how the property is actually being used.

Typical districts include:

- Residential
- Commercial
- Industrial
- Mixed Use
- Agricultural

Governments then review each neighborhood in their jurisdiction and decide which use zone, or zoning district, should apply to each piece of land. They prepare maps showing the extent of each zone, with borders drawn and dimensioned. In addition, zoning district designation is generally included on the tax records of each property.

Overlay Districts

In General

Overlay districts have no common border; they transcend base districts. Properties that fall within the overlay zone are not those within a certain boundary line, but those that meet certain criteria. They share a common trait but may be sporadically distributed around a town. Overlay districts may be designated for historic properties, as districts subject to design review, for enterprise zones, or as fire districts and floodplains. These last two are also established by some building codes (For information on floodplains, see pp. XX–XX).

Overlay Districts might include the following (examples in quotation marks are from the District of Columbia Municipal Regulations):

- Special Purpose (such as general Special Purpose or "Mixed Use Diplomatic District," "Uptown Arts–Mixed Use Overlay (Arts) District," and "Neighborhood Commercial Overlay District")
- Special Location (such as "Downtown Development District," "Pennsylvania Avenue Development District," "Capitol Interest District," and "Waterfront District")
- Special Incentive (such as "Hotel-Residential Incentive District")

Enterprise Zones

BACKGROUND
Enterprise zones are special properties whose owners are given tax incentives in exchange for bringing needed economic opportunities to a community. Enterprise zones are defunct buildings given new life housing businesses that offer employment and training. Only certain uses are allowed, generally of the light manufacturing and mercantile type.

Enterprise zones have been more successful in England, where the idea originated, than in the United States, where they have had very limited success. The problem derives from the costs of vacant buildings. In the United States, building shells are too expensive, since their owners would rather hold on to them than sell at a loss. Where properties can be acquired inexpensively, the idea tends to be effective.

Applicability
Enterprise zones are not places, they are designations. They aren't identified by borders on maps but are established by rules contained in tax codes and employment regulations.

Requirements
Rules explaining what qualified projects can do are found in appendixes to local zoning ordinances and pertain to intended use. For setback, density, parking, and other such rules, enterprise zone projects are governed by the same zoning rules that affect other projects in their district.

Fire Districts

Background
Fire districts, as established by building codes rather than zoning ordinances, restrict the density of construction and the types of occupancies in parts of a city that meet certain criteria. Cities can't afford the losses that result when urban fires spread from building to building. Such losses are more likely where there are large populations, a lot of combustible material, and a high percentage of buildings built before the use of codes.

Applicability
In the SBC, city blocks are included in a fire district if they are part of an area of at least two blocks where at least 40 percent of the land is occupied by buildings (50 percent if the average building height is under 2½ stories) and at least half of the occupied area is occupied by a long list of possible occupancies including business or storage occupancies, hotels, motels, theaters, nightclubs, and warehouses [SBC Appendix F].

The rules further specify that buildings located within 200 feet of larger fire districts (those that cover at least four blocks) also comply with the rules for fire districts, even though they don't meet the criteria for designation in their own right.

Requirements
Rules applicable to buildings in fire districts include [SBC91 Chapter 30]:

- General restrictions on permitted Construction Classification. Further, higher than usual fire ratings are required of structural components in certain occupancies and of exterior walls in buildings with heavy timber construction that are near property lines.
- Restrictions on permitted occupancies (no Group H).
- Limits on choice of roofing materials to those that meet a higher standard.

Common Issues and Rules

In addition to permitted use, other issues related to basic and overlay districts, and sometimes to developments governed by covenants, include the following:

- The density of development and the size of individual buildings allowed on properties (see Chapters 4 and 5)
- Sometimes, the particular design directions recommended for new development

Rules that pertain to all districts include provisions for:

- Broadly applicable activities such as recycling, screening, and landscaping
- Broadly applicable projects such as affordable housing, temporary uses, and day care
- Parking and loading (deliveries)
- Building and highway signage

Degrees of Permission

Individual zoning ordinances vary as to the way they grant usage permissions. Some processes are checklist-style whereby applicants are seen as either complying or not; others are much more deliberative. The two styles usually coexist in every ordinance, but the relative use of the two varies. For a side-by-side comparison of the differences between Permitted by Right and Conditional Use, see the matrix that follows.

Permitted Use "As of Right"

The simplest designation applies when a use is permitted by right. The project team looks for a description of its intended program in a list of uses permitted in the project's district. If it is found, projects can proceed with the next phase of development. With this designation, no further approvals are required.

Conditional Use

Conditional uses have no home district; they are not permitted as of right anywhere. Uses designated as conditional are allowed on a case-by-case basis, and only after in-depth review by the zoning authorities.

Special Exceptions

Projects whose uses aren't listed as Permitted or Conditional may be in jeopardy. Special exceptions are granted only when project teams can demonstrate that their proposed projects won't adversely affect existing districts through increased traffic, parking shortages, or other hardships. Arguments based on mitigation (see p. 68) are frequently used to justify special exceptions. Teams that are determined to proceed with special exceptions must be prepared to undertake substantial additional work and suffer extended delays (as discussed regarding reclassification at p. 71).

Accessory Uses and Buildings

In additional to primary uses for which approval is required, ordinances generally allow some quantity of accessory uses and buildings to share a property. These

Issue	Permitted Use Development "As of Right"	Conditional Use Development Requiring Quasijudicial Review
Staffing	Focus is on project review. Staff involvement is minimal.	Staff involvement is intensive due to public hearing requirements.
Permit Processing Time	Relatively short	Relatively long
Commission and Board Review	Not necessary	Can involve multiple and duplicate meetings/hearings
Environmental Sensitivity	Not generally adaptable to unique and site-specific conditions	Maximum flexibility and potential to require mitigation and conditions
Legal Exposure	Low	High
Community Involvement	None	Opportunity for community comment. Maximizes sensitivity of project to community values.
Flexibility	Limited to range defined by code	Maximum flexibility and discretion depending on type of permit
Other Agency Involvement	Not generally provided	Provided as part of review process
Special Study Requirements	Usually unnecessary	Potential to require any necessary study (e.g., environmental review)
Consistency of Review and Decision Making	Usually consistent but depends on staff experience and training. Standards applied uniformly. Emphasis on professional review.	Level of review and application of standards uneven based on changing community values and experience of review body. Potential for public pressure to prevail over evidence and for staff to act based on political considerations.
Application Fees	Based on administrative costs. Relatively low.	Based on administrative and hearing costs. Relatively high.
Quality of Building Design	Minimum standards	Higher standards possible
Consistency with Community Goals	Should meet minimum requirements established in comprehensive plan	Affords maximum flexibility and adaptability to requirements established in comprehensive plan

Permitted vs. Conditional Development. [Source: Table 8 from Lerable, Charles A., Preparing a Conventional Zoning Ordinance, PAS No. 460, American Planning Association. Used with permission.]

are defined as those that accompany and serve the primary uses and buildings. In some cases, quantity is no more clearly defined than "a reasonable amount." As accessories, these uses are subject to different rules. Generally these rules are less restrictive than would be required for independent uses or stand-alone buildings.

Accessory Uses

Accessory uses are uses that would normally require compliance with a different set of rules than the primary use. As a result of small size and minimal risk, they are allowed (see pp. 183–184). Examples in residential districts often include limited educational (tutoring a limited number of students), limited office (dentist's or doctor's office when the professional lives in the house), and limited multifamily residential (a limited number of boarders).

Accessory Structures

Accessory structures are detached, freestanding buildings that might normally require location on a separate lot. Due to their small size and programmatically supportive nature, they are allowed some relief. (For implications, see pp. XX–XX.) Examples include garages, sheds, and private stables serving residential buildings, and parking garages and day-care homes serving commercial buildings. In some cases zoning ordinances waive rules for accessory structures; in others they may provide a separate set of rules.

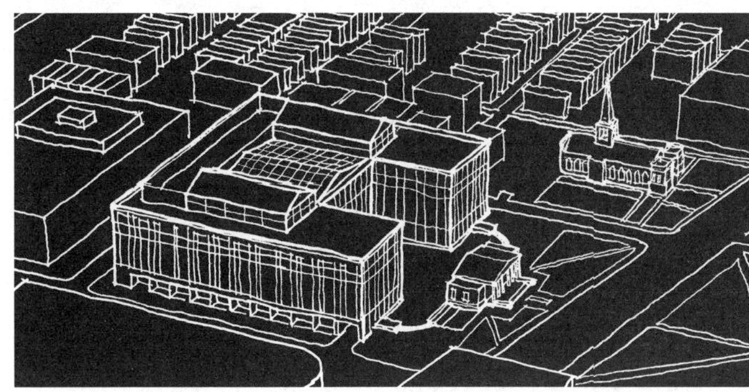

4 POSITION

Establishing Building Placement

CONTEXT

Why This Issue Is Important

Even before a project team decides where on the site to put a building, it needs to consult rules to find out what overall design approaches will be acceptable. From a regulatory perspective, site design is primarily a question of where to locate the building, both horizontally and vertically.

Horizontal Position

Horizontal position involves three related issues:

- Distance to property lines
- Distance to adjacent properties
- Distance to other site features

All three of these issues are strongly affected by the amount of space needed for vehicles, so project teams must also evaluate:

- the amount of ground-floor space required for parking, loading, and maneuvering.

For projects with substantial vehicular needs, the design of parking facilities is best done before, or at least along with, the earliest building design sketches. This is because our society is strongly affected (one might say "driven") by our commitment to personal transportation devices. The horse evolved into the car, not the train or bus. Moving and storing personal vehicles requires a tremendous amount of space.

Vertical Position

The position of buildings relative to the surrounding topography is also addressed by rules. Some aspects of vertical position are related to definitions of building height. Where determining total building height is related to where it intersects the ground rather than number of stories or height from foundation to peak (which it never is), vertical position becomes very important. Simply put, rules measure buildings as being of lesser height when their basements are pushed farther into the ground.(See pp. 138–139.)

Vertical position is also a question of inundation. Regulating the elevation of bottom-most structural elements is one way to minimize damage.

Why Regulatory Bodies Care

Horizontal Position

Whenever sites are larger than the buildings located on them, the positioning of the buildings becomes a critical decision, affecting many issues.

From the perspective of urban design embodied in zoning ordinances, maintaining the continuity of the "urban fabric" is stressed. Aligning new projects with existing buildings fosters a sense of stability; adjusting the width of the "urban canyon" can create a sense of either intimacy or monumentality. In addition, establishing setbacks can encourage a desired density, keeping buildings spaced at a comfortable distance. Last, traffic management plays a vital role in establishing a community's persona. Such management includes deciding whether parking lots are in front of or behind buildings, or whether streets will be "pedestrian friendly."

From the perspective of land use embodied in environmental regulation, portions of a site may be off-limits to construction for the reasons discussed previously (pp. 81–82).

From the safety perspective championed by building codes, spread of fire is a critical issue. It is affected by proximity, since distance is key to containment. It is also affected by the relative transparency and flammability of the enclosing walls. Walls that are relatively transparent to fire allow heat, flame, and burning embers to pass through them, while those that are flammable provide something for those flames and embers to ignite. These two issues, proximity and construction, are mutually dependent. Project teams that intend to use fairly open exterior wall designs incorporating large proportions of glass must consider the possibility of putting some distance between their building and others.

As for accessibility, the distance and impediments a handicapped person must traverse to get to a building must be carefully considered. Unless the project can support expensive and maintenance-intensive technology (lifts), grade changes can only be overcome with substantial distance (long ramps).

The remaining type of regulation addressed in this book, historic preservation, is mute on the subject of positioning.

The remaining public-domain issue influencing positioning is right-of-way. It can strongly limit design options, so the presence and exact location of such easements should precede any design work. (See pp. 84–85.)

Vertical Position

Lower buildings are more likely to be damaged during floods, and the federal government is generally expected to provide emergency relief funding to help people without flood insurance rebuild. Where conditions are not sufficiently extreme to justify forced abandonment of buildings, people can at least be asked to raise their buildings. Rules also contain provisions for owners who decide to simply move their buildings to higher ground.

Therefore, where flooding is likely, floodplain regulations strive to minimize damage either by setting minimum elevations for structural members relative to base flood elevations or by requiring special floodproofing measures below those elevations.

Rules address vertical position also because of its implications for urban design and for getting adequate light and ventilation to, and achieving egress from, subgrade levels. (See p. 190.)

Where to Find the Requirements

Each set of rules must be consulted individually, as each has a different perspective to consider.

- Rules for setbacks and parking requirements are found in zoning ordinances, tied to zoning district, so district must be determined first, as noted previously (pp. 92–98).
- Rules affecting a project's distance to other buildings, involving such things as extent of openings permitted in exterior walls, are generally found in building codes and enforced by fire marshals. Use group must first be determined.
- Rules affecting building position out of concern for environmental conditions, such as the presence of wetlands, forests, landfills, and hazardous wastes, or site features, such as slopes, soils, and septic systems, are found in:
  - federal regulations (especially Title 40 of the Code of Federal Regulations: "Protection of Environment");
  - state and sometimes county regulations dealing with site preparation, drainage, and erosion control, including state health department controls on septic systems;
  - building codes (especially in appendixes for such localized concerns as floodplains and fire districts);
  - standards published by such organizations as ASTM (for site assessments) and FEMA.

Regarding the jurisdiction of the rules, project teams should check at federal, state, and local (county or city) levels of government.

DEALING WITH PROPERTY LINES

Designing for Consistency of the Urban Fabric, a Zoning Ordinance Issue.

SETBACKS

Context

A setback is, conceptually, a construction-free zone bordering, but just within, a site's property lines. This area is sometimes called a setback zone. In residential terms, it is this concept that establishes minimum distances for acceptable front, back, and side yards.

The use of setback rules can be traced to the "Twelve Tables" of Roman building law from 450 B.C.

They required a 2½-foot setback around every house for repair access and fire separation. Rome burned partly as a result of loose enforcement of this law, attributed to the need to quickly house a swelling population. Then, as now, codes and their enforcement were made more rigorous following the fire. (Hindsight is always twenty-twenty.)

Current setback rules are intended to:

- Inhibit the spread of fire, still.
- Establish consistency of look, since they put all buildings in the same district in the same envelope within their sites.
- Establish minimal degrees of visual and acoustic separation, and therefore privacy.
- Influence the disposition of surface parking on a property, since most planning rules allow parking within setback zones. See "Parking: Private Vehicles" later in this chapter.
- By leaving a margin, help prevent buildings on one property from partially occupying another. Such overlaps, when they do happen, can cause bureaucratic and financial complications for the property owners.
- In unusually progressive ordinances, help ensure that individual buildings will be sufficiently far from neighboring properties to avoid casting shadows on them, maintaining the neighbor's access to sunlight.

Although often thought of as such, setbacks are not necessary to maintaining low densities, since there are other regulatory methods, such as lot coverage (see pp. 133–137). and floor area ratio requirements (see pp. 132–133), that more directly do that. This realization may be useful in arguing for relaxation of setback requirements.

Last, distance to property lines is only indirectly related to fire separation distance. Fire separation distance requirements seek to reduce the ability of fires to spread from building to building by ensuring that buildings aren't too close together for the fire resistance of their enclosing walls and roofs. For this issue, the real question is distance to adjacent properties, explored in the next section.

Which Side?

Setbacks are generally defined for front, side, and rear yards, each based on different intended goals.

Front Setback

Background

As front is generally defined as the side facing the street, this is the minimum distance allowed between a proposed building and the property line adjacent to the street. Different municipalities use any number of formulas for calculating this distance, including fixed and variable formulas.

Where Is It Measured From?

In many jurisdictions, such as Pittsburgh, Pennsylvania, setback is measured not from the property line but from the centerline of the street. This allows for future street widening without recalculating setbacks.

In some jurisdictions, such as Washington, D.C., front setbacks are not required due to a trick of subdivision. Property lines are required by the local government to be placed the desired setback distance from the street. Since, in effect, the property owner does not own any front yard property, the issue of building on it is moot.

Special Considerations

Since greater setbacks translate to wider public spaces, the front setback is the one that most affects the public domain. Premiums or bonuses are generally granted for exceeding this setback, even though they are uncommon for setbacks along the other edges. Greater setbacks are also associated with reduced interaction among neighbors and reduced supervision of sidewalk and street activity by residents.

Implications of Requiring a Front Setback

By requiring a front setback, regardless of the distance required, it is hoped that all the facades on any particular street will align.

One obvious implication is that the desired look will be achieved so long as distances are maintained. Another implication is that streets with consistent setbacks are better environments or create better communities. Both are well-established urban design concepts.

In commercial terms, front setbacks are partially responsible for the difference in look between urban areas, where buildings start at the edge of the sidewalk and cars park behind them, and suburban areas, where buildings are separated from roads by wide seas of cars.

Implications of the Size of the Front Setback

Residentially, narrow front setbacks are often championed by supporters of "the new urbanism" as helping to foster interaction, and thus increased communica-

Zero-lot-line suburban residential development. Photo by K. Wyllie.

With front yards like this, rear yards take on renewed importance. Photo by K. Wyllie.

tion, among neighbors. New urbanists claim that this leads to a stronger sense of community and reduced crime rates. Conversely, wide front setbacks are often seen as symbolic of affluence and are valued for their ability to acoustically and visually distance houses from public streets. These totally divergent attitudes fall precisely along the lines of pedestrian-dominant or vehicle-dominant planning goals. (See pp. 119–120.)

Rear Setback

Background

Currently, residential rear setbacks provide space primarily for the residents' recreational activities. In this role, it has replaced the front yard, since the front yard has been given over to the car. It has also replaced the community park in neighborhoods whose residents value privacy above community. For past generations, it provided supplemental housework space for such chores as hanging the wash to dry or for raising needed food stocks, such as vegetables, chickens, or goats.

Commercially, rear setbacks mandate space for services, such as trash removal and parking. If properly anticipated, such setbacks can provide convenient areas into which HVAC systems can discharge stale air.

Where Is It Measured From?

Generally, rear yards are measured as the least distance between the rear property line and the wall nearest it, with limited exceptions made for cornices, trims, and other building components that project past the wall.

Where existing structures predate current regulations, rules may consider setback to start at some distance above grade. For example, where older one- and two-story structures are common, rear yards may be measured twenty feet above grade. This allows existing buildings to remain but requires any proposed third-floor additions to observe the new setback rules.

Some districts use distance-to-centerline formulas for the lower stories, and distance-to-property-line formulas for upper floors.

Implications

Much of the reasoning for rear setbacks is related to the notion of alley access. Where properties are serviced from the street, rear setbacks serve primarily to maintain privacy distances between buildings. For smaller residential properties, they can put caps on rearward expansion of buildings and encourage compact multistory additions over sprawling single-story ones.

Where properties are steeply sloped, dropping in grade from front yard to rear, first-floor back yard decks may be several feet in the air, and higher as they approach the rear property line. With small lots, if such decks encroach on the rear setback, they may be too tall to be exempted from setback rules. (See pp. 105–107.)

Side Setbacks

Background

Side setbacks face "internal" property lines, those adjacent to other private properties. They are critical for providing space for maintenance of exterior walls, providing some degree of acoustic and visual separation between properties, and for reducing the likelihood that fires will be able to travel from building to building.

The distance required for side setbacks is generally structured as a total of both side setbacks. Additionally, a minimum is generally listed for one side (at the design team's choice), indirectly setting the other side as whatever is left over from the total. This still leaves project teams a variety of options regarding the widths of the two setbacks. For example, a rule may require 40 ft of side setback divided between the two sides, while further specifying a minimum of 15 ft per side. In this case, minimum side yards of 15 ft and 25 ft would work, as would 18 ft and 22 ft, or 20 ft each.

Obviously, a formula of this type would not be used to regulate sites intended to allow duplex projects (known in some regions as semidetached), where side yards would be appropriate, but for one side only. Duplex developments require either zero-lot-line rules for side yards (which would also allow town-house developments) or side yard rules that require no minimum for the smaller side yard.

IMPLICATIONS

The implications listed for rear setbacks, above, also generally apply to side setbacks.

Side setback requirements are generally smaller than front and rear setbacks but may still be too big for some projects. Neotraditional towns apply for zoning relief or develop their own PUD rules before starting to build.

Rights-of-way can achieve some of the results of setbacks. In some older cities, side yards are found at the ground-floor level only. For instance, row houses in Baltimore's Fells Point area and in San Francisco are often continuous on upper floors but separated by three-foot-wide passages at grade. Where fire separation and privacy are of less concern, such passages provide the needed access to rear yards and alleys, so are established as rights-of-way rather than setbacks.

Corner Lots

For a corner lot it is important to determine the side that the ordinance considers to be the front. Is a corner lot considered to have two front setbacks? Which property line is then considered to be the rear?

Different zoning ordinances take very different attitudes toward this issue, generally starting by defining *front*. *Front* is variously defined as the setback:

- That the main entrance door faces. This provision lets the project team decide.
- Adjacent to the street that is listed as the property's address in the tax rolls.
- Facing a street—which means there could be more than one setback. Some zoning ordinances that carry this condition compensate by requiring sides not facing streets to simply follow side setback requirements, waiving rear setback requirements.

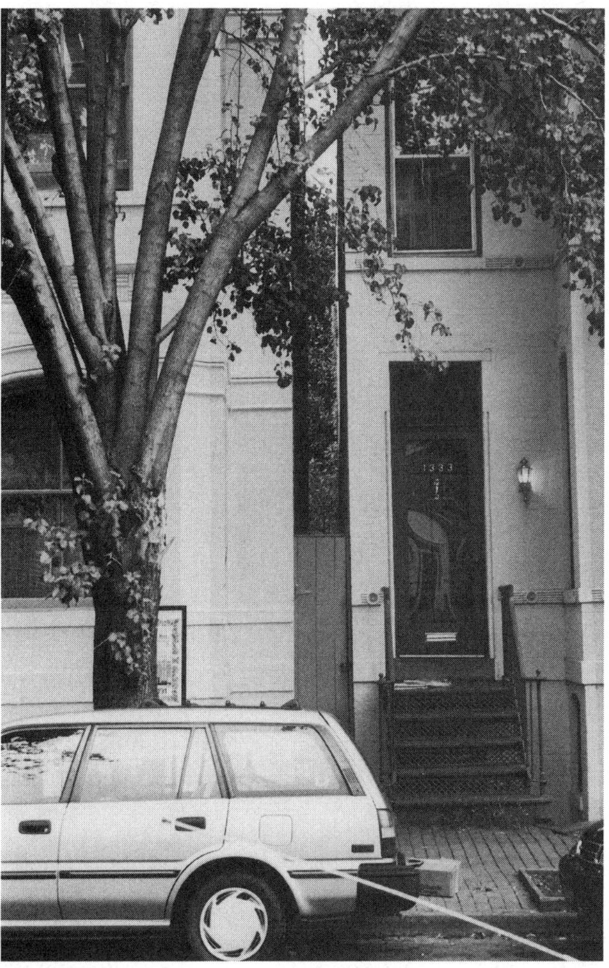

A town-house separation in the Georgetown area of Washington, D.C. Is it worth the bother? Photo by the author.

Where *front* is defined by the location of the front door, carefully choose the facade in which to locate that door.

Distance

No Distance

BACKGROUND

In many zoning districts, including most commercial ones, no setbacks are required or even allowed. This is known as zero-lot-line development.

IMPLICATIONS

Zero-lot-line development suggests that views and natural light and air are necessary only along the one or two property edges at each floor that are adjacent to public rights of way, at least where the height allowance of adjacent properties is the same or higher.

The density thus created enhances the urban feel, the critical mass suggested by a crowded sidewalk, and helps create a sense of community.

Optimizing Zero-Lot-Line Development Rules

Zero-lot-line rules pose implications for the future development of adjacent sites. Windows may be placed in walls that sit on property lines if they are above the maximum height allowed the adjoining site. Windows may be located below the height allowance if they are not required for egress, light, or ventilation, since they might someday have to be removed and infilled if the adjoining properties are ever built to their maximum envelopes.

Fixed vs. Variable Formulas

Background

In some cases, rules list fixed distances. They are generally expressed as an exact distance from building face to property line. In other cases, the setback distance is a formula that includes other factors. The height of the building is a key factor and may itself be subject to formulaic determination. See also the following section on rules pertaining to height. (For graphic depictions of the following formulas, see p. 280.)

Requirements

Fixed

Traditional. With a traditional fixed format, a setback may be listed as "30 ft," meaning that no part of the building may be within 30 ft of the property line. The simplicity of this type of formula makes it easy to understand the available range of design options. For low-density development, it tends to work well. It does, however, create a vertical, as in boxy, zoning envelope. When used on narrow streets bordered by tall buildings, fixed setbacks can reduce sunlight and view to meager levels and raise wind speeds as air is channeled into the relatively narrow spaces between buildings.

Building Restriction Lines. Building restriction lines are a type of fixed setback. Where used, they establish a common area of consistent setback, providing a mechanism for the future exercise of eminent domain.

These streets in historic Genoa, Italy, are extreme in the degree to which they block light and air from adjacent habitations. Photo by the author.

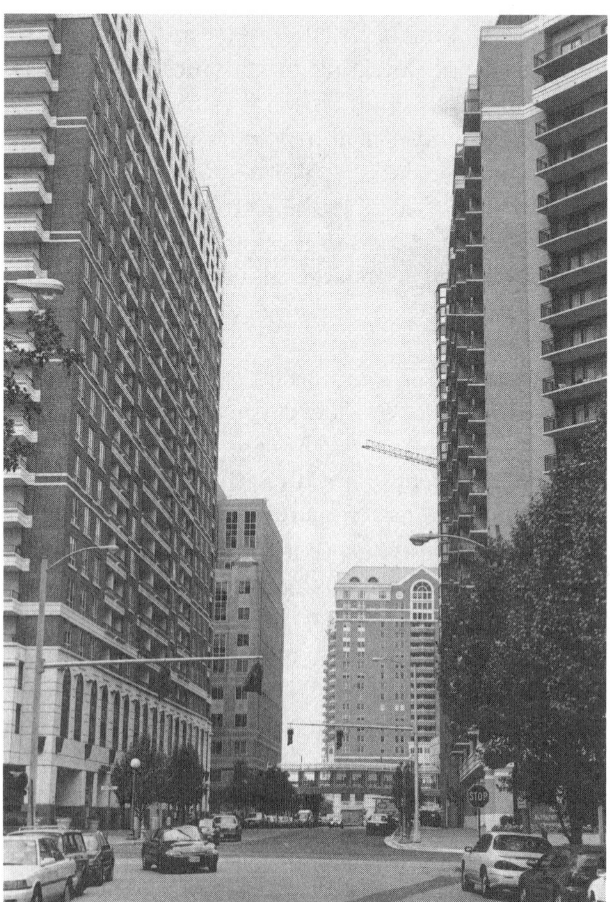

Streets in some American cities let sun hit sidewalk almost as rarely as those in Genoa. Although they don't preclude ventilation, such canyons tend to raise ground-level wind speeds. Photo by the author.

Variable with Cap

There are two parts to this type of formula:

- Calculating the amount allowed by using some form of variable formula.
- Setting a maximum cap on the amount calculated.

The maximum cap portion is fairly straightforward, like the fixed setback. A cap clause might say "... but no less than 30 ft." The variable formula portion of the calculation is more complex, determining setback incrementally for every unit of building height. It can take the form of either a step variable or a linear variable.

Step Variables. A step variable accumulates setback increases only at established trigger points. For instance, a setback may be listed as "30 ft plus 5 additional ft for every 25 ft of height beyond 40 ft." Translated, it allows buildings to start with a setback of as little as 30 ft from grade to a point 40 ft above grade. Five more feet are applied to the portion of the building that is between 40 ft and 65 ft high, for a total setback of 35 ft. The portion of the building between 65 ft and 90 ft above grade is subject to a setback of 40 ft. This 5-ft increase for every added 25 ft of height continues until there is no more building height or no more site width.

When first used, formulas like this led to the "wedding cake" layering seen in older buildings in New York City. The added cost of stepped facades has more recently encouraged developers to abandon that design approach in favor of simpler facades that don't maximize the available building envelope. Mies van der Rohe showed the way, breaking with the obvious solution when he located the entire exterior wall of the Seagrams building at the maximum required setback for its height.

Linear Variables The alternative to step variables are linear variables by which setbacks are continuously adjusted. There are two ways of calculating them:

1. By a ratio of number of inches of setback per foot of building height.
2. By a percentage of building height.

Linear variables require setback increases to be imposed for every increase in height, not just once in a while. A typical ratio formula might establish setback as "2 inches for every foot of building height" [DCMR87]. The same requirement, listed as a percentage formula, would establish setback as "16.6 percent of the height of the building."

Such formulas are somewhat complex. They are also incomplete. Although it is not generally stated, such formulas are generally intended to be recalculated wherever the height changes, rather than once based on the building's maximum height. Finally, they call into question the definition of *height.* (See pp. 138–139.)

As for the effect of these rules on project management, complex formulas slightly increase the complexity of the design process and may have some minor impact on the time required to review and weigh possible design options.

OPTIMIZING FORMULAS

Work with what you're given. Project teams can't optimize any of these formulas per se, but they can explore the implications of step and continuous formulas. While this takes time, it often points out greater design options than does following the obvious possibilities. (Additional ideas for generating and analyzing setback requirements with zoning envelopes are listed and illustrated at p. 280.)

Encroachments

"Encroachments" are occupied spaces, regardless of size, extending horizontally beyond a setback or property line from the primary mass of a building. (Unoccupied horizontal or vertical "projections" are explored at pp. 145–146.) Encroachments can give project teams some truly valuable juggle space. Setback encroachments and property line encroachments are permitted under quite different conditions and for different reasons.

Construction within the Setback Zone

BACKGROUND

Some amount of construction is generally allowed in setback zones.

- Regarding use, low walls generally up to 4 ft high, patios, surface parking, and accessory structures such as sheds and garages (see pp. 97–98.) are frequently permitted. The height limit on walls is intended to facilitate fire department access to buildings. Firefighters can cross barriers with their gear and while pulling hoses if the barriers are sufficiently low.
- Regarding construction quantity, some ordinances specify maximum amounts of the total setback zone that can be occupied by permitted types of construction [DCMR87 2502].

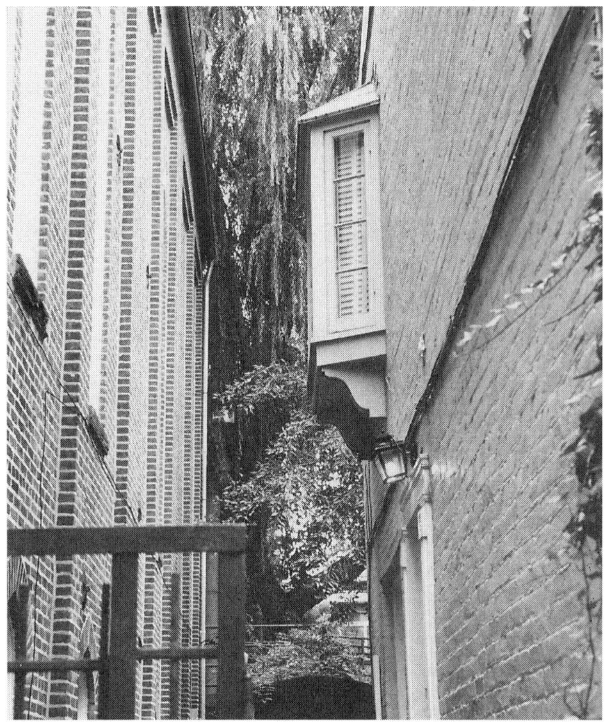

Within limits, chimneys, cornices, bay windows, and similar elements are generally permitted to encroach on required setback zones. Photo by the author.

Optimizing Setback Encroachment Rules

Since flush or low elements are generally unrestricted, finish grading may be raised to reduce the apparent height of fences and patios. It may reduce the maximum height of the building as well, depending on how that is calculated. Alternatively, it may be worth trying to qualify encroaching construction under rules for accessory structures.

Construction Beyond the Property Line

Building codes generally allow above- and below-grade extensions beyond the property line. This issue is regulated by the model codes' Chapter 32 regarding construction in the public right-of-way. In many communities, however, zoning ordinances do not permit such encroachment.

Below-Grade Encroachments

Background

One might think of below-grade encroachments as basements that extend under lawns or sidewalks. They come in two types—areaways and vaults.

Areaways are below-grade spaces that are unenclosed. Obviously they have walls to hold back the soil, but they generally have gratings for roofs and are climatically uncontrolled, meaning unheated and not waterproofed. They may have dirt floors, but more often have concrete floors and are fitted with drains for rainwater. Such spaces are generally used to bring fresh air to, or stale air from, below-grade mechanical rooms. They can also be very convenient as a path for removing damaged basement mechanical equipment for servicing or replacement.

Vaults are below-grade occupied spaces, so are climate controlled. In older cities, they are frequently found as receiving rooms for deliveries, used in conjunction with sidewalk freight lifts.

Issues

To a large degree, use of off-property areaways and vaults is a question of ownership and control of land use. Although they are underground, they are somewhat visible. Vaults may go unnoticed by passersby, but areaways blow hot air and trip pedestrians' heels, while sidewalk lifts pose their own hazards. Further, if they develop settlement problems or water leaks, repairs and maintenance tend to temporarily preclude sidewalks from use.

Requirements

Reference: IBC 3202.1

Occupied spaces enclosed by vaults must meet the same general building code requirements as any occupied spaces, including rules for exiting, light and ventilation, accessibility, and ceiling heights.

The structural slabs above vaults must bear the same loads required by public facilities manuals for any lawns, sidewalks, or roads not underlain by vaults.

The uneven cover panels on sidewalk freight lifts and vault access doors can pose a hazard to pedestrians. Photo by the author.

The one building code rule specific to below-grade encroachments prohibits any part of the structure of the main building envelope from being located beyond the property line [IBC 3202.1.1].

ABOVE-GRADE ENCROACHMENTS

Background

Above-grade property line encroachments, when permitted by zoning, can include overhanging rooms, bay windows, and balconies. These are all major building elements. When they occur above the pedestrian level of activity, smaller elements such as cornices, steeples and spires, and building signage are considered "projections" rather than encroachments (and are thus discussed in Chapter 5). When they occur near the ground, they can easily interfere with, or encroach on, human activity, so are discussed here.

Requirements

Encroachments Must Not Interfere with At-Grade Public Activities. Sidewalks are generally held to be public property. With few exceptions, when above-grade projections are allowed, they must leave the sidewalk to the public. The same concept also makes it easier for fire-fighting equipment to get close to a building. These goals are accomplished by requiring a minimum height above grade for encroachments.

As with other issues, zoning ordinances vary from place to place. The following rules are therefore derived from building codes. The omission of zoning rules here means only that readers must check with their local rules, not that there are none.

For building codes, minimum height is generally between 8 ft and 15 ft. This effectively forces creative structural design approaches such as cantilevers and stays, since any columns used to support the permitted encroachments would themselves be prohibited encroachments. An exception is allowed for columns supporting marquees, awnings, or signs. Pedestrian bridges attached to facades avoid the problem, since they can be supported at the other ends of their spans.

Allowed Elements. Within the minimum height above grade, building codes establish two subzones:

- a lower zone where encroachments might interact with people [IBC 3202.2], and
- a higher one, beyond reach, which might interfere with fire trucks or regular servicing handled from the ground, such as window washing [IBC 3202.3].

Flagpoles, awnings, marquees, and other encroachments are permitted when they don't interfere with activities in the public way. Photo by K. Wyllie.

For the lower zone, model building codes permit columns, other architectural trim, and steps, but only within a maximum depth that differs for each element. Even so, the deepest allowed is one foot. In addition to fixed elements deeper than this, moving elements that encroach, such as doors and casement windows, are prohibited. Recess those entries! Despite this requirement, building codes allow a break during the winter: temporary vestibules of limited encroachment are allowed for up to seven months of the year.

For the higher zone, permitted elements include awnings, marquees, architectural features, and such mechanical equipment as through-wall or window air conditioners. Encroachment is allowed on a sliding scale to a maximum of four feet, depending on height.

Maximum Extension. Beyond the higher zone, there are no limits imposed by building codes [IBC 3302.3.3]. However, zoning or other codes may continue to affect this area.

Optimizing Above-Grade Encroachment Rules

Sometimes the best way to achieve an encroachment that doesn't encroach is to recess the facade from which it extends. Since most rules are concerned with encroachment across property lines rather than facade extensions in general, moving a wall away from a property line allows room for extensions. Of course, this also tends to reduce the buildable area, so it must be weighed against other project goals.

Cones of Vision

Background

In residential districts, where sidewalks tend to be narrower and children may be present, some ordinances try to help drivers see around corners by restricting construction at intersections. Some provide similar requirements for commercial and mercantile districts in response to the high volume of pedestrian and vehicular traffic they generate. These areas of restricted development may be referred to as vision clearance areas, cones of vision, or some similar term. Provisions require the triangular areas at the tips of each corner property on an intersection to remain open at windshield height for a certain distance from the corner. Although cones of vision are mandated as a way to deal with vehicles, they are related to property lines and thus are included here.

Requirements

As always with zoning requirements, requirements vary. They generally establish two things: the location of the cone of vision and the types of improvements prohibited within it.

As for the location of the cone, there are only a few methods for setting it. Vertically, some provisions require that the cone extend from grade to the sky. Others are more generous, establishing a "window" by listing two dimensions, referenced to curb height, between which limited construction is permitted. This latter requirement allows cantilevered space to overhang the cone of vision.

Horizontally, a corner property's cone of vision is established as follows: Every property on an intersection is flanked by two street centerlines and two right-of-way lines. Ordinances vary as to which of these sets of lines are used to establish the position of the cone. Whichever it is, each ordinance sets a minimum distance measured along the lines from where they intersect. The two points established by this measurement, when connected to each other, form the diagonal edge of the cone of vision.

The restrictions on improvements in a cone of vision can be strict. Ordinances may list both architectural and landscape elements such as walls and plants. Building to the corner while providing visibility through the use of large strategically placed windows is generally not permitted.

DEALING WITH ADJACENT BUILDINGS

Designing for fire spread, a building code issue.

Context

This is the issue over which building codes were first written, following those disastrous turn-of-the-nineteenth-century urban fires. It is still of critical importance, as demonstrated by the millions of dollars in property loss caused by uncontrolled spread of West Coast forest and brush fires during exceptionally dry summers, the spread of fires during such urban riots as the one that followed the Rodney King verdict in Los Angeles, and the multiple blocks of Philadelphia that burned down following the police raid on "Move" in 1985.

Fires can't spread if there are adequate fire-resistive barriers in their path. Unfortunately, most barriers that can contain fires also inhibit occupant circulation, visibility, and air movement. One barrier that doesn't have this drawback is vacated horizontal distances. This is because hot air rises, and flames generally travel upward. Fires do not travel significant distances horizontally except when carried by wind-blown embers or when consuming fuels, such as building materials, that spread horizontally from areas already burning. In an analogous situation, forest fires are fought by clear-cutting treeless swaths completely around them. With no fuel, the fires cannot spread. Regulating the distance between buildings and regulating the fire resistance of exterior walls and roofs help control the spread of fires.

The main problem with using distance to provide separation is twofold:

- On small properties, once the unbuilt zone is set aside, there may not be enough land available to fit the proposed building.
- Land is expensive. The investment team pays a price in lost revenues for any land left undeveloped. This argument can actually be used to justify the costs involved in building fire-resistant separations, since compared with the cost of the equivalent physical distance, such construction is cheap.

Fire Separation Distance

A good place to start is the definition of *distance.* Fire separation distance measures, if indirectly, how far one building is from another, since buildings are a fire's sources of fuel. Let's look at the logic behind the different possibilities for defining fire separation distance:

The shortest distance between two adjacent buildings. Unfortunately, shortest distance formulas could lead to buildings with insufficient separation distance, or to a loss of property value. To illustrate, let's suppose a new library was being positioned fifty feet from a neighboring apartment building and twenty-five feet from the common property line. According to most building codes, it would be sufficiently far from the apartments to need no fire-resistance rating. Since no rating would be required, the project team would most likely decide to build it with an unrated wall. If the apartment building was subsequently torn down to make way for a new one, one of two situations would occur:

- If the new apartment building was built closer to the property line than the old one (assuming that zoning ordinances allowed it), the library, with its unrated wall, could be endangered. The owners of the library could be forced to refit the library with a new exterior wall before construction on the new apartments started.
- If a portion of the apartment building site had to be left off-limits to the new apartment building because of the nearness of the unrated library, the value of the site would suffer. The owners of the apartment building might have grounds for suing the owners of the library for the value of the unbuildable land.

For these reasons, this approach is not used by any of the model building codes.

The shortest distance between a building and a property line. Using this approach, one is assured that regardless of what adjacent property owners decide to do, minimum separation distance will be maintained. This is the approach taken by all three model building codes for interior lot lines. An interior lot line is a property line that borders other properties rather than adjacent streets.

In some cases, however, it would provide more distance than necessary. This happens when the adjacent land is permanently set aside as unbuildable. In such cases, the closest adjacent building is separated at least by the width of the intervening unbuilt land. This is true for the side of the property facing a street, since no building could be closer than the other side of the street. For these conditions, distance can be calculated to the centerline of the unbuilt land, as noted below.

The shortest distance between a building and the center of some unbuilt land. Using this approach, distance is defined from the building face to the centerline of an unbuilt strip of land. This is the approach taken by all three model building codes for exterior lot lines. An exterior lot line is a property line that borders streets, parks, or other land permanently precluded from construction. Since most properties are wider than the maximum separation distances required by codes, this approach effectively allows for unrated exterior walls even in zero-lot-line situations. It explains the common acceptance of glass curtain walls in street facades.

When an owner decides to build several buildings on one piece of land, this approach is also appropriate. Since there is no property lines separating adjacent buildings, model codes allow the project team to establish an "imaginary line between two buildings" and set their distances from that. Since a single entity (the owner) controls both the setting of this line and the property's development, the intent of the distance is assured [BOCA93 702.1: fire separation distance].

TRANSPARENCY OF THE EXTERIOR SHELL

Background

Even when distances are minimal or nonexistent, a fire-resistant wrapper protecting a building can prevent the spread of fire. The exterior walls and roof can form such a wrapper. Conversely, the more porous exterior walls are to heat or particulate matter, such as burning embers and sparks, the more likely they are to transmit fire. Porosity can occur in three forms:

- openings, such as windows and doors
- fire-susceptible walls, such as wood-framed construction or glass curtain walls
- fire-susceptible roofs

Fortunately, in all three cases, the farther those assemblies are from other buildings, the less likely it is that they will be exposed to heat and particulates. Codes consider this in their prescriptive regulation of exterior assemblies and openings.

This issue does not relate only to the distance separating unrelated buildings. It can also pertain to the distance between exterior walls of the same building when the spaces those walls enclose require a rated separation. For example, a team planning a building with assembly space and mercantile space in two adjoining perpendicular wings would have to carefully design the portions of each exterior wall nearest the inside corners, since fires could easily spread from one wing of the building to another at such locations. The same concern applies to internal courtyards.

Requirements

Walls, Basic Calculation

Rules mandate minimum wall ratings, in hours, based on the distance to adjacent construction and the occupancy of the enclosed space. This hourly rating applies to the entire wall and all of its parts but can be used flexibly by taking advantage of exceptions. A key exception is allowed for protected and unprotected openings constituting no more than a prescribed maximum percentage of the wall (see next section).

According to BOCA, most occupancies can be built with unrated exterior walls where a minimum distance of 10 ft is provided, measured as noted above. Exceptions are for factory, hazard, mercantile, storage, and utility uses. Even for the most dangerous condition, uses susceptible to flammability hazard, projects can be designed with unrated walls so long as there is at least 30 ft of separation. When insufficient distance is provided, ratings as high as 4 hours may be required, with 1-hour reductions allowed for sprinkled buildings [BOCA93 705.2: exterior wall fire resistance ratings].

In the One and Two Family Dwelling Code, a distance of 3 ft is the minimum for unrated walls and wall openings. Below that, no openings are permitted, and 1-hour fire resistance is required [CABO92 R-202].

Walls, Openings Allowed

Concept

The need for fire-resistive exterior walls does not mean that all buildings must exclude daylight and views. Such an exclusion would be in direct conflict with the need for light and air, not to mention what it would do to the pleasantness of spaces enclosed. Windows, doors, and louvers, known technically as openings, are permitted whether bare or protected, either left open or filled by glass or other materials. An opening is simply a breach in a fire-rated exterior wall. As such, it may allow a fire to spread from neighboring buildings. **Unprotected openings,** as the term suggests, are permanently unrated. They are therefore simpler and cheaper than protected openings. **Protected openings** are openings that are unrated under normal conditions, just like unprotected openings. However, in case of a fire, their fire resistance is automatically, and without electrical assistance, raised. It must be done automatically so that people aren't required to make the system work, and because fires often cause power failures.

Sizing

While both protected and unprotected openings may be used in any project, the amount that can be used varies depending on distance to adjacent buildings. Amounts are given as the percentage of the total wall that is occupied by openings. For example, a north facade that is 10 ft x 100 ft, or 1000 SF and allowed to have 30 percent protected openings, can have a maximum aggregate protected window and door area of 300 SF [BOCA93 705.3:maximum area of exterior wall openings; BOCA93 706].

Strategies

Rather than reducing the number or size of exterior wall openings, you could alter the ability of such openings to transmit fire. Get as close as you want by protecting the integrity of the enclosure at openings. Commonly, this is done by protecting such openings in one of several ways:

- Providing drop-down shutters. Shutters are often concealed in ceilings above openings and can be made of a variety of fire-resistant materials. They will normally close if opened and are made automatic by holding them open with fusible links that melt when exposed to heat (see p. 254).
- Using glazing with intumescent glass that foams up when heated, losing its transparency but becoming fire resistant.
- Providing closely spaced sprinkler heads adjacent to the opening. (See pp. 215–218 for information on smoke and fire control.)

Roofing

In most situations, some degree of fire resistance is required of roofing materials. Quality is evaluated as Classes A, B, and C, designations that are paired with different construction classifications. Codes even allow the use of materials that have no documented fire resistance when:

- used in buildings that qualify for the lowest construction classifications (BOCA 5B) and are a substantial distance (30 ft) from adjacent buildings.
- used in detached residential projects (R-3) that are a moderate distance (6 ft) from adjacent buildings. [BOCA93 1506.1.4: nonclassified roof coverings].

Optimizing Transparency of the Exterior Shell

Get as close as you want by:

- Adhering to stated limits on the amount of exterior wall openings.
- Modifying the percentage counted as opening by increasing the fire-resistance characteristics of more of the wall. One way to do this is to change portions of the wall to exterior wall assemblies with higher fire-resistance ratings; another is by use of protected openings.
- Providing sprinkler systems to protect enclosure assemblies, allowing for an increased percentage of opening.
- Reducing the fire exposure and preparing an equivalency, typically granted through a variance interpretation or a board of appeals hearing.
- Demonstrating that the rules are unnecessarily conservative for your project. For example, open garages must be open for purposes of ventilation. When adjacent to office buildings, the facing walls may be required to carry ratings. By pointing out that your office is designed to be fully sprinklered, and that open garages are classified in the S-2 (low-hazard storage) use group, you may be able to get the shell transparency the garage needs without increasing the distance between the office and the garage.

Dealing with Site Features

The conditions explored in this section often affect only part of a property. They may, however, either singly or in combination, affect entire properties. When the latter situation is encountered, any development may be infeasible. Properties are rendered unusable, as explored previously (see pp. 88–92).

CASE STUDY

Fairfax County, Virginia, has a fairly extensive mechanism for environmental regulation in place. Its Department of Environmental Management (DEM) requires site permits before issuing building permits. The site permit process involves reviewing proposals for compliance with rules found in the county code regarding:

- Soil suitability
- Excavation design
- Grading, drainage, and erosion control
- Site clearing (removal of vegetation, including trees)

In addition, the county has a Problem Soils Ordinance, which requires testing, and follow-up inspection when poor soils are present.

Since the middle 1980s, Fairfax County has operated an Engineers and Surveyors Institute to educate professionals about their regulations. This effort is part of a plan to certify Designated Plans Examiners who are allowed to use an expedited permitting process by virtue of their expertise. The program has recently been extended to architects.

Those interested in following or learning from these efforts can visit the Fairfax County Department of Environmental Management at www.co.fairfax.va.us/gov/dem on the Internet.

Where these conditions affect only a portion of a property, careful positioning of construction within the property can minimize the restrictive effects of these conditions.

Such conditions include floodplains, wetlands or habitats, steep slopes, soil conditions, placement of septic systems, solar access, and archaeological finds. Even landfills, if the property is sufficiently large, may be contained and localized to a limited portion of a property. In such cases, these conditions may be thought of as site features to be designed around, rather than as "deal breakers." Some site features are rarely so global as to rule out an entire property. Taking the optimistic tack, these conditions are being explored here in the hope that project teams will more commonly proceed with projects through careful positioning than be forced to abandon them.

Finding requirements for the site issues described below may require some hunting. Project teams may need to check several sources, from the local to the federal level.

FLOODPLAINS

Background

Floodplains are low-lying areas that flood periodically when adjacent river banks or lake shores are unable to contain heavy rains or melting snows.

Rules and maps may refer to 10-year, 50-year, or 100-year storm lines, lines that trace the boundaries of areas expected to be inundated by the floodwaters of such storms. The reference to time does not mean that a bad storm two years ago will preclude another this year. It means only that in any given year, there is a 1-in-10, -50, or -100 chance that a storm of such intensity will happen. Changing weather patterns over the last few decades, influenced by global warming, are indicating that such odds may need to be revised, meaning that areas designated as floodplains may need to be expanded.

Federal law excludes new projects constructed in designated floodplains from receiving aid or assistance through federal flood relief programs. In addition, many zoning ordinances prohibit new construction in floodplains, and some communities have resisted the temptation to rebuild flood-caused damage. Still, there are many communities in which occupied buildings continue to be found in floodplains. For this reason, all model codes have rules governing construction in floodplains, as does the Federal Emergency Management Agency (FEMA).

Applicability

Floodplains are identified on Flood Hazard Boundary Maps (FHBMs) in engineering reports and maps prepared by the Federal Insurance Administrator. The administrator prepares these maps because the federal government carries much of the financial risk of flooding through its National Flood Insurance Program (NFIP). These documents are available through local government offices. Different parts of a floodplain carry different restrictions depending on their potential to be damaged by floods. These maps designate several levels of hazard:

- Areas of Special Flood Hazard, or Zone A. These are areas within the designated (10- or 100-year) flood zone.
- Areas of Shallow Flooding, or Zone AO or ZO. These are a subset of Zone A areas. Shallow flood areas anticipate one to three feet of flooding, and have no clear channel, making it harder, or impossible, to predict the path floodwaters will take.
- Coastal High Hazard Areas, or Zone V, VE or V1-30. These are areas of high-velocity flooding including hurricane-driven waves.
- Floodways. These are channel and adjacent land that must be kept clear of construction to allow efficient drainage of floodwaters.

All development is prohibited in floodways, but properties in the literal middle ground, low enough to be damaged but not so low as to require abandonment, are allowed to proceed when designed in accordance with floodplain rules.

Requirements

In the model building codes, floodplain rules require that subdivisions are laid out in such a way that every property includes "adequate buildable area outside the regulatory floodway" [IBC A31-500.2.2]. Many such rules were developed by the SBCCI and published in its *Standard for Floodplain Management.* With code unification, many of these provisions are finding their way into the IBC.

Federal law [33 USC 2318] denies inclusion under federal flood damage reduction projects, such as loan programs, to "any new or substantially improved structure (other than a structure necessary for conducting a water-dependent activity) built in the 100-year floodplain with a first floor elevation less than the 100-year flood elevation." Where whole counties are at least 50 percent below the hundred-year floodplain, the limit is dropped to properties built in the 10-year floodplain. Programs include the National Flood Insurance Act of 1968 [42 USC 4001 et seq.].

When construction is permitted or allowed to remain, some otherwise permitted uses are restricted, such as [IBC App. 31-100.1]:

- those uses that may expose people or property to damage;
- those uses that may intensify the flooding or such flood effects as erosion and water speed.

Even when properties susceptible to flood damage carry no use restrictions, they are required to be built to withstand flooding. (See p. 167.) The elevation of lowest structural members is generally set by and indicated on Flood Insurance Rate Maps (FIRMs).

Optimizing Floodplain Rules

What other kind of development is possible for sites that prohibit any form of occupiable construction? Use is generally limited to water-related facilities, such as docks and piers. Consigning such properties to use as recreational facilities or gardens may be the only option possible. When only part of a site is within a floodplain, careful positioning can make a property usable.

WETLANDS

Background

Wetlands include areas with high water tables that are permanently soggy but rarely flooded, and floodplains that dry out but are flooded again on a regular schedule. One might say that they are the interface between land and sea. Wetlands are defined in the Clean Water Act [33 USC 1251 n.18] as "those areas that are inundated or saturated by surface or ground water at frequency and duration sufficient to support . . . vegetation typically adapted to life in saturated soil conditions."

In terms only distantly connected to the construction industry, they are home to a wide range of plant and animal species. Much more directly critical to human needs is that wetlands stabilize coastal areas, slowing erosion at oceans, rivers, and ponds. In addition, they are our planet's filtering agents, processing waste products into nutrients. Control over wetlands is a key issue in the property rights movement, whose advocates are understandably upset to find out that they won't be allowed to develop properties they own as they see fit.

The ecological complexity of wetlands has made a mockery of attempts to artificially create them. Still, that hasn't prevented many projects from gaining permission to develop wetlands on the promise of building "constructed wetlands" elsewhere.

Applicability

Although development of wetlands is prohibited by federal law, many property owners have had trouble determining if their properties are considered to be wetlands by those same federal laws. Congress has rewritten the laws defining wetlands multiple times in the last decade or two and will probably continue to do so in response to ongoing scientific research and the fluctuating political clout of those on either side of the wetlands debate. The latest versions can be found in the *Federal Register* even before they are published in the *Code of Federal Regulations.*

Requirements

Issues arising out of wetlands protection include:

- Prohibition of landfill. One of the best ways to destroy a wetlands is to bury it, to fill it in [33 USC 1344, Section 404].
- Prohibitions against discharging of hazardous substances, especially oil [40 CFR 117: *Determination of Reportable Quantities for Hazardous Substances,* and 40 CFR 110]
- Prohibition of impermeable surfaces. Asphalt and concrete parking lots prevent the infiltration of rainwater, so interfere with wetlands functions.
- Prohibition of trenching, which although it increases usefulness to humans by drying out the top few feet of soil, hurts wetlands by increasing the speed with which water migrates across the site.

A study conducted at Lawrence Livermore National Labs summarizes the requirements very effectively[24]:

The most wide-reaching wetlands program is administered by the U.S. Army Corps of Engineers, which regulates discharges of dredged or fill material to, and placement of structures in, wetlands. This program requires parties wishing to develop wetlands to obtain a permit. Before granting a permit, the Corps may require measures to be taken to mitigate the impact to wetlands. Special policies (e.g., requiring floodplain/wetlands assessments) also apply that further limit the extent to which federal projects may destroy or degrade wetlands.

The U.S. Army Corps of Engineers derives the regulatory authority to administer its wetlands protection program from Section 404 of the Clean Water Act. The program is overseen by the EPA, and other agencies such as the Fish and Wildlife Service and California State Water Resources Control Board may make comments that affect the permitting decision. It is important to note that Section 404 wetlands permitting is time-consuming, typically taking six to eight months (and sometimes longer) to complete.

Certain types of projects do not require individual Section 404 permits. Projects such as maintaining drainage ditches, maintaining currently serviceable structures, and very small projects involving less than 0.1 acre of wetlands are typically eligible for general permits (sometimes called "nationwide permits"), which bypass the individual permit application process. However, it is prudent (and sometimes required) to notify the Corps before commencement of the project and receive official confirmation that an individual Section 404 permit is not required. Even if a project qualifies for a general permit, it will still be required to obtain water quality certification from the Regional Water Quality Control Board before proceeding. Also, other state and local programs that may affect wetlands, such as the streambed alteration agreement program administered by the Department of Fish and Game, may impose additional requirements on a project.

In addition, all DOE actions must comply with the DOE policy regarding wetlands as stated in 10 CFR 1022: "Compliance with Floodplain-Wetlands Environmental Review Requirements." The DOE wetlands policy was put in place to comply with Executive Orders 11990 and 11988, which require federal agencies to consider adverse effects on wetlands and floodplains during decision making. For certain proposed actions that will affect floodplains-wetlands, the DOE must prepare a Floodplain-Wetlands Assessment that includes an evaluation of the effect of the proposed action on these resources and an evaluation of less-harmful alternatives to the action. Typically, the DOE's wetlands policy is satisfied by incorporating the Floodplain-Wetlands Assessment into documents prepared in compliance with the National Environmental Policy Act.

Optimizing Wetlands Rules

A middle ground may be possible in reclaiming wetlands damaged by earlier development or misguided flood control efforts. In fact, there is a growing cadre of "wetlands mitigation bankers" who rehabilitate damaged wetlands and then sell sponsorships to developers who own natural wetlands in more desirable locations. The developers then use their sponsorship of the reclamation efforts as credits in obtaining permission to develop (generally meaning destroy) the healthy wetlands on their better-located properties.

Wetlands banking has some impressive supporters, including the EPA, Congress, and the Clinton administration. A recent article quotes EPA scientists and enforcement officers as saying that "consolidated restoration efforts... lead to 'more thoughtfully designed wetlands' that are easier to monitor... 'Mitigation banks are a potential win-win. Banks can retain wetlands... They also reduce delays for builders who can replace wetlands by going to a bank and buying credit.'"[25]

In contrast, this is all based on the assumption that saving one wetland on the verge of destruction is the equivalent of destroying another one, even though when all is done, only one reclaimed wetland remains where there were once two healthy ones. Further, the wetlands mitigation has not been followed long enough to know if it has lasting effects. "The federal government hasn't set requirements . . . or standard practices that wetlands developers can expect to adhere. Until it does, 'the risk associated with these ventures is great.'"[26]

Soil Conditions

Background and Rules

Soils are classified into sixteen grades based on particle size and particle type. These two characteristics in turn affect the structural stability of the soil and the speed with which water drains through it. Those two characteristics in turn affect the amount of lateral load imposed by the soil on retaining walls, including foundations and the soil's stability under seismic forces.

The cost of replacing "bad" soils with "good" soils is enough to make any other alternative look good by comparison. Still, "bad" soils can increase project costs substantially by forcing higher strength structures and more elaborate waterproofing and dewatering methods.

With respect to hydrostatic pressure, codes mandate the amount of lateral load a design is required to resist, based on the types of soils present. The difference in the load required in response to well-draining soils (30 lb) and poorly draining soils (60 lb) is 100 percent [IBC 1610]. This is sufficient to justify a reasonable degree of soils investigation when property is sufficiently large to allow a choice of building locations and has a variety of soil types in different areas. Where margins are particularly slim, it could justify rejecting a site from consideration.

Few variables and no exceptions are listed with these requirements. Design lateral loads are specified as measured in pounds per square foot per foot of depth, so increase arithmetically as the depth of the excavation increases. In addition, the loads noted above are increased to 60 and 100 lbs/ft/ft for many larger scale projects.

As for seismic performance, soils are classified as Types A through F (see p. 85) to indicate their degree of stability under the kinds of loads imposed by earthquakes.

Optimizing Rules Related to Soil Conditions

Several options suggest themselves, some or all of which may not be feasible:

- Take additional soil borings in an attempt to find areas of the site with better soils.
- Minimize the depth of the excavation, either by reducing the floor-to-floor heights of basement space or by raising the elevation of basement space as much as possible.
- Amend or replace soils with better-draining soils.

Rights of Passage

Background

Rights-of-way and easements are areas within properties whose control is shared between the owner and designated others. They may be set aside for either of two reasons:

- Rights-of-way: To grant people other than the landowner a right-of-way. This is useful when certain people need to get across a property and don't want to violate trespass laws when they do.
- Utility Easements: To allow for maintenance, repair, and expansion of buried pipes, cables, water lines, storm sewers, or other utilities as needed.

In that they are set-asides, they have something in common with setbacks. Unlike setbacks, however, they follow a designated corridor or the path of utility lines, rather than property lines. Where possible, utilities are located in public rights-of-way, such as roads. They may stray, however, occupying bands along property lines or slicing right through properties.

Rights-of-Way

Rights-of-way are circulation paths, usually but not necessarily at grade. They may be granted to fire departments so fire trucks can be driven across private property to reach burning buildings. Rights-of-way needed by government entities are mandated by right of eminent domain. With private parties, they are extended by individual landowners at their option and recorded in deeds. Once extended, they may only be withdrawn by the granter. As an example, rights-of-way may be granted to nearby residents so they may cross large estates on their way to a private beach. Within the right-of-way, construction may be allowed above and below the path, so long as a "virtual tunnel" remains.

Utility Easements

Many properties contain no easements. For those that do, locations and widths of easements are recorded on plats and surveys.

Easement rules prohibit construction that interferes with access. In general, this means that all permanent construction is prohibited in the easement. In particular, elements of the project required for occupancy must be kept out of easements, as they would likely need to be demolished if access became necessary.

For specific regulations, check state and federal codes. For federal properties, see 43 USC Chapter 22: "Rights of Way and Other Easements in Public Lands."

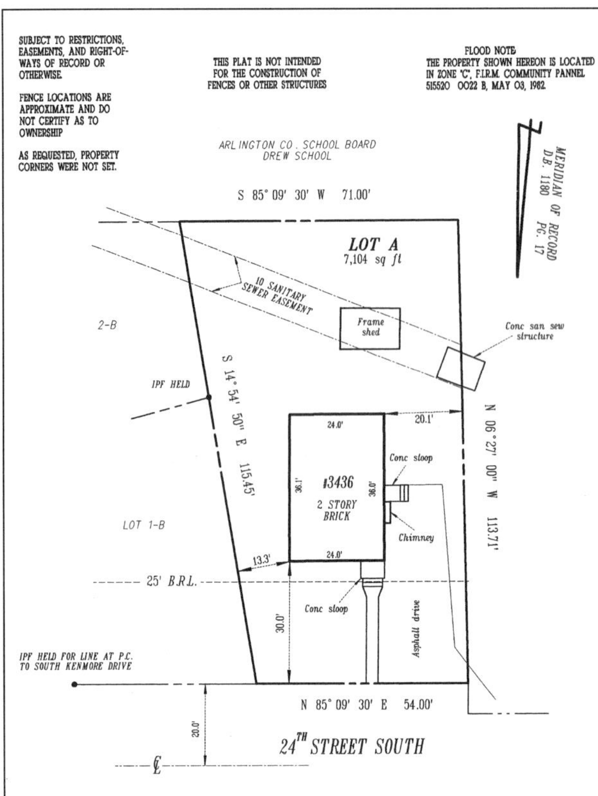

A plat, indicating positions of utility rights-of-way. © AVW & Associates.

ARCHAEOLOGICAL FINDS

Background

Some sites are home to significant archaeological resources. Particularly in older urban areas, virgin sites are rare. Before a project team makes too many plans, an archaeological survey of the site can establish the value of any foundations or artifacts found on the site. A common approach in cases where land values make it infeasible to cancel a project no matter how significant its historic value is to delay construction. This allows for excavations to document the discovery and for removal of significant, movable elements.

Applicability

Almost any site may hold artifacts of archaeological value. The problem is, few sites are known, and it's close to impossible to tell by cursory analysis of a property. Finds are usually discovered during excavation operations, when it is too late to do anything but deal with it.

Requirements

The Secretary of the Interior's Standards for Rehabilitation [#8] require that historic materials "affected by a project shall be protected and preserved." By stipulating that "mitigation measures shall be undertaken" in situations where resources "must be disturbed," they accept the notion that it is not always possible to avoid disturbing such sites, and attempt to provide a way of compensating.

Other rules related to archaeological finds are found in 43 CFR 7 (archaeological resources protections afforded by the Department of the Interior) and 33 CFR 325 (historic property protection procedures mandated by the U.S. Army Corps of Engineers).

Clearly, the implications for large buildings in our cities' central business districts are significant. In situations where the conflicting interests are resolved well, some amount of public goodwill is gained at the cost of often significant delays.

Optimization of Archaeological Finds Rules

With a proper head start triggered by sensitized project teams, "mitigation measures" can be completed before design work is completed.

Other solutions suggest themselves as well:

- Buildings can be carefully positioned to avoid sensitive areas of the site, so long as lot coverage is expected to be less than 100 percent and the evidence is concentrated.
- Decentralized (short span) structural systems can result in smaller footings. Even though more footings will be needed, they will exert less pressure on soils that may still contain artifacts.
- Centralized (long span) structural systems may be able to bridge across sensitive areas, leaving them undisturbed for ongoing exploration.

OTHER SITE FEATURES

Forests and Habitats

From a regulatory perspective, forests and habitats are areas where older growth woodlands or rare species are being endangered by development. As with wetlands, legislators are continually modifying their definitions of which forests and species are in need of protection.

CASE STUDY

During excavations for the new federal office building in lower Manhattan, New York, an African burial ground was discovered containing the remains of some of the first slaves brought to this country. The entire project was suspended while archaeologists carefully unearthed the remains and artifacts. The entire site was designated as historic by the New York City Landmarks Preservation Commission on February 25, 1993. Some remains were removed to allow construction to proceed.

Also as with wetlands, mitigation has become a major management method for forests and habitats. Developers pay to repair damage done to other properties in exchange for permission to damage proposed properties.

Steep Slopes

Development is often limited in areas where surface slopes exceed stated maximums. Regulators' concerns are those of possible soil instability and excessive erosion. Media coverage of mudslide damage too often includes images of buildings moved off their foundations. Some local codes limit construction on steep sites. Others prefer to approve permits on a case-by-case basis, allowing projects to proceed when clever design of footings and sensitive use of site features raise the chances for a good building to acceptable levels.

Where rules limit construction on steep slopes, *steep* is commonly defined as slopes exceeding 5 percent. This is relatively flat, where any instability in soils or problematic erosion could be compensated for relatively easily through proper engineering. Rules like this bear some responsibility for the overwhelming tendency of developers to level any site before developing it.

Solar Access

If no part of a site is ever in the shadow of another building, one would say that it has complete solar access. Energy awareness, with its emphasis on daylighting, passive heating and cooling, and photovoltaic electrical generation, is bringing new interest in solar access. Still, codes have not yet been needed to address this issue directly. They already do it indirectly, through lot occupancy and setback rules. Typical Floor Area Ratios required of suburban "flex" space (industrial park one-story-plus-mezzanine projects) are in the 0.25 to 0.40 range. With such low densities, solar access rules are hardly needed.

Septic Systems

State and local health department rules require that minimum distances be maintained between new or existing septic fields and new construction due to concerns for hygiene as well as soil instability. They also generally regulate the locations of septic fields relative to floodplains, to reduce the likelihood of system backup when water reaches flood stage. For low-lying sites, this double regulation may eliminate the possibility of fitting floodplain, septic field, and building on one site.

POSITIONING FOR PEDESTRIAN ACCESS

Accommodating strollers, wheelchairs, bicycles, and delivery carts (hand trucks).

Keeping People Away

Occasionally a project comes along where one of the design goals is keeping pedestrians away. This might be the case to secure against entry (an embassy or high school), to secure against damage (a historic property), or to protect people from injury (a zoo building or an airport). In some cases it is important to keep pedestrians away only from certain portions of a building (loading docks or mechanical exhaust grilles).

Where projects that need to keep people away are scattered, key strategies include positioning to allow wide expanses of open ground that make concealment difficult, or erecting physical barriers such as railings and moats. Shrubs can also be effective. There do not appear to be any rules addressing this issue. Vandalism and injury do occasionally follow close encounters with buildings, but perhaps drafters have felt that rules would have little effect on undesirable behavior.

A better strategy is not to scatter projects that need to keep people away. Zoning ordinances can "enclave" like facilities by carefully crafting permitted use lists. Higher hazard industrial projects can be banished to areas of lower density development. High-security projects like embassies and museums can be "enclaved" where there are enough such projects to justify doing so, perhaps in special-use districts.

Letting People Get Close

Meeting Them Halfway

Some projects strive to encourage pedestrian interaction. Restaurants put tables on sidewalks to entice pedestrians to become customers. Retailers do it just as vigorously with sidewalk sales. Oftentimes such activities happen spontaneously and without official approval. Generally it isn't until such activities generate complaints that affected localities adopt ordinances specifically addressing sidewalk cafés and other annexations that spill over property lines to occupy public rights-of-way.

Broadcasting a Welcome

Signs are generally scaled for vehicles, but buildings also encourage pedestrians to get close with display windows.

Getting in Their Faces

Zoning guidelines that discourage or prohibit front yard setbacks have the effect of mandating closeness. This is one of the key distinctions between urban and suburban environments.

Ramping

Background

Sloped sidewalks for even small changes in grade from property lines to interior floors can consume significant amounts of horizontal space. Ramps are long. Where a code-compliant stair ascends at 32°, a ramp rises at under 3°. The increased number of landings required for ramps makes them even less efficient. Between slope and landings, ramps occupy almost thirteen times the space of stairs.

Many accessibility rules seem oriented toward wheelchairs. This is not necessarily the same as being oriented toward people who, given current technology, depend on wheelchairs for mobility. Still, there is little functional distinction between the two concepts. Hundreds of patents for alleged stair-climbing people carriers have been granted, but no successful devices have yet reached the market. And although still many years away from human application, recent laboratory experiments are indicating potential for treatments that regenerate severed spinal cords. Until these are adapted for people, or radically different walking aids are developed, we will continue to need ramps.

Requirements

Allowable slope depends on distance. Wheelchair ramps are generally limited to a maximum slope, expressed as a percentage of vertical distance (rise) over horizontal distance (run). This is not only to ease the effort required to ascend the ramp, but also to control speed on descending. The shorter the ramp is, the steeper it is allowed to be, up to a maximum slope (generally set at 8 percent). This is because difficulty and danger increase as length increases, since handicapped people need more time to negotiate longer ramps. A curb cut is an example of a steep but short ramp.

Uninterrupted length is also limited, since weak people (or attendants pushing heavy wheelchairs) need to rest between exertions, as well as due to the danger posed by a runaway wheelchair. For this reason, most rules require level landings to be placed on a regular basis throughout long ramps, usually defined as those over 30 ft long.

No matter what design approach is taken, ramps are relatively long. The steepest ramp generally permitted, sloping at a ratio of 1:12, would need four runs plus three landings, for a total length of 135 ft to ascend a fairly tight 10 ft floor-to-floor height.

Optimizing Ramping Rules

On a sloped site, let grade be your ramp, and provide exterior doors on all levels. Obviously this will not work for most ten-story buildings, even in San Francisco.

Consider taking advantage of the sculptural aspect of ramps to further other design goals. Long, winding circulatory elements do a lot to enliven spaces. They're not just for the handicapped anymore!

Provide elevators. Even for exterior use, special low-maintenance, weather-resistant models are available.

Elevators are expensive, but so is the cost of devoting valuable building space to a function that doesn't directly contribute to the project's program.

POSITIONING FOR VEHICULAR ACCESS

Fire vehicles, cars, trucks, and motorcycles.

Context

Vehicular access is primarily in the domain of planning rules, the guardians of community character and available amenity. Poorly planned parking and circulation leads to traffic congestion. It also leads to confusion over whether people or cars are the intended beneficiary of societal effort.

"Slowing the Traffic." © Roger K. Lewis.

Roads

Required Dimensions

Background

APPLICABILITY

Private. Where a development site remains undivided, on-site roads are privately owned. Responsibility for design, construction, and maintenance rests with the investment team rather than the government. Private on-site roads serve as driveways, frontage roads, and service streets.

Public. Where developments are subdivided into lots, the ground occupied by roads connecting the lots is usually deeded to the local government. Through this mechanism, roads are designed and built by developers before transferring title, and are thereafter maintained by the government. This is particularly advantageous to property owners in areas with high snowfall accumulations. Requirements for public roads are more demanding. (See pp. 32–33.)

HISTORY

In many suburbs, zoning ordinances were first written in the 1950s. Governments took an active role in providing housing for families starting out on the American dream following World War II, including assistance with such suburban infrastructure as roads. Design guidelines for streets were developed to promote easy flow of traffic rather than residential character. Twelve-foot-wide lanes and long lines of sight at intersections increased the speed with which cars could safely get to and from residents' houses. Level of Service (LOS), as measured by clocks and traffic counters, has been the prime determinant of successful street design for many decades. This situation could be seen alternatively as the ascendancy of drivers over residents, or as an admission that designing for safety minimizes the danger posed by rushed drivers.

Requirements

Public Facilities Manuals generally specify minimum roadway widths for one-way and two-way traffic, and the lane widths of multilane roads. Widths are generally specified as edge to edge dimensions, inclusive of any shoulder requirements. Although the radius required at corners generally has minor impact on overall design, manuals frequently specify minimum radii for inside corners at intersections. Ordinances usually also specify that vehicular curb cuts be kept a minimum distance from any intersection to avoid conflicts between through traffic and merging traffic.

Optimizing Road Size Rules

Proponents of "new urbanism" are challenging the assumptions of Public Facilities Manuals and working to have these concepts of suburban design excellence reconsidered. They are returning to narrower lanes, median strips, and alleys. In the neotraditional town planning of Seaside, Florida, by architects Duany Plater-Zyberk, driveways were listed as "parking lots" to take advantage of reduced requirements. They complied with requirements for aisles in parking lots; they simply didn't have parking stalls, an element not required by the applicable rules.

Many communities are fighting to assert their agendas on existing roads through a combination of planning innovations and tactical maneuvering, including:

- minimizing cut-through commuter traffic: redesignating two-way roads for one-way traffic during rush hour;
- reducing effective street width: rededicating lanes adjacent to curbs for on-street parking;
- forcing reductions in travel speed: constructing traffic circles and speed bumps in existing roads, deliberately procrastinating in repairing potholes, and parking retired police vehicles at strategic locations.

Grading

Background

Grade for roads, as noted for pedestrian ramps, is generally measured as a percentage of rise over run. A road that slopes 5 ft vertically for every 100 ft horizontally has what is referred to as a 5 percent grade. Public Facilities Manuals specify maximum grades for roads, including dedicated indoor roads such as ramps in parking garages. Often fairly steep grades are allowed so long as the grade is achieved in successive segments to prevent "bottoming out" of vehicles against the road surface.

Requirements

A typical requirement allows grades of up to 8% slope. Where the road is part of a parking lot, the maximum is lower, as noted in the section on parking, below.

Optimizing Road Grade Rules

Grades of up to 16 percent are generally allowed if done in 10 ft long (minimum) sections of 8 percent slope (maximum) each.

Fire Department Access

Background

Fire trucks must be able to get sufficiently close to burning buildings on all sides (including internal exterior courts) to rescue trapped occupants and to fight fires. For small properties, trucks can reach buildings from the street. For larger properties, provisions must be made to ensure access. Lawns are soft. Fire trucks are heavy and tend to involve lots of water, which makes the ground softer. In addition, some trucks, such as hook and ladder engines, use stabilizing outriggers to transfer off-center loads generated by side-stretching ladders to the ground. Fire trucks can't do their jobs unless suitable roads extend sufficiently close to those buildings for trucks to get where they're needed. Civil engineers and architects should review options for access very early in the site planning process for large properties.

Requirements

Distance and Arrangement

For low buildings, where hoses can reach every surface as long as the trucks carrying them can get close enough, roads as needed must be provided. The distance needed to accomplish this is a function of typical hose lengths combined with stream length, which in turn is a function of water pressure. Distances of about 300 ft are typical, which not coincidentally is the distance between fire hydrants on city streets. For taller buildings, trucks must be able to get sufficiently close, within about 75 ft for truck-mounted ladders to reach.

With multistory buildings, requirements are eased if a ladder truck can be used to increase access to the

CASE STUDY

An unusual accommodation was made during the design of the MCI Arena in Washington, D.C. The facility was intended to handle circus performances, and upon checking, the design team discovered that elephants panic and stampede when the surfaces on which they walk have grades of more than about 12 percent. A truck ramp leading to the below-grade loading dock was correspondingly designed for that maximum. Now, that's information design teams aren't likely to find in codes.

lower floors, both for fighting fires and for evacuating trapped occupants. This can pose particular difficulties for courtyard designs, since a truck simply cannot reach courtyard facades unless a wide opening is left in at least one of the courtyard walls.

DIMENSIONS

Model building codes require fire department access roads to be 18 ft wide [IBC 913.5]. Local ordinances are likely to differ, with 20 ft being a common requirement.

Roads are typically required to allow for 30 ft turning radii, although New York requires 35 ft. Again, it is prudent to ask.

QUALITY OF ROAD

Rules specify minimum bearing capacity and drainage required of fire department access paths. Thankfully, like most rules, there are creative ways to provide the needed capacity without turning the approach to a building into a parking lot. Some of these are noted below.

Optimizing Fire Department Access Rules

Paved roads are not the only way to keep fire trucks from losing traction and sinking in soft ground. Paved areas within ladder distance of exterior walls are problematic because:

- They may conflict with the intended landscaping design. The aesthetic impact of a building even partially surrounded with paving is not always appropriate.
- They affect a building's energy loads in two ways:
  1. On sunny days, paved surfaces get hot and retain their heat, which radiates into adjacent buildings. This is particularly true of dark paving such as asphalt. Paved surfaces can affect any facade including north-facing ones so long as unshaded portions of the paving are relatively close to exterior walls.
  2. Paved surfaces adjacent to east-, south-, or west-facing facades reflect much more solar radiation against a building than does a landscaped surface. This is particularly true of light-colored paving such as concrete. Energy-conscious project teams can use fire department access roads as part of an integrated approach to passive solar design.
- Like all roads, they create water management problems since the amount of permeable ground available to absorb rain is reduced.

To avoid these problems while retaining the required fire department access, many manufacturers now make fairly invisible ground reinforcement grids of concrete or plastic that distribute truck loads while maintaining permeability and allowing grass to grow.

A final issue concerns two-sided access to projects. Developers of dense urban town-house projects sometimes meet market demand for decks by erecting them in shallow backyards, narrowing the open space at grade. Due to the size of larger fire trucks, this arrangement may impede fire truck, but not fire department, access. These designs are sometimes approved when hydrants are available for both facades of each unit, since firefighters have access to water once they can carry their hoses to the rear facades by hand.

PARKING: PRIVATE VEHICLES

On- or Off-Site?

Background

Zoning ordinances require most projects to include on-premises parking facilities. Technically, such facilities are referred to as "on-site."

Requirements

What constitutes on-site? Codes generally require mandated parking to be within the property lines. Parking available on the street in front of a building cannot be counted in fulfillment of parking requirements.

Parking tends to take up so much room that design options for a building might be unnecessarily limited. Back when the architects' licensing exam included a twelve-hour comprehensive design project, would-be architects taking it were sometimes advised to design the parking lot first, then fit the building into whatever space was left. Since buildings are usually the reason for doing projects in the first place, this kind of situation is a significant problem.

Some rules allow on-site parking to be located outside the property line if it is permanently, legally, and contractually available within a certain maximum distance (generally about 200 ft) of the project. This allows, for example, a proposed church to transfer much of its parking requirement to an existing office building garage if the garage would normally be empty on Sundays when the church draws the most cars. The transfer would be allowed only upon negotiation of a contract obligating the office building to make the spaces available to the church at the required times.

Optimizing Off-Site Parking Rules

When sites are sufficiently large to accommodate all requirements through surface parking, no optimization strategies are necessary. When available parking exists within the distance limit, and the ordinances allow for its use, teams can start negotiating use rights. For the vast majority of urban projects and many suburban projects with limited sites, the economic implications of providing a parking structure will need to be considered. Although large surface parking lots carry their own problems (see p. 121), they are inexpensive and relatively easy to maintain. Parking structures raise levels of complexity in significant ways.

Location within the Site

Background

For parking that remains on-site, a layout must be developed. Where will it fit? As with other vehicular issues, the conflict between well-designed urban spaces and transportation realities is difficult to resolve. The issue is primarily of concern for projects with high parking requirements.

Requirements

Almost all zoning ordinances allow parking within setback zones. Since parking can be there and the building can't, it usually winds up there. This is a powerful reason for carefully distinguishing front setback distances from side and rear distances.

Also of concern are restrictions on locations of curb cuts, since they provide entry to and egress from parking lots. Curb cuts must be certain minimum distances from intersections. For corner properties, such distances encourage parking lot sprawl as a way to get two entries.

Optimizing Rules Related to Location within the Site

Surface parking is almost always the parking strategy of first resort. It beats underground parking and parking structures because it is cheap and can easily be demolished when the time comes to put a "real project" on a site. When located within the setback zone, it does not occupy land that could be used for the building. It does, however, consume a lot of real estate and is harder to conceal than enclosed parking.

When the directive is for surface parking in the setback zone, strategies are limited. Screen walls must be sufficiently minimal to be permitted within the setback. Berms are generally unrestricted and may also provide a way to deal with spoil, the dirt excavated to make room for the foundation. They are fairly wide, though, and can't easily be traversed.

When enclosed parking can be justified, considerations for its location are generally the considerations for placing any building of similar use group.

A common suburban parking strategy—the front setback. Photo by K. Wyllie.

A two-story parking structure of the type generally prohibited in a setback zone. Photo by K. Wyllie.

Required Quantities

Background

Many developers choose to program their projects for more than the minimum number of parking spaces. Adequate parking spaces, whether paid for or included in leases without additional charge as marketing incentives, are a real asset when trying to attract prospective tenants. Retail projects that suffer from inadequate parking have a very difficult time attracting customers. Still, one must be aware of rules, if for no other reason than to be able to explain the extent to which a particular project exceeds them.

On the other hand, there are projects whose feasibility can be pushed beyond reach by excessive parking requirements. A case in point is the fact that most residents of SROs (Single Room Occupancy residences) do not own cars, either because they can't afford them or because they are transient. Despite this, most zoning ordinances categorize SROs as apartment buildings, which require at least one parking stall per apartment. Since parking spaces consume about 270 SF each, and land costs are high in most of the urban areas in which SROs are located, rules can have an enormous and inappropriate effect on the affordability of the apartments.

Another situation in which high parking requirements can make a project infeasible is that of existing properties, especially if historic in nature. If the existing buildings occupy most of the site and there are no nearby parking lots available, minimizing required quantities may be critical.

Finally, since ordinances require certain minimum amounts of parking relative to the sizes of the projects they serve, it is important to understand the required method for calculating project size.

Requirements

Parking quantities are dependent on both zoning district and building size. A set of rules may count building size of a retail project by dimensional area (square feet), of a multifamily residential project by the number of apartments, in a theater by number of seats, or in a hospital by number of beds. Quantities may be specified in a two-tier format, often independently citing parking for employees and for shoppers, or for tenants and for visitors.

Optimizing Quantity of Parking Rules

Project teams can always try to get relief by pleading hardship. If it frequently worked, few projects would include required quantities. Other options are available:

- Use mass transit: When a project is accessible to public transportation in the form of buses and rail lines, teams may be able to argue that many occupants will leave their cars elsewhere. Nearby subways stations are particularly convincing for business and mercantile uses. In a strategy that has worked for many major cities around the world, taxicabs and bicycles have evolved to serve this market.
- Establish private ride shares: Make long-term arrangements for private ride share programs such as shuttle buses.
- Offer home delivery: In Manhattan, few people use cars, even if they own them. They shop, buy merchandise, and leave-empty handed. Businesses deliver everything to customers who can't take it with them.

Somewhere in these possibilities is a strategy for most projects with entrepreneurial project teams.

CASE STUDY

A Home Depot building supply store, proposed for the site of a former Sears in Arlington, Virginia, was successfully defeated by local residents partly by arguing that few customers would get there via the nearby subway system. Even though a station was only a few hundred feet away, the store was unable to demonstrate that sufficient numbers of customers buying sheets of plywood and kitchen cabinets would use it.

CASE STUDY

Stein & Company, an imaginative Chicago-area developer, used a creative strategy to minimize parking and maximize tenant amenity in developing 203 N. LaSalle Street, a high-rise office building on a highly unusual site in Chicago's Loop. The building is a convenient subway ride from O'Hare airport and is immediately adjacent to two elevated routes and one subterranean route, with another nearby. Stein & Company developed the project with two floors dedicated to airline ticket counters. Mr. Stein then arranged the private lease of subway cars to bring luggage from the building to O'Hare Airport twice a day. This gives businesspeople with afternoon flights the option of commuting by subway and dropping their luggage at the airline counter on their way to work. Trains bring luggage to the airport, where it is loaded on the appropriate flights without further handling by the flier. Stein also arranged for incoming flight luggage pickup at his building. While the building includes a twelve-floor parking garage, this integration of transportation systems maximizes its availability to other users.

Required Sizes

Background

Sizes of parking spaces are related to vehicle type. Therefore quantity of parking is meaningless without establishing the types of vehicles expected. If 200 spaces are required, can a project team decide that 190 will be motorcycle or compact car spaces? What counts as a vehicle?

Minimum plan dimensions and vertical clearance are specified. Most zoning ordinances specify minimum stall lengths and widths, allowing a two-tier set of dimensions for full-size and compact cars, and specifying the maximum allowed percentage of the total that can use the smaller compact-car dimension.

Vertical dimensions become important in parking structures, where floor-to-floor space is often kept to a minimum. Clearance that is sufficient for cars and pedestrians is generally increased to allow vans, including ambulances, to gain access to part of every such structure.

Requirements

A typical requirement is a 9 ft x 19 ft full size stall, with up to 40 percent compact stalls at 8 ft x 16 ft. These sizes are for car stalls oriented perpendicularly to the roadway or aisle. Additionally, rules generally specify a special longer set of sizes for parallel-parked cars.

Optimizing Parking Stall Size Rules

Often the best strategy is to try to shift the ratio of compact to full-size stalls. Additionally, some ordinances permit motorcycle parking to count toward required parking spaces. Both strategies depend on documentation demonstrating the likely or actual composition of vehicles coming to the project.

Configurations

Rules tend to accept any stall configuration, including perpendicular, angled, and parallel, so long as it meets requirements for size and count.

Grading

Background

Rules influence the allowable pitch of the ground in parking lots. This is partially why parking lots in hilly areas look even more unnatural than usual.

A fairly typical loading dock, containing three raised berths with dock levelers for trailer rigs, and an overhead door enclosing either a van berth or a Dumpster. Photo by K. Wyllie.

Requirements

Rules generally look for grading in areas where cars are parked to stay in the 2 percent to 6 percent slope range. This is steep enough to allow for adequate storm runoff but shallow enough so that car doors can be controlled upon entering and exiting. When slopes exceed 6 percent, gravity tends to take over, making it much harder for a passenger to push a door open or prevent it from slamming into the side of an adjacent car.

LOADING: SERVICE VEHICLES

Background

Loading is the freight equivalent of parking, and berths are the equivalent of stalls. Retail and industrial projects need loading facilities to handle their merchandise; residential and office projects need them to move furniture and supplies in and out. Many buildings need them on occasion to service mechanical equipment. When projects have inadequate loading facilities, traffic problems become unmanageable, with double-parked 18-wheelers and panel trucks trying to make their deliveries.

Requirements

Quantities and sizes are related to the site's zoning district and the size of the project. For larger projects, rules will require a certain number of berths for vans (generally 12 ft x 25 ft) and a certain number for trailer trucks (generally 12 ft x 50 ft). Turning radii and headroom clearances must also reflect the needs of the required berths.

In urban areas, loading docks are increasingly being moved indoors. This is mostly in response to the need to maximize profits by building the maximum permitted under lot coverage provisions. When a building's footprint covers all available ground-floor space, there is little choice but to provide facilities indoors. Recent projects are even going to the extreme of providing vehicle elevators to bring delivery trucks to subbasement docks.

INTEGRATED STRATEGIES FOR POSITION

NOB HILL TAVERN

Albuquerque, New Mexico, Stephen Schreiber, Architect[27]

The project team for this new restaurant faced a difficult and contradictory set of rules. The 12,400-SF property had only one interior lot line, since it was a corner lot with an alley in the rear. Even the interior lot line was somewhat public since a private driveway ran along it. The site was essentially vacant, following the destruction of a previous building by fire. Positioning the new building on the site became a game of political compromise and ingenuity.

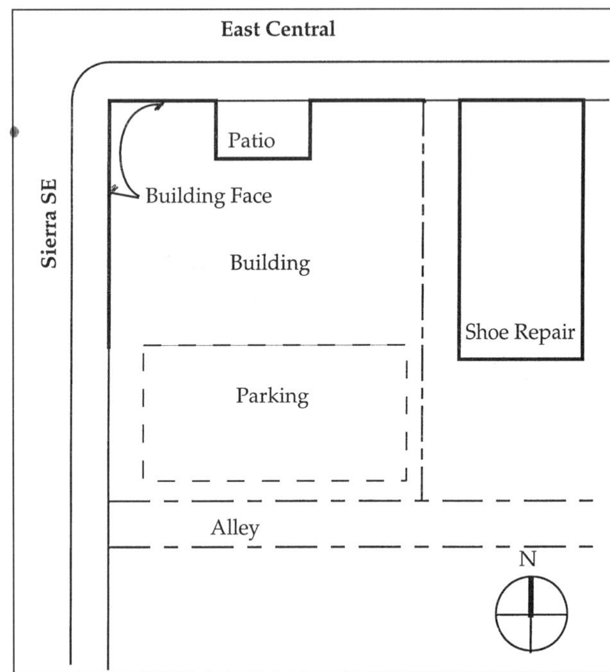

Applied sector plan guidelines. ©S. Schreiber.

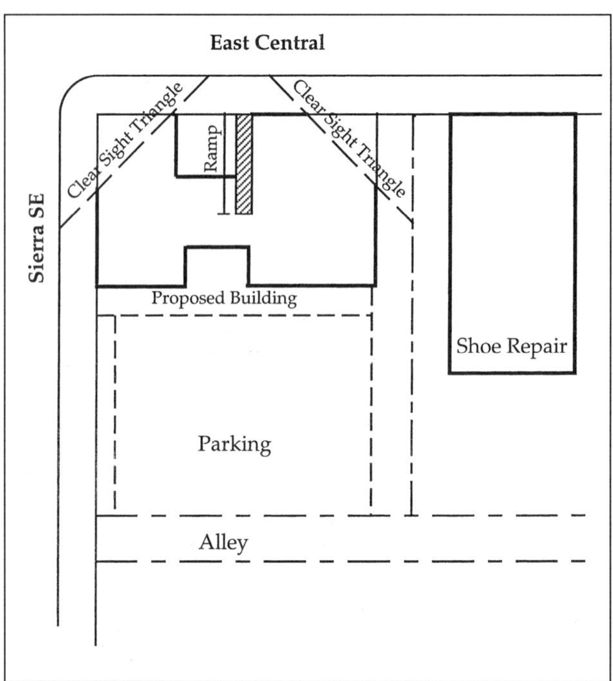

Applied ramp requirements and vision cones. ©S. Schreiber.

The regulatory landscape was crowded by:

- the Nob Hill "sector plan" (design guidelines), which favors infill construction, urban "street walls," and active storefronts, and seeks to reduce the impact of automobiles;
- the Albuquerque Comprehensive Plan, which seeks to raise developmental density so the city can grow without adding infrastructure;
- subdivision and zoning rules written and enforced by Albuquerque's planning and public works departments, oriented toward new subdivisions and mandating larger properties and wider streets;
- the usual complement of building codes and accessibility regulations.

The following contradictory goals were required or recommended by any number of applicable rules and essential program requirements:

- That buildings be constructed without setbacks to present a unified streetfront (sector plan).
- That parking be hidden from thoroughfares and accessed only from alleys (sector plan).
- That pedestrian plazas be included where possible (sector plan).
- That all entrances facing public rights-of-way be accessible, with ramps if necessary (building code).
- That the floor elevation be almost two feet higher than sidewalk grade since the property was marginally within a floodplain, making ramps necessary (zoning ordinance).
- That the existing alley be widened, occupying an 8 ft wide strip at the rear of the property (public works). The city eventually retreated from this demand when it realized that the additional paving would be useless due to existing power poles bordering the alley, which it was not prepared to move.
- That 25 ft vision cones be maintained at the street-street and the street-driveway intersections. Vision cones were not required at the street-alley or alley-driveway intersections (public works).
- That the tavern would have to serve one hundred customers (program), translating to 1,500 SF at 15 SF per occupant (building code), to make a profit. Additional area would be needed for entries, kitchen, toilets, and offices.
- That approximately 300 SF of parking lot [9 ft x 19 ft stall plus aisle] be provided for every 45 SF of floor area (one stall for every three occupants at 15 SF per occupant). This added up to 30,000 SF of parking for the intended 100 customers, not counting

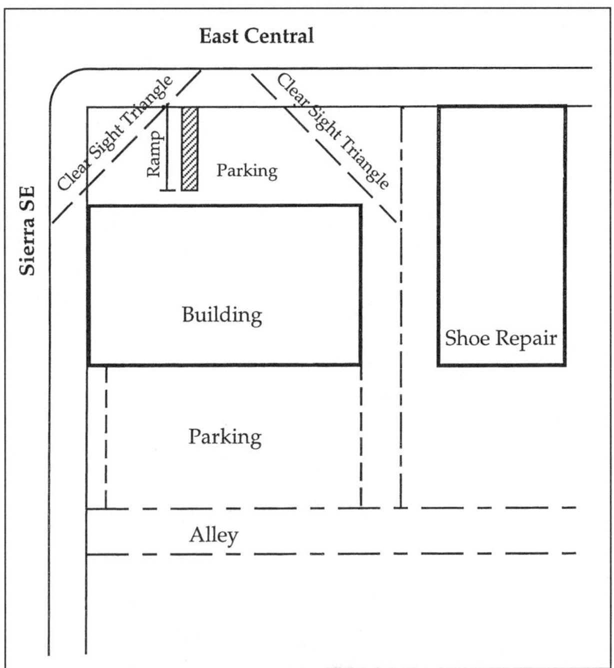

A typical suburban response. ©S. Schreiber.

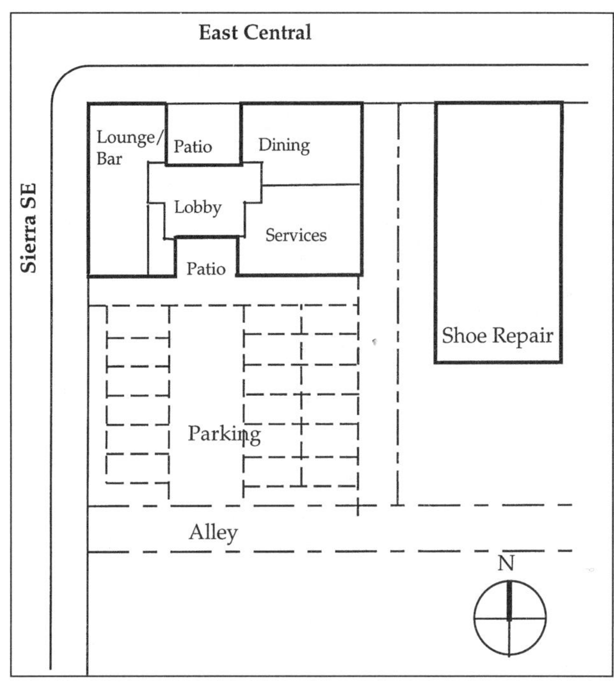

The response as built. ©S. Schreiber.

employee parking (a combination of building code and zoning ordinance).

Conflicts and their resolutions included:

1. Maintaining the vision cones while keeping the zero-lot-line approach, mutually exclusive goals. The project took advantage of exceptions permitted for certain intersections, among which was the intersection at this property. If they hadn't asked, they wouldn't have found out.
2. Providing the required parking, since feasibility studies had demonstrated that the tavern would have to serve one hundred customers (translating to 1,500 SF) to make a profit. With the addition of entries, kitchen, toilets, and offices, the site wasn't big enough for both building and parking. The ordinance allowed reductions in parking count upon agreement by the owners of adjacent properties, neighborhood associations, and the zoning hearing administrator. Since all parties thought that this project would help maintain the vitality of the neighborhood, they all agreed, following a public hearing, to a reduction.
3. Making the building accessible from the street while keeping the zero-lot-line approach. Only two solutions semi-worked:
   - Finding room for a ramp from the sidewalk, even if it meant canceling a hoped-for entry patio. The grade change could be accomplished at the off-street facade by regrading.
   - Providing no entries on the street, allowing entry only by way of the off-street facades.

In the end, the team went with the first approach.

In this case, much of the problem came from Albuquerque's history of suburban (vehicular) ap-proaches to development. The architect, Stephen Schreiber, said it best:

The numerous constraints, designed for suburban areas, are a major reason why so much land in older neighborhoods (such as the center city) is left vacant.

Albuquerque remains a suburb "in search of a city." Only the most optimistic and persistent property owners and architects can maneuver through the multiple contradicting codes for even seemingly simple jobs. It is easier to develop a new parcel of land on the edge of the city, regrade where necessary, plunk the building in the middle of the property, and surround it with parking.

Still, the project team's success illustrates the value of code optimization made possible by familiarity.

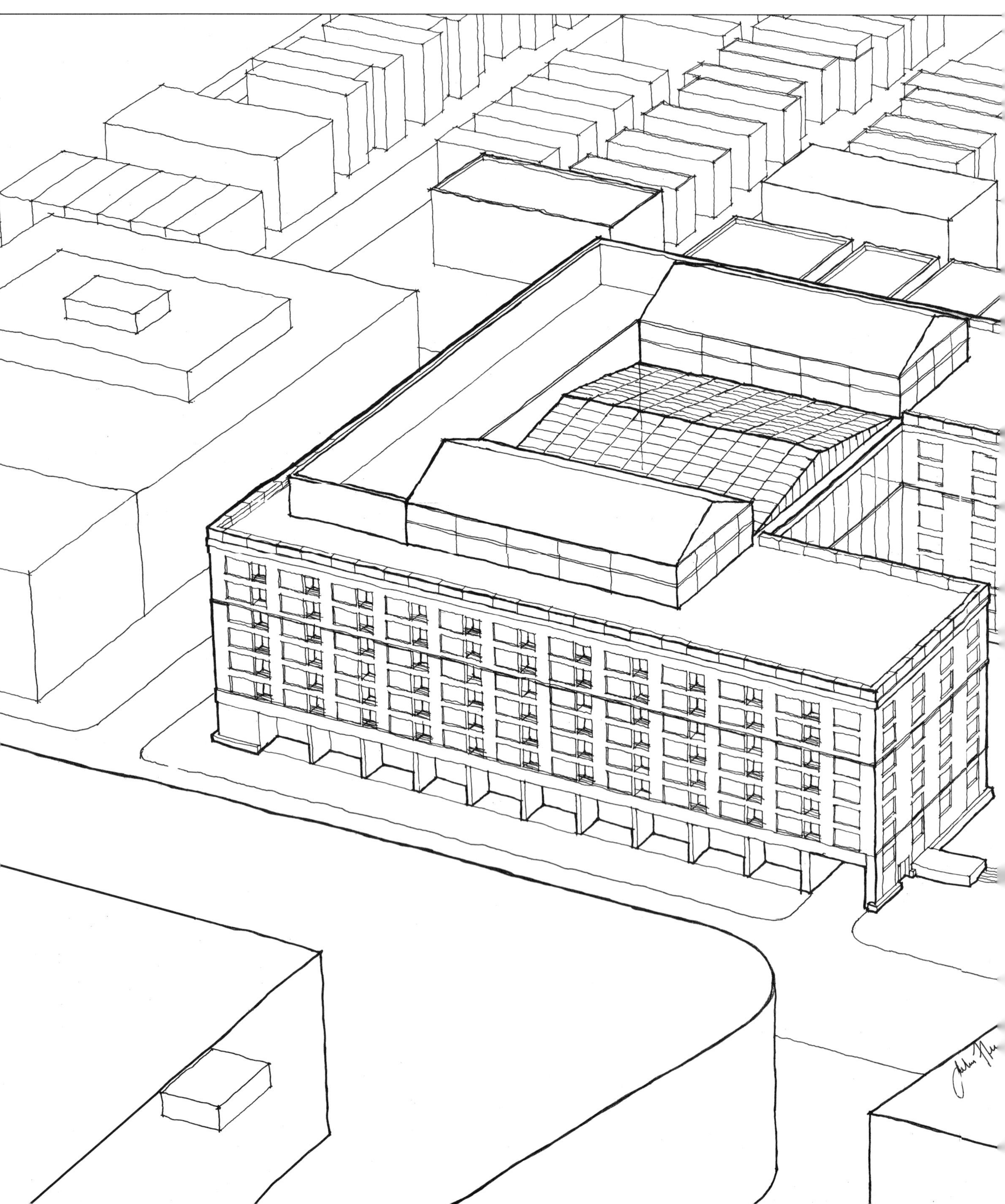

PART 

URBAN DESIGN

While project teams are usually concerned primarily with their project's issues, many of the design choices they make affect the community beyond their project's property lines. Designs that are sensitive to urban design issues strengthen the communities in which they are located. Rules influence the design of a building's shell and therefore the exterior of a building. This section discusses the issues involved in the following design questions:

- How big should it be, fulfilling aesthetic intent without adversely affecting available infrastructure or density, and without becoming unsafe?
- What should exterior elevations look like, fulfilling aesthetic intent while admitting sufficient light and air, allowing egress, and achieving other functional ends?
- What materials should be used, fulfilling aesthetic intent while being sufficiently durable, fire resistant, energy efficient, and well crafted?

5

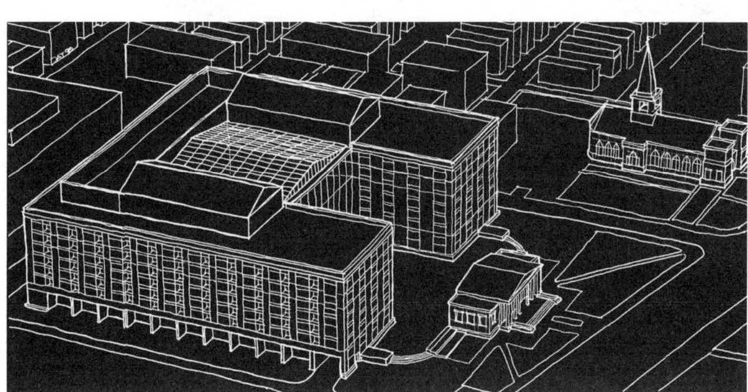

MASSING

Establishing Massing

CONTEXT

Why This Issue Is Important

In terms of urban design, one of the worst aesthetic offenses committed by bad projects is that of being insensitively out of scale. For individual buildings, the offense of being too large for safe exiting of their occupants during emergencies, if it were allowed, would be similarly problematic. Even before schematic design is begun, rules must be consulted to find out what overall massing is acceptable. How high is reasonable? How big? In what configuration?

Why Regulatory Bodies Care

The effect of a city's zoning ordinance on the massing of its buildings very strongly influences, if it doesn't outright establish, the identity of the place. A quick glance at the differences between the look of some well-known examples makes this clear:

- New York, with its buildings of almost unlimited height moderated by stepped setbacks to allow sunlight to penetrate to street level
- Chicago, with canyon-like streets bordered by equally tall but zero-lot-line high-rises
- Washington, D.C., with its dogged adherence to mid-rise blocks

But looks aren't everything. Perhaps more critically from a functional and environmental perspective, limiting project size allows urban infrastructure and building structure to grow in a coordinated way. The bigger the projects are, the more people tend to occupy them. The transportation needs of those people must be accommodated by available roads, parking, and mass transit. Their wastes must be accommodated by available sewer lines, trash removal, and recycling. More police patrols and mail delivery are needed. Massing rules attempt to keep the demands made by projects from exceeding the carrying capacity of the infrastructure. As additional infrastructure becomes available, FARs can be adjusted through the hearings and votes that constitute the regular political process.

Building codes have equally crucial effects on massing. Buildings must be able to provide their occupants with adequate light and air, and sometimes artificial sources won't do. This tends to establish maximum distances from exterior walls. Buildings must be able to remain safe during a fire long enough for occupants to safely exit and for firefighters to safely extinguish the blaze. This tends to relate the allowed size of a building to the fire resistance of its construction.

Where to Find the Requirements

In Zoning Ordinances

Requirements for use, parking, and massing are often found in three separate parts of an ordinance. Don't give up if you don't find massing requirements right away. Unless you find a place where it says "no requirement," you probably haven't looked hard enough. If it isn't found in a chapter devoted to the site's particular zoning district, it will be in a chapter devoted to massing as a regulatory issue.

In Building Codes

In all the model building codes, massing requirements are brought together in matrices that stipulate allowable area per floor, allowable number of stories, and allowable height of a project when given the Occupancy and Construction Classifications (see the definitions listed in Chapter 2).

These matrices also provide links, through remarks or footnoted cross-references, to exceptions provided by the codes in exchange for including various design options such as sprinkler systems. Be relentless in pursuing these noted exceptions—follow them to the end of the trail.

Such matrices serve as a sort of "home base" for a journey through the codes, as they are the origination point for many code provisions. Get to know them well.

ESTABLISHING TOTAL SIZE

Context

Rules influence the total size of projects—the aggregate sum of all the individual floors. Zoning ordinances do it through formulas that relate building size to property size.

However, total size does not always mean total size—some kinds of spaces are not counted toward the total. It is critical to understand, thoroughly, which spaces are included and which are not. Anything not counted can be added on an unlimited basis, although subject to the constraints of other provisions.

Interestingly, building codes do not directly address total size except by addressing the factors that add up to total size—area per floor and building height. For this reason, the issues discussed in this section refer only to zoning issues.

Zoning ordinances do not all agree on the appropriate method for relating building size to site size. Some use floor area ratios and others use unit ratios. As noted below, each suggests certain project types. While this may be of interest to those who write codes, project teams have no choice but to work with local ordinances as written, with whichever method of calculation such ordinances contain.

AREA OF BUILDING

Background

Floor Area Ratio, or FAR, is the ratio of the maximum total floor area a project may contain compared with the total area of the site. FAR simply sets the upper limit—it need not be reached. For example, an FAR of 2 would allow a project enclosing a maximum of 50,000 SF on a 25,000 SF site.

The area of the site is not subject to debate; one need only consult the survey, or plat. This seems simple enough, until one asks how floor area is to be counted. Is it rentable area, occupiable area, or gross area? Are columns, exterior walls, mechanical equipment rooms, elevator shafts, or volume spaces included? This is fertile ground for regulatory optimization.

The answer to these questions lies in the chapter of definitions included in every regulation, listed as floor area for FAR calculation, or some similar heading. The definitions listed simply as floor area do not apply. The FAR definitions spell out the inclusions and exclusions that can then be applied to the project program.

Requirements

FAR formulas generally exclude:

- Enclosed garages
- Mechanical equipment rooms and chases
- Spaces under sloped ceilings with headroom distances of less than 5 ft. This is a definition with optimization potential.

FAR formulas generally include:

- All occupiable space
- Columns and exterior walls

Implications

FAR controls only the amount of building built, measured in floor area. It does not control the number of occupants, the program density (how tightly spaces are packed), or the quality of the permitted area.

In this respect, FAR can favor more luxurious projects—if a project team can only build so much and no more, it may as well make it fairly expensive. This maximizes the project's revenues and corresponding profits. Expensive projects generally have higher profit margins.

On the other hand, FAR can also favor maximizing occupant load. This can be seen in two quite contradictory ways. The more people (in an office or residential project) or equipment (in the case of a factory) one can squeeze into the same area,

- the more rent the investment team can justify charging (although they'll use it for the increased maintenance caused by increased usage);

- the less area one needs to build per occupant, and therefore the less rent one must charge per occupant, making occupancy more affordable.

Optimizing FAR Rules

Although FAR is a fixed ratio, it fortunately isn't too fixed. Most ordinances permit allowable FARs to be stretched in exchange for including certain public amenities in the project. Premiums are frequently given for:

- Setbacks beyond the district requirements.
- Special pedestrian spaces such as plazas and arcades.
- TDRs. When a site can take advantage of transferable development rights, a whole range of possibilities opens. (See pp. 89–90.)

NUMBER OF BUILT UNITS

Background

Another way zoning ordinances have of controlling total project size is by restricting the number of units a project on a given site can contain. Units might be defined as beds (in hospitals or prisons), classrooms, apartments, or any other convenient measure. In this way, the ordinances attempt to regulate how many people might be using the facility, something that does not necessarily correspond with area.

Requirements

Counting

Most ordinances that use this type of rule have breakdowns by size, but they aren't broken down very much. For example, an ordinance may allow one efficiency apartment per 75 SF of site, and one multiroom apartment per 125 SF of site, regardless of how many rooms the multiroom unit has.

If it isn't a unit, it isn't counted. If zoning ordinances restrict a prison project to a maximum number of beds, the project team is free to design as many classrooms, workshops, and other facilities as it feels are appropriate, without restriction.

Implications

With "number of unit" rules, ordinances don't concern themselves with how large or small each unit is; they're attempting to limit the number of occupants living, working, or institutionalized there, perhaps in an attempt to avoid overloading the available infrastructure. Clearly this type of requirement favors larger units with better amenities, since investors can only increase revenues by increasing income per unit, not by increasing number of units.

Optimizing "Number of Unit" Rules

Combine Units

The only way to optimize this requirement is to modify the way units are counted. Most project teams would want to increase the number of units permitted, so if they could get approval to count two units as one, they could achieve some planning flexibility.

As an example, let's say a developer is trying to put 150 low- to moderate-income apartments on a site whose area limits the number of residential units to 125. The project team needs to develop 150 to make the project financially feasible, but rules prohibit this. If financial returns per unit can't be raised, combining units might present a possibility. The project could adhere to the 125-unit maximum, but each unit might include an "in-law suite," permitting a family to share rent with extended family members who would otherwise have to rent their own unit.

Maximize Accessory Facilities

Find ways to minimize the number of counted units while maximizing the uncounted units. For example, where an ambulatory surgical center is restricted by number of operating rooms, the project might benefit from additional examination rooms—spaces that would not be counted toward total size.

OPTIMIZING TOTAL SIZE RULES

Size isn't everything, but it helps. Get there by use of creative classification and creative counting.

ESTABLISHING SIZE PER FLOOR

Context

Before the advent of effective artificial lighting and ventilation systems, floor size was of major concern, since proximity to windows was directly related to maintaining health. With the rise in concern over Sick Building Syndrome, it has resurfaced as an issue.

Additionally, as the saying goes, "the bigger they are, the harder they fall." Bigger buildings are heavier, may have longer spans, and hold more people, so tend to create bigger disasters unless built of safer construction than smaller buildings.

From a design perspective, there is no floor size too large to be evacuated safely. Even though egress is easier when people don't have too far to travel, it is certainly possible to design large floors so long as they include closely spaced exits.

Lot Coverage

A zoning ordinance issue.

Background

Lot coverage regulates the amount of the site that can be covered by a building, expressed as a percentage of total site area. A portion of the site is generally considered to be covered if enclosed space occupies it or hovers over it. For simple designs, lot coverage is the area of the site divided by the area of the building footprint. Where occupied overhangs extend beyond the footprint, they must be added to the area of the footprint.

Requirements

Obviously, the maximum lot coverage allowable is 100 percent. This would be considered to be a "zero lot line" situation. However, just because a particular site may carry a 100 percent coverage allowance does not mean that there are no conflicting setback requirements. Make sure to check both areas of the regulation.

Like other requirements, the percentage of lot coverage allowed varies by district, and can range substantially.

Optimizing Lot Coverage Rules

Not every built or paved area is considered coverage. Generally only occupied ones are, leaving parking areas and patios out of the definition. This may not be true when the parking or patio is atop an unexcavated retained area raised substantially above grade.

Reducing the size of the ground floor, setting it back relative to the upper floor plans, doesn't change any issues, because coverage is measured based on the maximum footprint even if it doesn't occur in contact with the ground. Think of lot coverage as the area of ground that all parts of the building, in combination, would shield from sunlight if the sun were directly overhead.

Allowable Area

A building code issue.

Background

Allowable floor area provisions set maximums on the amount of space buildings may contain. Allowable area is limited so as to maintain a balance between:

- the number of occupants,
- the potential for fire based on use group,
- the fire resistance of the construction.

Different codes do it in different ways. In some, areas are listed by floor and get progressively smaller as buildings get taller, except for the unlimited area allowance permitted the most fire-resistive construction types [BOCA93 506.4]. In others, areas are listed as aggregate totals of the whole building, regardless of number of stories [UBC91 505(b)].

These provisions can be optimized based on compartmentation. (See pp. 180–181.)

Requirements

Basis

Allowable area generally depends on construction classification, because classification is chosen based on the relationship between three factors:

- Use group, important because some uses are inherently more dangerous than others
- Budget and schedule, important because the degree of fire resistance the construction achieves depends on how much time and money are invested in the effort
- Desired area and height

Since use group is known and fairly fixed very early in the project, project teams can determine allowable area (and height, which is discussed in the next section) if they can assess budget and schedule and thereby decide on construction classification. Clearly this is not only a complex decision but one with very far reaching implications.

Requirements

Each model code contains a matrix that relates allowable area to use group and construction classification. Although the form of each code's matrix varies, the intent is equivalent. When you find the matrix, you will find the requirements.

- In the draft IBC, the matrix is noted as Table 503. It is interesting in that it varies height in stories and area per floor by use group, but not height in feet, which remains constant for all use groups.
- In the "big three" model American codes, since 1993, the matrix is noted as Table 500 or 503. In older editions, see BOCA Table 602, UBC Tables 5-C and 5-D, and SBC Table 400.
- CABO, since it is written specifically for small structures, assumes an Unprotected Combustible construction classification and a residential use group.

The values in these tables must be modified for multistory buildings [BOCA93 506.4]. Exiting from one- and two-story buildings is fairly efficient. Occupants of the first floor can exit directly outside, and occupants of the second floor have exclusive use of the stairs. Beyond that, the taller a building is, the longer it takes to evacuate. In response, values for allowable floor area are modified for taller buildings, except, of course, those of highest construction classifications.

For example, the base values specified in BOCA establish maximum floor areas for one- and two-story buildings only. Table 503 permits some projects of Type 1A and 1B construction to be of unlimited height in recognition of their inherent fire resistance. For the rest, the maximum height allowed is nine stories. An additional floor can be added by including sprinklering, bringing the total to ten. For these limited-height projects, BOCA has a schedule of percentage reductions that apply to every floor of buildings three or more stories high [BOCA93 Table 506.4]. It is mathematically linear, changing by the same amount for each floor added. It doesn't need and therefore doesn't have any values for projects over ten floors. Using these formulas, a building five stories high of Type 2A construction is limited to a floor area 15 percent smaller than (meaning 85 percent of) the area allowed by Table 503. The UBC takes the opposite approach, giving allowed

Table 506.4
Reduction of Area Limitations

Number of stories	Type of construction		
	1A & 1B	2A	2B, 2C, 3A, 3B, 4, 5A, 5B
1	None	None	None
2	None	None	None
3	None	5%	20%
4	None	10%	20%
5	None	15%	30%
6	None	20%	40%
7	None	25%	50%
8	None	30%	60%
9	None	35%	70%
10	None	40%	80%

Reduction of Area Limitations, BOCA93 Table 506.4.

areas for tall buildings and providing multipliers to adjust them for lower buildings. In its current form, the draft IBC is moot on the topic, although it allows maximum heights of eleven stories plus one for sprinklering [IBC 504].

Implications

This system is based on the notion that dangerous uses justify tighter restrictions on allowable area, since such restrictions keep the balance between amount of people or goods at risk and the likelihood of loss.

It is also based on the notion that the more fire resistant the construction is (the "higher" its construction classification), the more area per floor a project can justify.

Both these notions have been developed through observation of building fires and the extent of losses incurred. Is there room for adjustment? Or would that be, quite literally, playing with fire?

Optimizing Allowable Floor Area Rules

The major model codes include options for increasing, as well as requirements for decreasing, allowable floor area due to the following project conditions or features:

- Compartmentation: Listed floor areas refer to area allowed within a single building or fire area. Where it doesn't overly compromise design intent, compartmentalizing a design into several buildings or fire areas is an option well worth considering for the other freedoms it brings. (See pp. 184–185.)

THE BOCA NATIONAL BUILDING CODE/1996

Table 503
HEIGHT AND AREA LIMITATIONS OF BUILDINGS

Height limitations of buildings (shown in upper figure as stories and feet above grade plane)[m], and area limitations of one- or two-story buildings facing on one street or public space not less than 30 feet wide (shown in lower figure as area in square feet per floor[m]). See Note a.

Use Group	Note a	Type of construction									
		Noncombustible					Noncombustible/Combustible			Combustible	
		Type 1		Type 2			Type 3		Type 4	Type 5	
		Protected Note b		Protected		Unprotected	Protected	Unprotected	Heavy timber	Protected	Unprotected
		1A	1B	2A	2B	2C	3A	3B	4	5A	5B
A-1 Assembly, theaters		Not limited	Not limited	5 St. 65′ 19,950	3 St. 40′ 13,125	2 St. 30′ 8,400	3 St. 40′ 11,550	2 St. 30′ 8,400	3 St. 40′ 12,600	1 St. 20′ 8,925	1 St. 20′ 4,200
A-2 Assembly, nightclubs and similar uses		Not limited	Not limited 7,200	3 St. 40′ 5,700	2 St. 30′ 3,750	1 St. 20′ 2,400	2 St. 30′ 3,300	1 St. 20′ 2,400	2 St. 30′ 3,600	1 St. 20′ 2,550	1 St. 20′ 1,200
A-3 Assembly	Lecture halls, recreation centers, terminals, restaurants other than nightclubs	Not limited	Not limited	5 St. 65′ 19,950	3 St. 40′ 13,125	2 St. 30′ 8,400	3 St. 40′ 11,550	2 St. 30′ 8,400	3 St. 40′ 12,600	1 St. 20′ 8,925	1 St. 20′ 4,200
A-4 Assembly, churches	Note c	Not limited	Not limited	5 St. 65′ 34,200	3 St. 40′ 22,500	2 St. 30′ 14,400	3 St. 40′ 19,800	2 St. 30′ 14,400	3 St. 40′ 21,600	1 St. 20′ 15,300	1 St. 20′ 7,200
B Business		Not limited	Not limited	7 St. 85′ 34,200	5 St. 65′ 22,500	3 St. 40′ 14,400	4 St. 50′ 19,800	3 St. 40′ 14,400	5 St. 65′ 21,600	3 St. 40′ 15,300	2 St. 30′ 7,200
E Educational	Note c	Not limited	Not limited	5 St. 65′ 34,200	3 St. 40′ 22,500	2 St. 30′ 14,400	3 St. 40′ 19,800	2 St. 30′ 14,400	3 St. 40′ 21,600	1 St. 20′ 15,300 Note d	1 St. 20′ 7,200 Note d
F-1 Factory and industrial, moderate		Not limited	Not limited	6 St. 75′ 22,800	4 St. 50′ 15,000	2 St. 30′ 9,600	3 St. 40′ 13,200	2 St. 30′ 9,600	4 St. 50′ 14,400	2 St. 30′ 10,200	1 St. 20′ 4,800
F-2 Factory and industrial, low	Note h	Not limited	Not limited	7 St. 85′ 34,200	5 St. 65′ 22,500	3 St. 40′ 14,400	4 St. 50′ 19,800	3 St. 40′ 14,400	5 St. 65′ 21,600	3 St. 40′ 15,300	2 St. 30′ 7,200
H-1 High hazard, detonation hazards	Notes e, i, k, l	1 St. 20′ 16,800	1 St. 20′ 14,400	1 St. 20′ 11,400	1 St. 20′ 7,500	1 St. 20′ 4,800	1 St. 20′ 6,600	1 St. 20′ 4,800	1 St. 20′ 7,200	1 St. 20′ 5,100	Not permitted
H-2 High hazard, deflagration hazards	Notes e, i, j, l	5 St. 65′ 16,800	3 St. 40′ 14,400	3 St. 40′ 11,400	2 St. 30′ 7,500	1 St. 20′ 4,800	2 St. 30′ 6,600	1 St. 20′ 4,800	2 St. 30′ 7,200	1 St. 20′ 5,100	Not permitted
H-3 High hazard, physical hazards	Notes e, l	7 St. 85′ 33,600	7 St. 85′ 28,800	6 St. 75′ 22,800	4 St. 50′ 15,000	2 St. 30′ 9,600	3 St. 40′ 13,200	2 St. 30′ 9,600	4 St. 50′ 14,400	2 St. 30′ 10,200	1 St 20′ 4,800
H-4 High hazard, health hazards	Notes e, l	7 St. 85′ Not limited	7 St. 85′ Not limited	7 St. 85′ 34,200	5 St. 65′ 22,500	3 St. 40′ 14,400	4 St. 50′ 19,800	3 St. 40′ 14,400	5 St. 65′ 21,600	3 St. 40′ 15,300	2 St. 30′ 7,200
I-1 Institutional, residential care		Not limited	Not limited	9 St. 100′ 19,950	4 St. 50′ 13,125	3 St. 40′ 8,400	4 St. 50′ 11,550	3 St. 40′ 8,400	4 St. 50′ 12,600	3 St. 40′ 8,925	2 St. 35′ 4,200
I-2 Institutional, incapacitated		Not limited	Not limited	4 St. 50′ 17,100	2 St. 30′ 11,250	1 St. 20′ 7,200	1 St. 20′ 9,900	Not permitted	1 St. 20′ 10,800	1 St. 20′ 7,650	Not permitted
I-3 Institutional, restrained		Not limited	Not limited	4 St. 50′ 14,250	2 St. 30′ 9,375	1 St. 20′ 6,000	2 St. 30′ 8,250	1 St. 20′ 6,000	2 St. 30′ 9,000	1 St. 20′ 6,375	Not permitted
M Mercantile		Not limited	Not limited	6 St. 75′ 22,800	4 St. 50′ 15,000	2 St. 30′ 9,600	3 St. 40′ 13,200	2 St. 30′ 9,600	4 St. 50′ 14,400	2 St. 30′ 10,200	1 St. 20′ 4,800
R-1 Residential, hotels		Not limited	Not limited	9 St. 100′ 22,800	4 St. 50′ 15,000	3 St. 40′ 9,600	4 St. 50′ 13,200	3 St. 40′ 9,600	4 St. 50′ 14,400	3 St. 40′ 10,200	2 St. 35′ 4,800
R-2 Residential, multiple-family		Not limited	Not limited	9 St. 100′ 22,800	4 St. 50′ 15,000 Note f	3 St. 40′ 9,600	4 St. 50′ 13,200 Note f	3 St. 40′ 9,600	4 St. 50′ 14,400	3 St. 40′ 10,200	2 St. 35′ 4,800
R-3 Residential, one- and two-family and multiple single-family		Not limited	Not limited	4 St. 50′ 22,800	4 St. 50′ 15,000	3 St. 40′ 9,600	4 St. 50′ 13,200	3 St. 40′ 9,600	4 St. 50′ 14,400	3 St. 40′ 10,200	2 St. 35′ 4,800
S-1 Storage, moderate		Not limited	Not limited	5 St. 65′ 19,950	4 St. 50′ 13,125	2 St. 30′ 8,400	3 St. 40′ 11,550	2 St. 30′ 8,400	4 St. 50′ 12,600	2 St. 30′ 8,925	1 St. 20′ 4,200
S-2 Storage, low	Note g	Not limited	Not limited	7 St. 85′ 34,200	5 St. 65′ 22,500	3 St. 40′ 14,400	4 St. 50′ 19,800	3 St. 40′ 14,400	5 St. 65′ 21,600	3 St. 40′ 15,300	2 St. 30′ 7,200
U Utility, miscellaneous		Not limited	Not limited	5 St. 65′ 19,950	4 St. 50′ 13,125	2 St. 30′ 8,400	3 St. 40′ 11,550	2 St. 30′ 8,400	4 St. 50′ 12,600	2 St. 30′ 8,925	1 St. 20′ 4,200

Note a. See the following sections for general exceptions to Table 503:
Section 504.2 Allowable height increase due to automatic sprinkler system installation.
Section 506.2 Allowable area increase due to street frontage.
Section 506.3 Allowable area increase due to automatic sprinkler system installation.
Section 506.4 Allowable area reduction for multistory buildings.
Section 507.0 Unlimited area one-story buildings.

Note b. Buildings of Type 1 construction permitted to be of unlimited tabular heights and areas are not subject to special requirements that allow increased heights and areas for other types of construction (see Section 503.1.4).

Note c. For height exceptions for auditoriums in occupancies in Use Groups A-4 and E, see Section 504.3.

Note d. For height exceptions for day care centers in buildings of Type 5 construction, see Section 504.4.

Note e. For exceptions to height and area limitations for buildings with occupancies in Use Group H, see Chapter 4 governing the specific use groups.

Note f. For exceptions to height of buildings with occupancies in Use Group R-2 of Types 2B and 3A construction, see Sections 504.6 and 504.7.

Note g. For height and area exceptions for open parking structures, see Section 406.0.

Note h. For exceptions to height and area limitations for special industrial occupancies, see Section 503.1.1.

Note i. Occupancies in Use Groups H-1 and H-2 shall not be permitted below grade.

Note j. Rooms and areas of Use Group H-2 containing pyrophoric materials shall not be permitted in buildings of Type 3, 4, or 5 construction.

Note k. Occupancies in Use Group H-1 are required to be detached one-story buildings (see Section 707.1.1).

Note l. For exceptions to height for buildings with occupancies in Use Group H, see Section 504.5.

Note m. 1 foot = 304.8 mm; 1 square foot = 0.093 m^2.

A typical Construction Classification Table [BOCA96 Table 503]. Notice the universal exceptions for height and area listed as "Note a." Familiarity with them is essential to design optimization. Although the corresponding tables in the UBC, SBC, and IBC are different in form and requirements, they are similar in concept. © BOCA.

- Automatic sprinkler (fire suppression) systems: Codes allow, within certain limits, floor area increases of 100 to 200 percent [both IBC and BOCA93: 506.3].
- Street frontage: Increased areas are allowed for designs where a fairly large minimum percentage (25 percent) of the exterior walls face either streets or open spaces of specified minimum widths directly accessible from streets. The ease of fire department access possible with this configuration justifies the exception [IBC 506.2]. BOCA makes a word problem of it, allowing an extra 2 percent of floor area for every 1 percent of total building perimeter above 25 percent that faces a street. The IBC makes a mathematical formula of it but as a result is able to fine-tune the calculation to provide a sliding adjustment for depth of unoccupied space.

  For rectangular buildings directly fronting a street, the minimum percentage isn't difficult to achieve—it requires only that one of the long sides of the rectangle face the street. A rectangular building on a corner lot always has 50 percent of its walls fronting the street, so is allowed a 50 percent increase in area (50 percent designed frontage—25 percent minimum frontage x 2).

  This exception gives a substantial advantage to any project whose street facade has lots of ins and outs. "C"- or "E"-shaped plans with their spines oriented away from the public way are virtual shoo-ins for surpassing the 25 percent minimum.

Unlimited area is allowed in three cases:

1. For freestanding buildings one story high, if sprinklered, since people can exit outdoors either directly from their rooms or through a length of hallway, but without first having to negotiate stairs. Stairs cannot be traversed as quickly as corridors, and multistoried projects introduce egress problems for the disabled. In some cases, such as schools and participant sports facilities (which are also exempt from sprinklering), primary activity rooms must exit directly to the outdoors [BOCA93 507]. (For further discussion of direct exiting, pp. 178–179.)
2. For most any building (except hazard groups and nightclubs) when all construction is classified as protected and noncombustible, as in BOCA Type 1 [BOCA93 507.1].
3. For high-volume, low-hazard industrial spaces such as fabrication shops [BOCA93 503.1.1].

ESTABLISHING TOTAL HEIGHT

Context

Domain

Total height, like total size, is both an urban design and community identity issue as well as a safety issue. Zoning ordinances aim for gradual transitions from low-rise areas to high-rise areas. This approach is intended to achieve a consistency of scale within any given neighborhood. Then there is the "down in front" syndrome, whereby owners of existing buildings with spectacular views of waterfronts, mountains, or other natural or cultural treasures try to legislate the prohibition of new buildings that might block their views. As for building codes, the primary issue is safety. Exiting from taller buildings takes longer, so such buildings must be designed with an extra degree of safety.

Background

The most crucial question with any requirement related to height, as in other issues we've already examined, is "How is it measured?" How does one determine where to hold each end of the metaphorical tape measure? As with any distance (as opposed to area) measurement, there are two ends to be defined. Zoning ordinances and building codes have very different ways of defining each, as noted below, corresponding to their differing end goals.

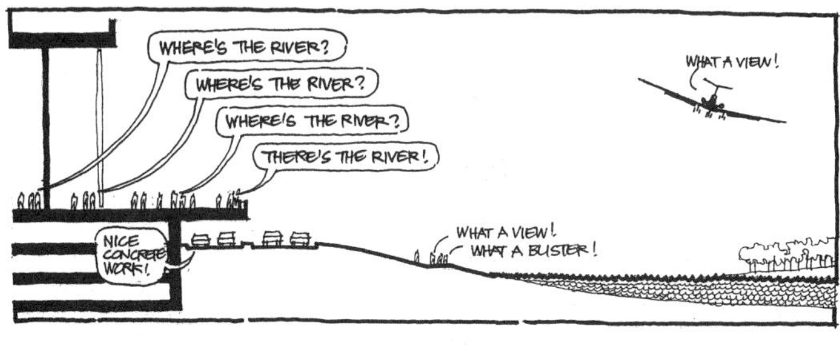

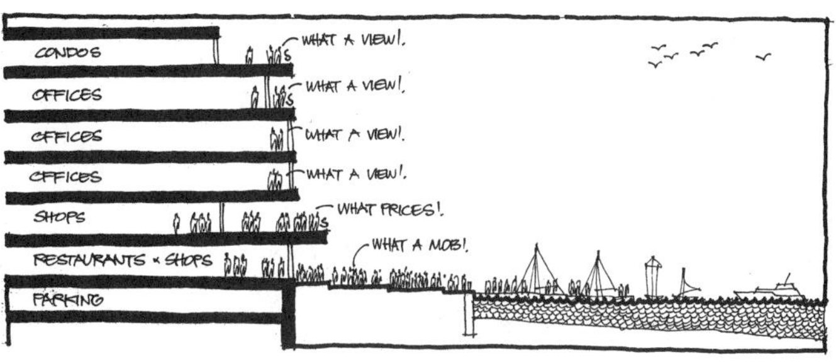

"What a View." © Roger K. Lewis.

Height by Zoning Ordinance

Background

As with all rules written locally, there are no universally consistent rules for defining height. The range of definitions is quite varied.

Where Is the Bottom of the Building?

Bottom is generally defined as grade level. The question then becomes one of establishing grade.

- In zoning ordinances, concerned as they are with identity and property value, grade is intended to establish a consistent look to a street facade. Grade may therefore be defined differently for large and small properties.

  Larger properties tend to create their own context, so grade can be established uniquely for the property. Large properties might be thought of as those whose buildings fit their properties with room to spare. Ordinances commonly quantify this as a ratio of building height to setback distance. Grade may be defined as the elevation of the ground in the middle of the front wall.

  Small properties must be more responsive to adjacent properties, so grade may be set as an elevation datum consistent across several properties. This may be achieved by referencing the elevation of the street curb nearest the center of the property. Where a building or group of buildings sit on a low terrace, the terrace elevation may be used. For districts with relatively low maximums, facade proportions (though not scale) may more closely approximate those found in larger properties, so grade at the center of a building's front wall is often used.
- In building codes, concerned as they are with safety and egress, definition of *grade* is usually related to the ability of firefighters to reach a building. The IBC and BOCA define grade ("*grade* plane") as the average elevation of the lowest spots within six feet of all exterior walls. Six feet may not be enough width for a fire engine, but it's enough room for a ladder. Beyond that distance, project teams are free to do as they wish with finished grades without affecting the official heights of their buildings.

Where Is the Top of the Building?

On flat roofs, which are rarely actually flat, but rather sloped to drains, the top might be considered to be the top of the parapet or the top of the gravel stop, or the high or low point of the roof. In the case of buildings with multiple roofs or multiple parapet heights, it is generally the "main roof" that defines height rather than roofs over mechanical penthouses or elevator overrides. With pitched roofs, the defining point is usually the ridge, the eave line, or the average height.

Perhaps few, and most likely no, ordinances define high point relative to an unenclosed piece of mechanical equipment, such as a cooling tower. While there may be cost savings, there are no regulatory incentives for leaving equipment exposed.

How Many Height Determinations Should Be Made Per Project?

Calculations are generally done once for an entire project, with each wall, wing, and individual building in the project expected to fit within the determined distance.

Requirements

The point of zoning ordinances limiting height is, again, maintaining community identity. To this end, heights may be specified in one of two ways.

Relative Heights

These are usually tied to grade at the base of the building, as the ordinances define that point. This favors consistency of height, regardless of the grade elevation at the project site. Buildings in such districts tend to rise and fall with the topography.

Absolute Heights

These are often tied to the heights of existing landmark buildings. For example, for many years Philadelphia held to a maximum height matching the top of its city hall. In the current market, most landmark buildings are significantly shorter than new projects need to be for commercial feasibility. The few cities with such restrictions periodically face pressure from developers to ease restrictions.

For hilly cities, this method argues for locating taller buildings on lower lying sites. It also gives cities a "flat-top" look.

Optimizing Zoning Ordinance Height Rules

In general, where height restrictions imposed by zoning are tighter than the heights allowed by building codes and supportable by the real estate market,

How high is this train depot? When a design is pushing the allowable height envelope, measurement points become critical. Is height measured to the top of the two-story exterior walls, the top of the dormer walls, the top of the "main" roof, or the top of the point on the cupola? Here the highest option is 200 percent of the lowest option. Photo by K. Wyllie.

Developers of projects in Washington, D.C., strain to keep their buildings within the fairly low maximum heights mandated by the D.C. zoning ordinance. Photo by the author.

designs tend to optimize floor-to-floor height. This explains why almost every building in downtown Washington, D.C., is made of posttensioned concrete: the material can shave a few inches off the thickness of every floor-ceiling assembly.

Adjusting Bottom Measurement Point

If Based on Elevation of Grade

Berm the base of the building. By raising grade, your building effectively becomes shorter. If the design calls for windows near grade, consider using areaways around them to retain the soil.

If Based on the Ground-Floor Elevation

Consider a split-level entry foyer. Since, one could argue, ground floor is the elevation of the level where one enters the building, this solution has the effect of lowering the building a half-story.

Adjusting Top Measurement Point

If Based on High Point of Roof

The high point is generally considered to refer to the "main roof." Roofs over mechanical penthouses are not main roofs. Since building codes limit the total area of all mechanical penthouses to one-third of the roof area (see "Squeezing into Allowable Height," later in this chapter), extensive mechanical systems may not fit.

Consider moving excess mechanical equipment onto the roof. Such equipment may be surrounded with screening walls, so long as it isn't roofed over. Since most mechanical equipment is available in weatherproof models, such exposure need not be a problem.

If Based on Eave Line or Average Height

Consider using a steep roof pitch, with or without dormers. If the ridge hasn't moved, the eaves will be lower than the same design with a reduced pitch. Mansard and gambrel roofs bring the added benefit of expanded interior volume.

Height by Building Code

Background

Conceptually, the same ideas discussed for allowable floor area apply to allowable height, and requirements for both are found in construction classification matrices.

Building code definitions of height are much more manageable, but also more strictly defined, than those in zoning ordinances. The only real question regards a building's safety. Therefore, height is defined by:

- The number of stories, since the more floors there are, the more occupants will need to egress during an emergency.
- Measured distance, since regardless of number of floors:
  1. Tall buildings are harder to reach, not only for the firefighters but also for the water in the standpipes, which require higher water pressures.
  2. It takes longer to exit from a tall building because the distance to safety is greater. When roofs are fitted with decks and occupied, buildings codes may interpret them as floors in their own right.
  3. Tall buildings contain more material than short buildings—materials with the potential to burn.
  4. Tall buildings are affected more by ground motion (earthquakes) [IBC Table 1613.4.1].
- There is another critical height-related measurement—the one that determines whether the building is considered to be a high-rise and is therefore required to meet the special provisions for high-rise projects. The only real issue is whether any occupied part of the building is beyond the reach of fire department ladders. The measurement is made from grade at the fire department access road to the floor level of the topmost occupied floor [IBC 403.1]. (See pp. 186–187.)

Number of Stories

Generally, every above-grade floor counts as a story, except those that qualify as mezzanines, penthouses, or deep basements. These and other exceptions are discussed below (see pp. 141–142).

Dimensional Height

To determine dimensional height, consider the same issues that apply for zoning requirements. The bottom of the height may be defined as the average elevation of grade at the exterior walls or when the ground is bermed around the building, the lowest point within 6 ft of the building. The high point may likely be set as the average height of the highest roof surface [BOCA93].

Optimizing Building Code Height Rules

The major model codes allow additional height and number of stories in exchange for inclusion of an automatic fire suppression system (sprinklers). Within certain limits, codes may allow an additional 20 ft of height and an entire additional floor [BOCA 504.2].

As with area, no limits are set for high-volume, low-hazard uses such as fabrication shops and power plants [BOCA93 503.1.1].

Although codes don't say so, general interpretation suggests that rooftop terraces complying with the one-third floor area rules can qualify either as penthouses (if directly accessible to exits and occupied spaces are allowed as penthouses) or as mezzanines (if open to a room on the floor below). When this interpretation holds, decks do not count as additional height or additional stories.

Height Due to Flight Paths

Background

Properties that are near airports are affected by Federal Aviation Administration (FAA) rules to minimize the danger low-flying aircraft pose to buildings and vice versa. Even when owners offer to accept full liability for the risk posed by aircraft, eminent domain is generally exercised. A legal defense based on occupants' prior knowledge of the risks may not work once they are manifest. Rules are not found in zoning or building codes, so they're easy to miss.

The primary issues are height and identification. Height is increasingly restricted as the distance from a property to an airport is reduced. Identification is a question of making the building visible to airplanes.

Applicability

Flight paths, and therefore the properties they affect, are indicated on zoning maps as "aircraft landing approach areas" or "aircraft navigational aid effect areas." Local ordinances require that projects in such areas be approved by the FAA before they will review the plans.

FAA notification is not required for buildings that are obviously shielded by other buildings, posing no danger. Buildings that are more exposed must notify if they have any of the following characteristics [14 CFR 77.13-15]:

- Height over 200 ft, measured from grade, even if not near an airport.
- Height passing a line drawn at a slope of 100:1 from a commercial runway when the project is within 20,000 ft (roughly 4 miles). The distance and angle are reduced for smaller airports and helipads.

When notification is required, project teams submit Form 7460-1. The FAA then studies the proposed design and issues recommendations for building massing and warning lights. Since the FAA averages 27,000 requests a year, projects should be submitted only once if possible.

Requirements

In general, buildings are considered to be obstructions to air safety [14 CFR 77 (C)] when higher than:

- 500 ft measured from grade at the building;
- 200 ft measured from grade at the airport if within 3 miles, 300 ft within 4, 400 ft within 5, and 500 ft within 6 miles;
- an imaginary plane calculated using rules found in 14 CFR 77.25-29.

Specific factors considered by the FAA in making its recommendations are beyond the scope of this book. The FAA has offices throughout the country, which can be found listed in the blue pages of local telephone directories. Ask questions sufficiently early in a project's development to make a difference.

Ultimately, allowable height changes very gradually with distance. Most sites are not large enough for building placement within the site to affect this requirement, so the property's placement relative to the airport must work for the scale of development proposed.

See below for the more minor issue of building-mounted warning lights (p. 166).

Squeezing into Allowable Height

Background

Just because a certain number of stories or a certain measured height are legally allowed doesn't mean that they are functionally, technologically, or aesthetically appropriate. Once height has been optimized as much as possible, project teams must find ways to make the allotment work.

For the specifics of strategies listed below, see the relevant sections of this book.

When Number of Floors Is More Restrictive Than Maximum Height

Not all floors need be counted. Codes allow two major "freebies" on height: deep basements and partial floors used as mezzanines or penthouses.

Basements

While all floors are defined as "stories," only "stories above grade" count toward the total number of stories permitted. The codes turn a blind eye to any below-grade basement levels. Such "free" levels are effectively defined as any whose ceilings aren't sufficiently far above grade [IBC 502.1]. Specifically, a basement is considered to be below grade when the height of the finished floor immediately above it fulfills all three of the following rules:

- **Average height:** The floor above is less than 6 ft above the average grade measured 6 ft from the building.
- **Majority height:** The floor above is less than 6 ft above grade for over half of the building's perimeter.
- **Localized height:** The floor above is nowhere more than 12 ft above grade.

This may be a terrific boon for so-called "underground buildings," those that improve energy efficiency and ease of maintenance through earth berming. Further, even traditional buildings can put space that far below grade to good use, since artificial light and ventilation are allowed for most occupancies. (See also p. 190.)

Partial Floors: Mezzanines and Penthouses

Background

Neither mezzanines nor penthouses, as defined by codes, are counted as separate stories in terms of allowable height [BOCA93 505.1]. They can therefore be used to advantage in effectively reducing the height of a project. Take note of the definitions of such spaces used by codes; they apply to only some of the spaces we might commonly think of as mezzanines or penthouses.

Top floors of buildings do not necessarily conform to the definitions of "penthouse" used by model codes.

Model codes put typical top floors in the same category as other floors. Special provisions are allowed, however, for top floors used to protect vertical shaft openings or to house machinery and equipment related, for instance, to elevators and HVAC systems.

Model codes agree that penthouses are distinguished from top floors by the fact that they enclose no more than one-third the area of the floor immediately below them. Some codes make no mention of occupancy; others accept only unoccupied spaces. Some put no restriction on height; others do. The IBC is likely to limit penthouse height and exclude occupied spaces.

REQUIREMENTS

Mezzanines. Maximum size: By definition, mezzanines must remain below a maximum proportion (one-third, with an exception of two-thirds for certain industrial projects) of the area of the floors immediately below them.

Maximum number: There are no restriction on the number of mezzanines a building or floor can contain. Every floor and every room can have them. There are only restrictions on the amount of space they occupy per floor: the total area of all mezzanines must fall within the one-third rule [BOCA93 505.2].

An additional floor qualifying as a mezzanine. Note the lack of enclosure, the total containment within the space below, and the relatively minimal size. Photo by the author.

Penthouses Maximum size: A penthouse must remain below a maximum proportion (one-third) of the area of the floor immediately below it. Unlike mezzanines, they are not expected to have any particular relationship to the room layout of the floor below.

Maximum number: There are no restrictions on the number of penthouses a building can contain. There are only restrictions on the amount of space they occupy per floor: the total area of all penthouses must fall within the one-third rule [BOCA93 1502.1].

STRATEGIES

For Mezzanines

Use them to add space to a project without adding stories. This is likely to be beneficial only for projects facing a maximum story height. Most projects classified as protected noncombustible construction have no limits on height anyway, depending on occupancy.

For Penthouses

Unenclosed mechanical space is not counted toward penthouse area. Enclosure is defined as fire-rated walls and a roof. When a cooling tower is roof mounted and concealed by screen walls, it isn't counted as a penthouse. Therefore, one-third of the roof area remains available for use as elevator machine rooms and enclosed mechanical equipment.

If one-third of the area isn't sufficient to house the needed equipment, some may be able to be located elsewhere, such as in a basement mechanical equipment room. If the required area exceeds the one-third allowed, the penthouse would be counted as an additional floor for purposes of determining total building height.

As with mezzanines, penthouse designation is likely to be beneficial only for projects facing a maximum story height. Most projects classified as protected noncombustible construction have no limits on height anyway, depending on occupancy.

When Maximum Height Is More Restrictive Than Number of Floors

Reducing Headroom [BOCA93 708]

Less headroom contributes to tighter floor-to-floor dimensions, which make for a shorter building. Keep

A rooftop mechanical room that qualifies as a penthouse. The screened but unroofed area adjacent to the penthouse does not count as a floor or toward FAR. Photo by F. Banogan.

in mind that each small optimization made applies to every floor in the building. Even very small reductions can be significant on larger projects.

Another strategy, but one that applies only to the topmost floor of a project, is designing accessory spaces with clearances that would be unacceptably low for occupiable spaces. One can take advantage of limited headroom space under sloped roofs (attic spaces) [BOCA93 1110 and 1111]. Although lower spaces can't be used for occupied program elements, they can be used for mechanical chases and storage rooms. (See pp. xx–xx for discussion of ceiling heights required of occupiable spaces.)

Reducing Thickness of Floor/Ceiling Assemblies

The thinner the floor/ceiling sandwiches can be, the more space you will be able to pack in a fixed height limitation. Acceptable assemblies are generally limited to those that have been tested and documented by UL, Underwriters Laboratories, based on required fire-resistance needs [BOCA93 Table 602]). These tested assemblies form the core of the construction details from which buildings are built. To this core can be added other materials and components, such as equipment, fixtures, trim, and finishes. The thickness of the total assemblies is dependent on the following factors:

Structural Depth

The height required of slabs, beams, and trusses is a function of the materials used, the way they are combined into assemblies, the design spans (how far apart the columns are), and the severity of anticipated live loads (for which minimums are set by code).

Thickness of Infrastructure

Common infrastructural elements include, in order from largest to smallest:

- Mechanical ducts, which are the total thickness of the duct itself, reinforcing angles (except on flex duct), and insulation, if any. Keep in mind that some system components, such as VAV (Variable Air Volume) boxes, can be quite deep.
- Light fixtures. Recessed fixtures add to depth, while surface or pendant fixtures may not, since the latter are generally not considered as encroachments to minimum headroom requirements. With wall sconces, table lamps, and torchères, floor assembly depth is even less of an issue.
- Piping for water, including not only pipe size, but also size of hangers, and depth needed to achieve the minimum slope required for positive drainage. Depth needed to achieve required slopes depends on the horizontal distance pipes must go to reach vertical chases, something that is directly affected by design decisions within the team's control. Keep in mind that separate piping systems are needed for water supply, waste water, internal downspouts, and often sprinkler systems. Also, except for some forced-air and radiant electric heating systems, water for heating and cooling systems will be needed.

- Piping for gas. The primary differences between gas piping and water piping are that sizes are relatively small since clogs are not a concern, and that pipes do not require sloping for drainage. In addition to natural gas for heating and cooking equipment, laboratory, medical, and some manufacturing facilities may add oxygen, carbon dioxide, vacuum lines, and other piping to the list.
- Electrical conduit, and increasingly, data and communications wiring. The advent of "flat wiring" systems and infrared links is helping to minimize and eliminate even these systems.

The Interplay between These Components

How well can they share space? For example, if electrical circuiting can be placed within the corrugations of a metal deck, no additional depth need be added to the required structural depth. Codes can't help with this issue, but they won't stand in the way of a determined designer, either. Designers' abilities to efficiently fit components into limited spaces is limited only by their insight, the level of collaboration among the project design team members, and the amount of time they're willing and able to put into the effort.

Due to the extensive demands made of the floor/ceiling assembly, hospitals frequently ignore reduction of building height as a design goal and opt for interstitial plenum spaces, where construction and maintenance teams can actually walk around within this assembly.

Systems conflicts. © Roger K. Lewis.

Applying Rules

Most of these issues are only indirectly regulated by rules. Still, examination of these indirect influences may yield some design flexibility.

Building codes set minimum levels of fire resistance. Fire resistance limits the number of alternative construction assemblies from which designers can choose to those that have been tested and rated, such as those published in the *UL Fire Resistance Directory*. Each of these model assemblies carry minimum depth requirements based on the type of, number of, and spatial relationship between components in the assembly. This includes the added depth of sprayed fireproofing coatings.

Building codes regulate live load based on use group.

Building codes have a lot to say about whether an automatic fire suppression system should be included.

Ductwork can be eliminated through the use of pressurized plenums, either in program spaces or in exit access corridors. Such plenums are subject to a whole new set of requirements mandating fire-resistant ceilings and penetrations intended to control the potential of such plenums for spreading fire and smoke. (See pp. 244–245.)

Plumbing codes mandate minimum numbers of plumbing fixtures, thus indirectly setting minimum piping requirements.

ESTABLISHING SHAPE OF BUILDING

Context

Beyond setting guidelines for overall bulk, there are certain more finely scaled demands that rules may make of a project. Such finer matters may include influencing whether a project should be rectilinear, angular, or curved; whether it should be simple or have wings, turrets, spires, or bay windows; whether its lower floors should project or be recessed from its upper floors; and degree of symmetry, solidity, and transparency.

Rules don't get involved in these issues to cause headaches for the project team. Such issues affect structural soundness and community identity.

General Architectural Style

Background

Overall shape of a building is strongly associated with particular styles—the low horizontality of Prairie style, the verticality of Gothic, the foursquare solidity of Georgian, the lightness of Modern. This relationship between overall building shape and style must be established at the outset of a project. While it may not seem that rules would have much to say about an issue as personal as style, some elements of style are very much in keeping with rules' general role as public advocate. (See also the discussion of shape and style at pp. 265–267.)

Requirements

Style is of concern primarily to CC&Rs and historic preservation rules. Further, it isn't limited to residential work. Suburban developments occasionally develop style guidelines, such as the pseudo-Georgian buildings required in some of the tony suburbs found in mid-Atlantic states.

Advocates and detractors of neotraditional town planning both point to, alternatively, the shared sense of community or the suffocating hold on individual expression engendered by the styles such towns require.

For historic properties or those located in historic districts, project teams must defer to the dominant historic style of the property or district.

Optimizing Architectural Style Rules

For covenants, understanding the unwritten attitude behind the rules is key to optimization. Since CC&Rs are written and enforced locally by others vested in the community, attitudes can be discovered by talking to neighbors, by attending meetings of the homeowners' associations, and by reading community newsletters. Visibility can also make a difference. If project teams are perceived as active and interested in the community, proposed optimizations may be seen as less threatening.

For historic properties or in historic districts, project teams would do well to ask "What made the neighborhood worth preserving?" New work that sensitively responds to that question is likely to be appropriate.

"Projections": Ornamental Elements

Background

This issue differs from that of property line or setback encroachment. "Encroachments," as noted in Chapter 4 (pp. 105–107), are occupied spaces, regardless of size, extending horizontally from the primary mass of a building so as to cross a setback or property line. Conversely, "projections" are unoccupied, incidental, or ornamental elements (such as steeples on churches) that are embellishments on, rather than extensions of, buildings. Projections cross beyond the normally allowed massing envelope in any direction, vertical or horizontal. The reason for regulating projections is mostly structural. There have been some notable cases of projecting roof cornice failures, which sent dangerously large chunks of construction to the ground.

Requirements

Sometimes language can be quite loose. For example, DCMR Article 770.3, dealing with height of buildings in commercial districts, allows that

a spire, tower, dome, pinnacle, penthouse over an elevator shaft, ventilation shaft, chimney, smokestack, and fire sprinkler tank may be erected to a height in excess of that authorized in the district in which it is located.

Article 400.3 for residential districts goes further, adding "serving as an architectural embellishment" to the description. None of those key terms is defined in that ordinance. The determination of whether a particular building element would be defined as a "tower" or an "architectural embellishment," and therefore allowed to project beyond the permitted massing envelope, would rest with the code officials. Further, the amount of projection beyond the permitted massing is not quantified—at least not in the same article.

Some quantification can be found elsewhere, but only for horizontal projections. For example, DCMR Article 2502 allows cornices and eaves up to 2 ft deep, sills up to 6 in deep, and awnings up to 40 in deep, and even limits "self-contained room air conditioners" to a 2-ft projection. None of these limits restrains vertical projection beyond the allowable height.

Awnings, Canopies, and Marquees

Awnings and canopies that project, whether fixed or retractable, are discussed in Chapter 4, in the section dealing with property lines, since they are considered to be encroachments in the public way by virtue of their proximity to the ground (see p. 107).

Chimneys

Mechanical codes specify minimum heights that chimneys and vents must project above roofs. This is intended to prevent exhaust gases and embers from endangering adjacent areas of roofing or entering nearby windows. Such minimums are directly or indirectly based on roof pitch. Chimneys are generally required to be 2 ft taller than the highest roof elevation found within 10 horizontal ft of the chimney. Required vent height is measured at the vent and is fairly small for most common roof pitches but increases dramatically with steeper pitches. [IMC96 800].

Optimizing Projections Rules

Some latitude might be found in pushing the definition of *projection.* Confirm with building and zoning rules whether bay windows, deep cornices at roof lines, or theater marquees are limited, and if so, how such features are defined.

Since restrictions on horizontal projections are much more common than those on vertical projections, building designers might look to vertical projections as a way to distinguish their buildings from adjacent ones.

Roof Design

Background

Roof shape is a primary component of building shape; their interdependence suggests that roof and building shape are best designed together. Read simply, most rules have only a very limited concern with roof shape, aside from the style issues previously discussed.

However, since some shapes are inherently less likely to spread or pass fire, incorporating them in a design leaves more freedoms regarding materials selection, since fire resistance becomes a less critical quality. In most design projects, shape is determined before materials are selected, so they are generally not in conflict. However, in some cases, especially when a light or transparent look is desired, materials selection may take precedence, thereby forcing the shape. From a code perspective, where materials selection is not important as a formal design issue, shape is less likely to be of concern [BOCA93 Chapter 15].

However, not all rules are so ambivalent. Preserving historic character through a rehabilitation often depends on maintaining existing roof shapes. Some CC&Rs restrict materials selection, so through syllogistic connection, limit shape.

Design Concerns

Materials Selection

Codes specify minimum levels of material performance when some roof shapes are used. For example, BOCA requires that roofs pitched at more than 60°, such as mansards, be made of noncombustible materials and carry a 1- to 1½- hour fire resistance rating.

Codes also list minimum and maximum pitches for each potential roofing material, such as a 1/4 in per ft (roughly 2%) minimum and 3 in per ft (3/12) maximum for built-up roofing [BOCA93 1507.3].

Tent structures bring unusual pitch and material issues to the discussion. Even they can fulfill the requirements of any construction classification including the most stringent, depending on the height of the roofs above occupied floors, the membranes' combustibility, and the fire resistance of the frames supporting them [BOCA93 3103].

Again, although these considerations don't limit the design shape by themselves, they clearly peg shape to material selection.

Historic Fit

Historic preservation guidelines voice concern over shape. The *Standards for Rehabilitation* recommend "preserving roofs—and their . . . features—that are important in defining the overall historic character of the building" [*Standards for Rehabilitation and Guidelines,* 1990]. They specifically recognize the value of such roof shapes as hipped, gambrel, and mansard, and such features as cupolas and chimneys, but caution against trying to rebuild historic features that are completely missing without sufficient period documentation.

Minimizing Structure

Snow, wind, and the weight of the roof itself are the typical load components that must be considered when designing a roof and its supporting structure. Special load components, such as the landscaped roofs, heliports, and special occupancies, add further complexity to the issue. As with seismic concerns, the relationship of structure to shape should be considered along with the other more obvious architectural or marketing considerations during the early schematic design effort [BOCA93 1608, 1609, 1610, 1611, 1614].

Many of these issues do not vary with shape—rules mandate the use of minimum structural loads when doing calculations regardless of the actual design. Others are directly design-dependent.

Maximizing Interior Volume

Where egress strategies based on timed exiting reinforce other design decisions, design teams may want to let the high volumes influence roof shape.

Accommodating Nature

Rain

Roofs are generally designed to collect water rather than disperse it, and then to channel it to drains or gutters from which it can safely be conducted away from the building. Techniques for doing this depend on the shape of the roof.

SHAPE

On low-slope and flat roofs, water is usually channeled to internal (as opposed to perimeter) drains distributed across the roof area. This is a fairly invisible way to accomplish drainage, having no impact on facade design. It can, however, affect interior layout due to the presence of internal downspouts, which pass through the building from roof to grade.

The other option for low slope and flat roofs, and the standard for pitched roofs, is to direct water toward the edges, where either gutters or edge dams (parapets or gravel stops) pierced by scuppers (spouts) conduct water into downspouts. All of this is not only clearly visible to passersby on the ground below the building (unless it is built-in to the roof and walls) but may affect them if the runoff system clogs, if the weight of icicles winds up tearing the system from the building, or if the system in other ways fails to do its job properly.

SIZING

The magnitude of the issue is an important consideration. Drains, gutters, and downspouts must be sufficiently large to handle the amount of water that can be expected during a heavy rainfall. Plumbing codes include maps and tables that indicate the amount of rainfall to be expected, expressed as number of inches per hour, in a 100-year rain [IPC95 Appendix B]. Once

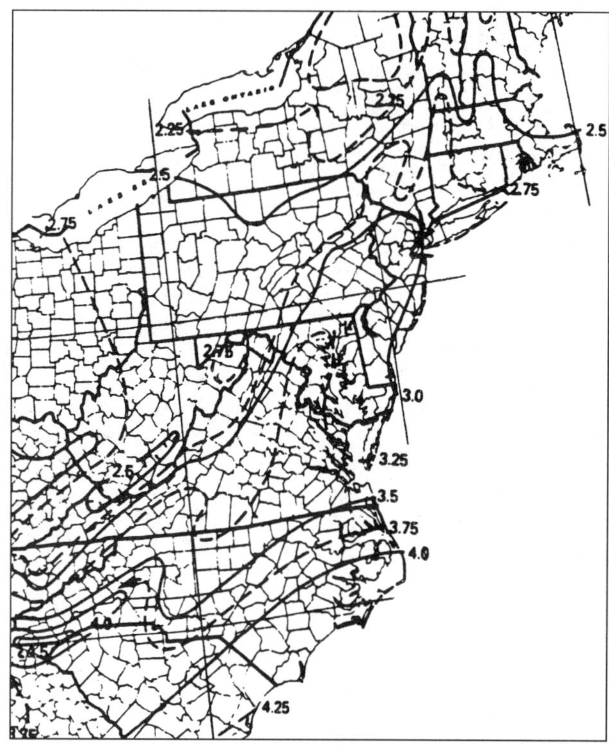

Detail from a rainfall map of the mid-Atlantic states, IPC95 Figure 1107.1. © ICC

the expected rainfall is known, it can be multiplied by the area of the roof. For example, if 3 in of rain per hour are expected to fall during a 100-year rain on a 100 SF roof, the drains, gutters, and downspouts must be sized to carry off 25 cu ft of water per hour (100 SF x .25 ft).

NO COLLECTION

Finally, there is the simple approach: letting water fall directly from the edge of the roof. If storm runoff is allowed to spill over onto the ground below, it can:

- increase hydrostatic pressure on foundations, causing structural problems;
- encourage leakage through the foundation wall;
- cause splashback of soil onto the base of the wall, staining it.

Snow or ice can:

- create hazards to passing pedestrians;
- lead to ice damming when found in conjunction with improper insulation and ventilation.

The model codes, including the CABO One and Two Family Code, are mute on this issue. Some state and

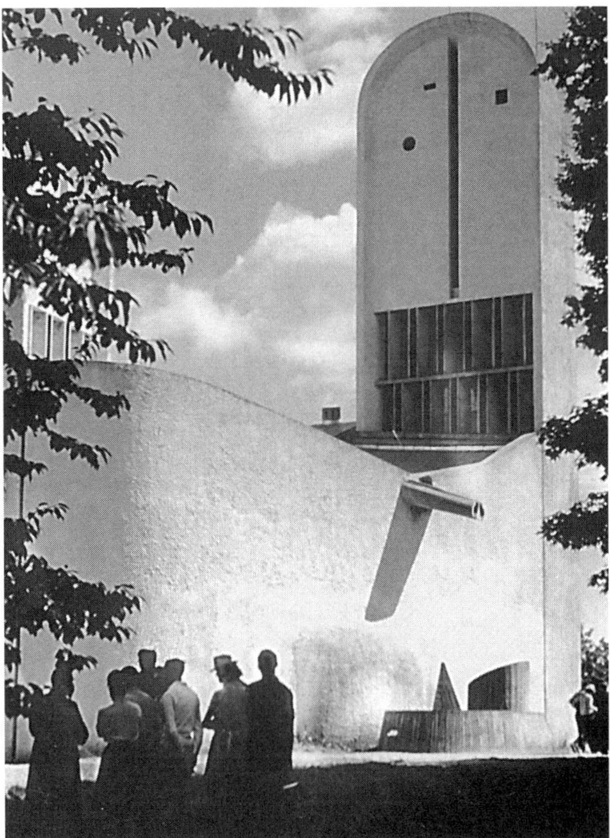

Playing with rainfall: the scupper at Notre Dame du Haut, Ronchamp, France. Le Corbusier, architect. Photo by J. O'Hear, III.

local codes allow for gutterless eaves when the roof extends a minimum distance from the exterior wall. The Massachusetts Residential Building Code sets 18 in as this minimum. Taking this approach, rainwater will hit the ground sufficiently far from the exterior wall to avoid problems.

Snow and Wind

Snow and wind were introduced earlier (see pp. 86–87). Assuming that neither has rendered a project infeasible, project teams must consider compensatory design. Most of the rules for designing for snow and wind originate in Section 6 and 7 of ASCE 7, "Minimum Design Loads for Buildings and Other Structures."

Since roof shape has a major role to play in managing snow and wind load (once a site is selected), it is important to have some understanding of the relationship between the shape of a roof and required increases in structural capacity. These increases compensate for snow drifts, for snow slides, and for the effects of wind, acting on its own and as an agent in determining patterns of snow accumulation.

Individual Components

Primary Elements

Flat and Low-Slope Roofs

Loads are generally required to be calculated for the worst-case scenario of flooding—the amount of water held by the roof when all primary drains are clogged, and only overflow drains or scuppers are operable. At 64 lb per cu ft, water is more than one-third the weight of concrete. This presents a strong argument for effective overflow design [BOCA93 1609.5].

The snow load formula for flat roofs (under 30°) differs from that for sloped roofs in that it allows for load reductions based on the amount of shelter provided by surrounding vegetation or buildings. On the other hand, it also mandates increases based on the criticality of the building's use.

Flat roofs also have the advantage of simpler snow load formulas based on "balanced" loading. These formulas factor out wind, and so assume even snow accumulations. This is not because some sites are never windy, but because the effect of wind is negligible for some roof shapes. Balanced loads are assumed for roofs that are either very flat (under 2.5°, or ½ in per ft), since wind pressures wouldn't differ much across a ridge, or very steep (over 70°), since snow would simply slide off.

Sloped Roofs

Snow load is calculated based on the pitch—the steeper the roof, the less snow, and therefore load, is assumed to accumulate. Codes are written with certain break points where required design loads change markedly.

As an example, BOCA93 1610.5 requires any roof sloped at less than 30° (a 7/12 roof) to assume a full snow load. In other words, a 6/12 roof gets no reduction in snow load over a flat roof, so has no advantage in column spacing or structural depth. Between 70°and 30°, the snow load required of structure is decreased geometrically, but not exponentially.

Sloped roofs also differ from flat roofs in that wind creates positive and negative pressures on different roof faces. This is referred to by codes as unbalanced load conditions. As a result of wind differentials, the design loads that would have been required for balanced snow loads are either increased or decreased using a multiplier.

Curved Roofs

Snow load can be ignored for parts of curved roofs, such as domes and barrel vaults, that pass a minimum pitch criterion (70°), since they are too steep to let snow accumulate.

Edge Elements

PARAPETS

The IBC requires parapets on all buildings unless they are listed as exceptions [IBC 704.11]. The most common exceptions are those granted when exterior walls don't need to be rated due to fire separation distance, and when roofs are pitched at least 4:12. Parapets must be of minimum height above the roof (IBC: 30 in, BOCA93: 32 in).

CORNICES

In the late nineteenth century, many high-rise buildings were designed with heavy, projecting cornices at the roof line. Strong winds and heavy snow loads have taken their toll over the years, in some cases causing the cornices to come crashing to the sidewalk. Many codes have responded by restricting the size (3 ft wide) and use of such cornices [BOCA 3203].

Roof Features

IN GENERAL

Snowdrifts and snowslides can be expected on lower portions of a roof and adjacent to such roof features as parapets, penthouse walls, and mechanical equipment. Several considerations are of note [BOCA96 1608.7 and .8]:

- Rules assume that snow won't drift more than 20 ft Therefore, one way to avoid the additional load is to design a gap of at least 20 ft between a higher roof and a lower roof.
- Rules assume that roof features less than 15 ft long won't block wind sufficiently to cause drifts. For longer features, drift height and depth are calculated as a function of the height of the feature, up to a maximum of 25 ft.
- When upper roofs are in a position to drop snow onto lower ones, loads calculated for drifts are increased by 40 percent.

LANDSCAPED ROOFS AND ROOFTOP PLANTERS

Interestingly, this is a give-and-take situation. Although the weight of all soils when completely saturated with water is expected to be considered as part of the dead load calculations, a reduced live load (20 PSF) is allowed based on the thinking that people and furniture will not be using a landscaped area as much as a paved area [BOCA93 1609.6.1].

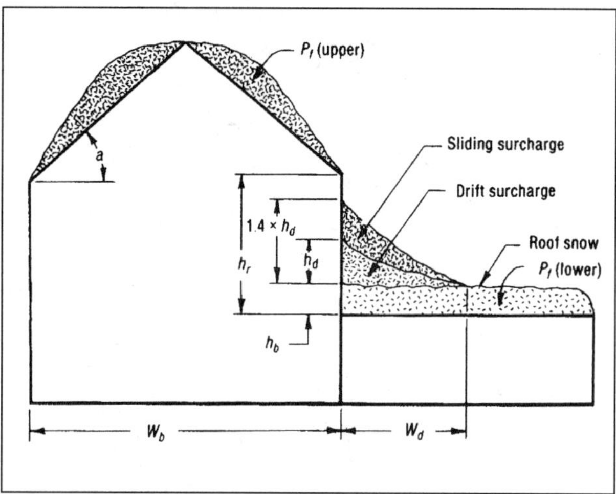

BOCA96 Figure 1608.8—"Additional Surcharge Due to Sliding Snow." © BOCA

HELIPORTS

Rules provide design guidance for helicopter landing pads including heliports and helistops (heliports minus refueling, maintenance, and storage facilities) [IBC-1ST Draft 412.4]. Although such landing pads may be located at grade, they may also be installed on roofs. In this latter position, such facilities can exert substantial static and dynamic forces on roof structures [IBC-1ST Draft 1605.5, adapted from ASCE 7]. In addition to the weight of helicopters and their passengers or cargo, one must also consider the momentum of a landing helicopter as it touches down. Building codes do not address the uplift on anchor points caused by wind buffeting a parked helicopter, apparently for lack of magnitude. Finally, allowing occupants access to a roof triggers egress requirements more typical of an occupiable floor than an unoccupiable roof [See IBC-1ST Draft 1007.2.2, and NFPA 418 (Roof-Top Heliport Construction and Protection), adopted by reference in BOCA93 1511.]

DORMERS

When deciding whether to locate a dormer side wall immediately above an exterior wall or to offset it by a few feet, note that the in-line wall must pass a higher fire resistance threshold [BOCA93 1510.6].

MISCELLANEOUS ROOF STRUCTURES

Towers, dormers, or other roof projections of average size and of strictly ornamental purpose are required to be only as fire resistive as the buildings to which they are attached. When they are overly tall (85 ft), large (200 SF), or fulfill a programmatic function, they are required to be of noncombustible construction (Type 1 or 2). Exceeding these limits can complicate otherwise simple projects [BOCA 1510.9].

Effect of Seismic Concerns on Form

Background

All contemporary model building codes require every building to address seismic loading as appropriate to its geologic location. Some regions not normally thought of as being "earthquake zones" are rated as surprisingly risky. Check before designing. Further, as geometry and load resistance are integrally connected, making design decisions without considering the impact of such decisions of seismic feasibility can be wasted effort.

Earthquakes shake the ground under buildings, primarily in a back-and-forth direction, but occasionally in an up-and-down motion. The Northridge, California, earthquake of 1994 was a notable example of this latter type. Some building shapes, those that are considered "irregular," have inherent problems withstanding these diverse kinds of movements and are therefore discouraged by building codes. In fact, codes deal separately with plan (horizontal) irregularity and sectional (vertical) irregularity.

Requirements

What Kinds of Designs are Likely to have Seismic Consequences?

Bilaterally symmetrical buildings whose mass and structure are distributed in a totally uniform and homogenous way (known as "regular" designs) respond to earthquake stresses in ways that are reasonably predictable. Predictability is good, since it allows us to be scientific about design. By following certain rules, we can be relatively assured of the structural sufficiency of a design. Almost no buildings are uniformly symmetrical and homogenous, however, and therefore almost all buildings are less than optimally predictable. Significant deviances from reasonably predictable design are classified according to type of design irregularity. Codes require each type to be handled with its own specialized set of rules [IBC 1613.4.2, BOCA96 1610.3.4].

This section describes each of the categories of design irregularity for which special rules exist. While it is unlikely that a design team would need to avoid certain designs simply because they trigger the related design rules, the team might reevaluate the benefits of only marginally irregular design features.

The terminology found in codes for each category of irregularity can be rather opaque. The following descriptions attempt to explain, in simpler terms, the essential characteristics of these categories. The terminology found in codes is given in parentheses.

Design and LFREs

Every building contains some elements that handle earthquake loads better than others. These may be vertical elements, such as walls and shafts, or horizontal ones, such as floors and roofs. They may be solid, as are shear walls, or skeletal, as are braces. However they are configured, they are capable of resisting lateral load better than other parts of the building, and it is on them the survival of a building depends in an earthquake. The descriptions below refer to such lateral-force-resisting elements as LFREs.

For a building as a whole to resist earthquakes, its LFREs must be related to one another with continuity and consistency so that they can perform as a system. In other words, an LFRE system, made up of individual LFREs arranged in a very particular relationship to each other and to the building as a whole, must be designed into buildings where seismic activity is a concern. It is most economical when structural (gravity-resisting) components are designed to do double duty as LFREs, but this is not a requirement. Potential LFREs must be identified and then designed to juggle the issues of balance, continuity, and consistency, as described below.

Not all potential LFREs in a conceptual design should be handled the same way. Design teams must identify not only those components that should act as LFREs but also those that could, but should not, act as LFREs (by detailing them to be isolated from adjacent structural components). The rules of irregularity described below are critical in determining which potential LFREs should be designed as actual LFREs.

Finally, elements may need to be added in order to complete the lateral force resistance system. In such cases, it may not be possible to do so without changing the design. Even project team members who have no expertise in structural engineering should be aware of these issues and their implications for a proposed design. Rules categorize irregular designs into ten basic types. For ease of comprehension, they are grouped here (though not in codes) as unbalanced, discontinuous, and inconsistent.

Unbalanced Designs

One might think of unbalanced designs as lopsided. In them, massive elements, such as bathroom and elevator cores, are not centered or distributed symmetrically in relation to the more open spaces, such as offices.

Design archetypes that trigger code provisions related to imbalance include:

- **Wing plans** (re-entrant corners): Buildings with re-entrant corners include those whose floor plans are bent into Ls, Es, Us, Ts, and the like. They fail when lateral loads are imposed because the building masses on either sides of a corner develop momentum independently, as they differ in orientation, even if not in design. Stresses would be relieved if the masses were not attached to each other, and are highest at the corner as a result. Buildings with longer wings are more irregular, with compensatory action required when wings extend more than 15 percent beyond the rest of their buildings.
- **Angle plans** (nonparallel systems): All buildings have primary horizontal axes – what one might call their lateral and longitudinal directions. Nonparallel systems occur in building designs when LFREs are at an angle to both of these axes. Failure occurs because LFREs can twist diagonally when the majority of the building bends under lateral loads. To avoid damage, they must move in synch with the majority of the structure.
- **Twist plans** (torsional irregularity): The concept underlying torsional irregularity is actually quite simple. A fundamental rule of physics holds that actions met by equal reactions are held in balance. Another rule says that the effect of a force acting on each part of an object is the same as the effect of a force acting on a single point at the center of an object. When the force is physical (gravity or lateral) the center of the object is the center of the object's mass. When the center of resistance doesn't match the center of the force, torsion results. Torsion makes things twist, or rotate. It is experienced, for example, when a bookcase is pushed near one end rather than in the middle. Objects, including buildings, experiencing torsional movement must be much stronger than those experiencing simpler linear movement. Balanced resistance depends on careful distribution of force-resisting elements.

When the object is a building, the action is exerted by earthquakes and the reaction is provided by LFREs. To the extent that the center of LFRE resistance fails to coincide with a building's overall center of mass, the building will tend to twist when exposed to lateral loads. Twist is not a question of symmetry (the centers of mass and resistance may or may not coincide in symmetrical designs) but one of balance.

Discontinuous Designs

One might think of discontinuous designs as having gaps. Gaps may take the form of holes, such as atriums, or offsets, such as overhangs. Because of these discontinuities, lateral loads can't take the most direct route to the ground. Loads are concentrated in materials that bridge gaps or serve as offset transitions, necessitating stronger construction details.

Design archetypes that trigger code provisions related to discontinuity include

- **Swiss-cheese plans** (diaphragm discontinuity): These occur when a large percentage of an LFRE consists of one or more openings. The rules are triggered when the cumulative area of all openings is more than half the size of the LFRE in which the openings are located.
- **Zigzag sections** (out-of-plane vertical element offsets): Here, LFRE walls shift back and forth (out of plane, like a folded ribbon) on adjacent floors. Since the LFREs are vertically discontinuous, little resistance is amassed.
- **Scatter sections** (in-plane discontinuity in vertical lateral-force-resisting elements): These occur in designs in which vertical LFREs shift from side to side (within the same plane) on adjacent floors, usually to work around large open design elements. The rules apply only when no part of an LFRE is continuous from the LFRE immediately above or below it. When there is some continuity, the design is considered to have wiggle sections, as described below.

Unbalanced Designs

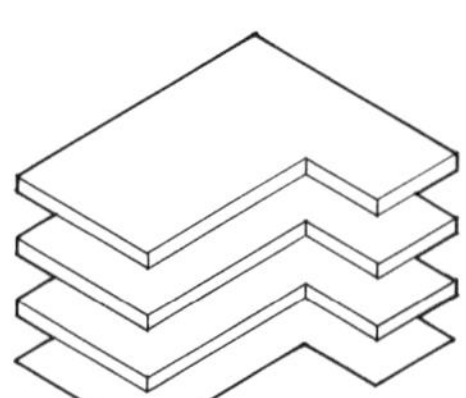

Wing Plan
(Re-Entrant Corners)

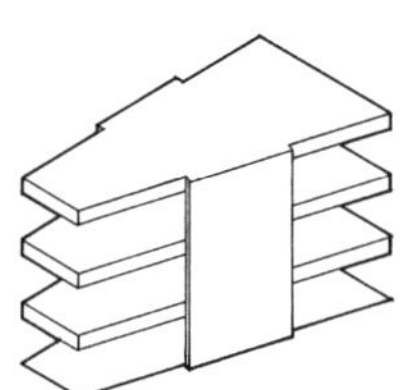

Angle Plan
(Non-Parallel Systems)

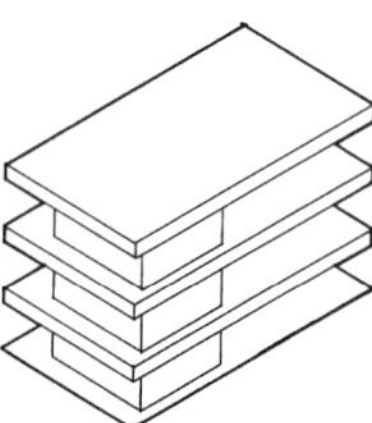

Twist Plan
(Torsional Irregularity)

Discontinuous Designs

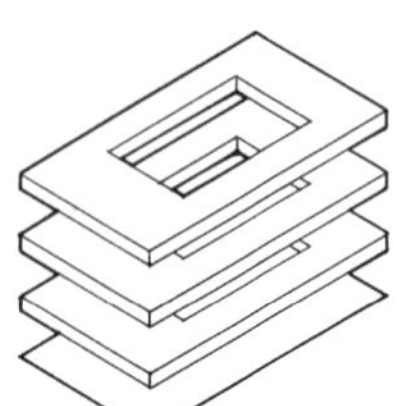

Swiss Cheese Plan
(Diaphragm Discontinuity)

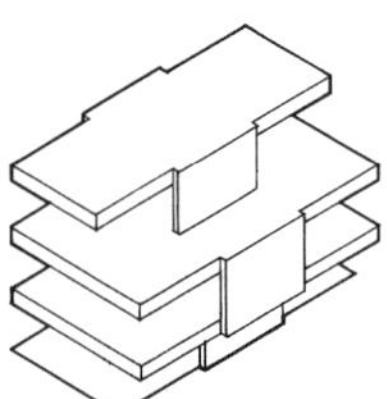

ZigZag Section
(Out of Plane Vertical Element Offsets)

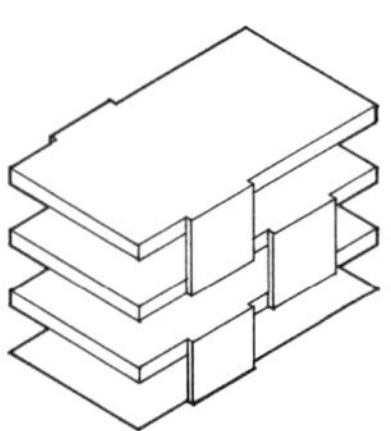

Scatter Section
(In-Plane Discontinuity in Vertical LFREs)

Inconsistent Designs

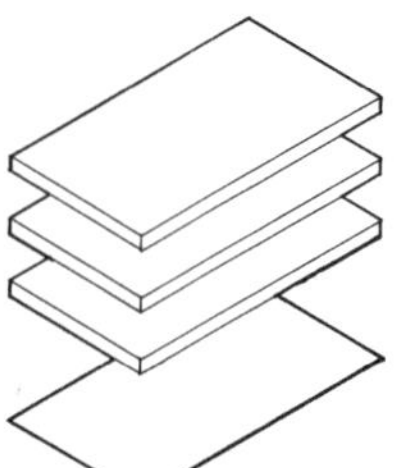

Wobble Section
(Stiffness Irregularity)

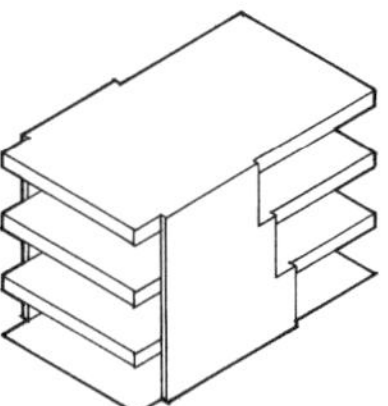

Wiggle Section
(Vertical Geometric Irregularity)

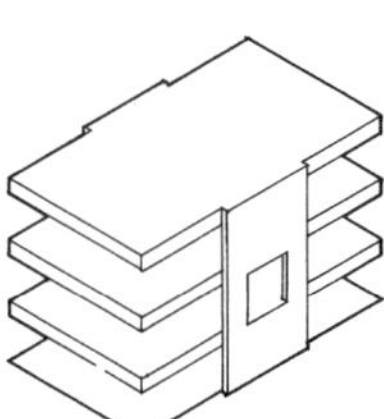

Weak Section
(Discontinuity in Capacity)

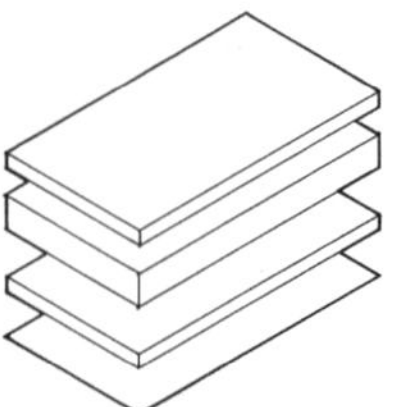

Rock & Feather Section
(Weight or Mass Irregularity)

Examples of unbalanced, discontinuous, and inconsistent massing.

Inconsistent Designs

One might think of inconsistent designs as lumpy. When some parts are more rigid and others are more flexible, it is more difficult to predict how they will behave when pushed.

Design archetypes that trigger code provisions related to inconsistency include:

- **Wobble sections** (stiffness irregularity): When LFREs on one floor are more flexible than those on floors above them, they can be squashed flat by those upper floors when lateral loads are applied. This can occur when an office or residential building with multiple walls and partitions is built over parking, since there are far fewer LFREs on the parking levels. It can also happen when a lower floor is taller than an upper one, so that its LFREs are more likely to buckle due to their higher slenderness ratios. Rules for wobble sections kick in when one floor has anything less than 70 percent of the stiffness of the adjacent floor, or less than 80 percent of the average of the adjacent three floors.
- **Wiggle sections** (vertical geometric irregularities): These occur when the load resistance (usually a function of the width) of a vertical LFRE varies from floor to floor. The cumulative ability of such a "wiggly" LFRE to resist load is limited by its narrowest dimension. When LFRE width varies by more than 30 percent, designs are required to comply with special rules.
- **Weak sections** (discontinuity in capacity): In buildings with weak stories, some LFRES differ significantly from others in their ability to carry lateral loads. This might occur when one floor, for the sake of an open plan, goes without some of the shear walls and columns that occur on other floors. The rules apply when the strength of adjacent LFREs differs by 20 percent or more.
- **Rock and feather sections** (weight or mass irregularity): These happen when adjacent floors, not counting roofs, have widely differing masses from each other. This creates an inertial imbalance, since lateral loads have a more profound effect on greater masses. Rules apply to differences greater than 150 percent.

What Is the Higher Standard That Such Designs Are Obligated to Meet?

For starters, designs with the irregularities described above require special structural analysis. This, of itself, clearly does not directly affect a design, but it does consume some of the design budget that might otherwise be spent developing more visible architectural or interiors work further. On the other hand, a decision regarding whether a design will or will not involve structural irregularity goes to the heart of many a bold design gesture [BOCA93 1612].

Aside from triggering analysis requirements, irregular designs are required to meet rules that other designs can ignore. Such rules are cited in IBC Tables 1613.4.2.1 and 1613.4.2.2.

Optimizing Seismic Behavior Rules

Seeing as how seismic consequences profoundly affect design work, and considering the unexpected regions that require buildings to address seismic loading, this is one issue that the entire project team should be aware of, if for no other reason than to avoid being caught unprepared. To optimize their efforts, project teams are well advised to contact code publishers regarding continuing education seminars and publications on the issue, and to bring structural engineers into the design team early in the process.

INTEGRATED STRATEGIES FOR MASSING

What strategies and attitudes deal effectively with the whole range of zoning, building, accessibility, and preservation issues affecting massing, including total size of building, size per floor, total height of building, and shape of building?

Look at the broadest possible context. Keep in mind the impact your project will have on adjacent properties in terms of visual impact, user load, and impact on infrastructure. To develop a convincing design, you need primarily to convince the project's neighbors that the new project will be a safe, polite, and undemanding addition to the community.

You don't really have to guess in trying to figure out what kind of project would be considered welcome. You do have to read the rules with this goal in mind. One very efficient way to learn more about these issues is to attend a public zoning or design review board hearing. You'll see, firsthand, the concerns, passions, and issues that inform such requests.

No single strategy will ever successfully address so many divergent demands, but it might be worth reviewing some of the most effective approaches to optimization.

- Carefully consider including public amenities in exchange for an agreement to boost permitted FAR.
- Use specified safety systems (sprinklering) whenever possible.
- When appropriate, develop assembly spaces on floors that have access to grade, to take advantage of direct exit bonuses.

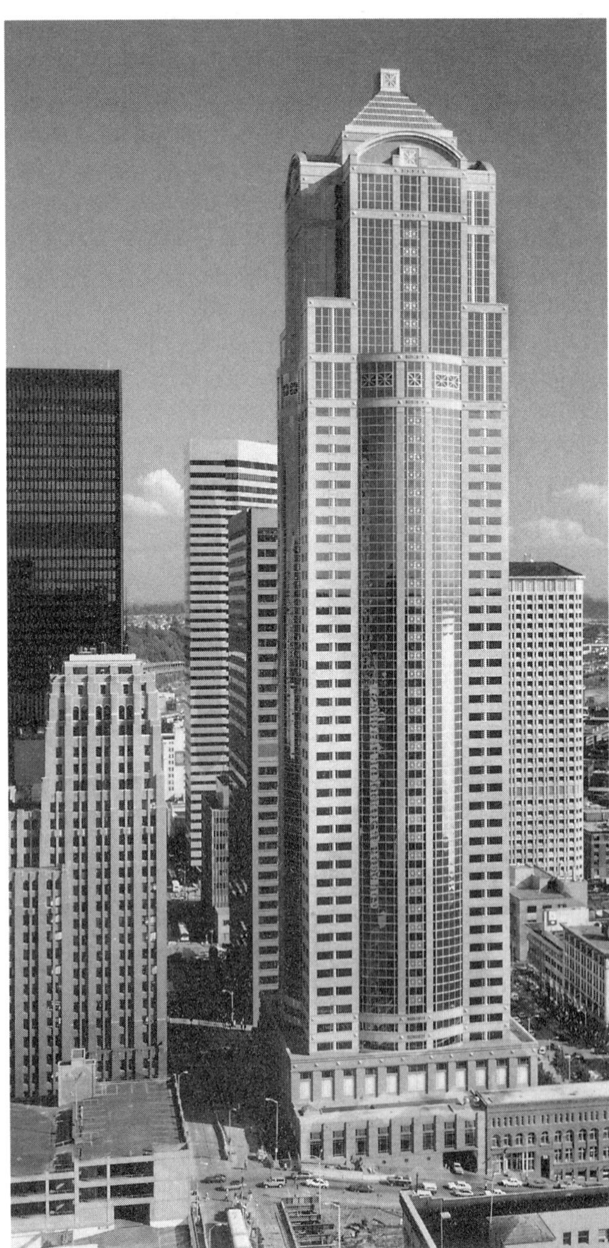

Washington Mutual Tower, in situ. Photo by Kohn Pedersen Fox, architects.

WASHINGTON MUTUAL TOWER

Seattle, Washington; Kohn Pedersen Fox, architects[28]

In an extremely efficient (some might say aggressive) demonstration of rules optimization, Seattle's Washington Mutual Tower, designed by architects Kohn Pedersen Fox, literally doubled its allowed density to one million square feet by taking advantage of just about every incentive offered by the rules. From the 27 stories allowed by right, it grew to 55 stories by adding:

- 13 stories for funding off-site housing efforts
- 2½ stories for including a public escalator to help people get up the steeply sloped site
- 2 stories for including an urban plaza in the project
- 2 stories in compensation for space occupied by the building's mechanical equipment
- 2 stories in compensation for the floor area on upper levels lost when the architects made them smaller to create a distinctive skyline profile
- 2 stories for including retail space in the project
- 1½ stories for including a public atrium in the project
- 1 story for space lost to a subway tunnel connection
- 1 story for including a ground-floor day-care center in the project
- ½ story for making a flowered terrace available for public use

To place this in context, one must understand that projects like this one do not necessarily change the density balance. When Seattle made all these incentives available, it simultaneously reduced the allowed density in the retail core from a Floor Area Ratio of 10 to 5 and the maximum height from 400 ft to 200 ft. Projects like the Washington Mutual Tower are not necessarily any bigger than they might have been before incentive zoning was implemented, but they certainly give city dwellers more of what they want in their buildings. One must wonder how the city is affected by projects whose teams are not so adept at using incentives, or who simply decide that they don't need them. Bear in mind that the alternative to incentive zoning might be thought of as "base-line" or "demand" zoning, where all projects are required to reflect community aspirations and incorporate community amenities.

6 ENCLOSURE

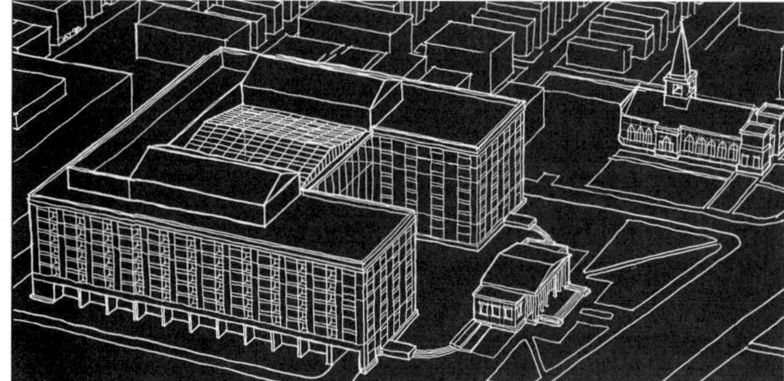

Schematic Design of Exterior Elevations and Roofs

IF A PROJECT HAS SUCCESSFULLY DEALT with the issues so far raised in this book, the project team has established that its intended use is acceptable and that its position on the site, its size, and its shape will most likely be approved. At this point, an urban massing model could be built from what is known of the design. What remain to be determined about the exterior of the building are materials selection and the subject we will now consider: the design of the exterior elevations, including window size and placement, the design of balconies and guardrails, and the shaping of the roof.

CONTEXT

Why This Issue Is Important

Enclosure Design Criteria

A building's enclosure provides most of a project's "signature," its public face. It does this on multiple scales including the distant view of the urban skyline, the view from across the street, and from the perspective of passersby only a few feet away. Designing for the distant view primarily requires a sensitivity to compositional, or formal, issues such as proportion, contrast, and color. In addition to all these issues, designing for direct contact requires a mastery of technical issues such as expansion and contraction, dealing with water- and weatherproofing, and energy performance, as well. Buildings that look as good close up as they do from a distance are generally those that are better designed and built. To the extent that construction quality is a concern of rules, they can be a useful guide to "close-up" facade design.

Further, facade design has implications for the interior of a project. In addition to establishing the view of the building from outside, the facades provide windows to the outside world from within the building, are the source of natural light and ventilation, and are an alternate means of escape from fire.

Criteria for Elevation Development

An exterior elevation is an architectural drawing that documents the design of a facade and portions of the

Massing model indicating positioning and bulk, but as yet lacking elevations and materiality. Photo by the author. Model by S. Jancy/K. Long.

roof visible to passersby. To the extent that this book parallels the design process, we will look at the design decisions related to elevation development, rather than simply look at facade or roof issues.

Elevations show not only how windows and materials are to be arranged on a facade but also the finer, more detailed elements such as jointing patterns for stone or precast concrete work and expansion joint locations for masonry and stucco.

We've already discussed some issues relevant to exterior elevations in the sections on positioning, insofar as elevations and distance to adjacent buildings are closely interdependent due to fire-spread considerations. Here we'll consider others, unrelated to positioning.

Why Regulatory Bodies Care

In the context of schematic design, the public concern for safety influences facades, fenestration, balcony design, and selection of materials. Before concluding schematic design work, a project team must consult rules to find out what overall construction is acceptable.

Where to Find the Requirements

Requirements related to exterior elevations are as diverse as the issue itself. Due to the public nature of facades, zoning ordinances, CC&Rs, and historic preservation guidelines, if relevant, should be consulted in addition to building codes. Also, projects that are designed in response to RFPs (Requests for Proposals) can often learn about the developer's design intentions through a careful reading of the RFP. When the developer is a governmental agency, such information is even more useful as it can help the design team understand where juggle space might be found for dealing with regulatory issues.

WALL AND WINDOW DESIGN

Context

Why This Issue Is Important

Basic facade design, focusing on the development of walls and windows, is the key element in establishing building identity. It's often what people remember most about a building. Before working out details, project teams must find ways to achieve overall design intentions for the mix of solid wall and dynamic opening of which exterior walls are made.

Alternate wall designs must be evaluated for their abilities to provide adequate but controlled access to light and air. They should be sufficiently durable and easy to maintain, not only due to inherent material properties but also due to the installation techniques associated with those materials. And, of course, the selected design will have a marked effect on fire spread.

Why Regulatory Bodies Care

Protection from the elements is arguably the most fundamental role of building projects. Controlling temperature, precipitation, wind, and solar gain are the mainstays of the health, safety, and welfare trilogy at the heart of many codes. In this regard, the walls and windows are arguably the most important points of control. They stand out even relative to the roof, which has relatively simple design parameters since it need not be operable. It's much harder to control weather when you sometimes have to or want to let it in.

The member of the project team most suited to managing these critical issues is the architect. Other design team members can resolve structural, infrastructural, landscaping, and interiors issues, but the weatherproofing "buck" stops at the architect.

Where to Find the Requirements

Project teams should review standards, such as those published by ASTM and ANSI, in addition to the usual coterie of codes. These documents are key to quality control in materials selection and workmanship. Where rules require achievement of specific quality levels, they usually refer to such standards, adopting them by reference.

When using codes, make use of the maps they contain to determine weather conditions in the vicinity of the project site. In particular, they can help quantify wind and snow conditions.

WALL ASSEMBLIES

Resistance to Weather

Background

Specifications could take the simple route to weather resistance through such performance statements as "If it's a wall, and it's in Chicago, it must be able to make it through the kinds of winters they have there." While that seems clear enough, it has little specificity. Would every assembly in every part of every project be required to meet the worst conditions possible? One rule could not adequately address the different design goals of an exposed exterior panel on the ninety-eighth floor of the Sears Tower and a mostly protected soffit (exterior ceiling) in the recessed ground-floor entry to the Ritz Carleton Hotel. Wind exposure can vary substantially from one face of a building to another. When wind is pushing (exerting positive pressure on) one face, it is sucking (exerting negative pressure on) the opposite face.

Fortunately, codes generally consider factors other than regional weather patterns, including the compass direction a given wall faces (accounting for the direction from which most storms approach) and the degree of shelter afforded by other parts of the building, such as overhanging roofs. As an example, code sections on masonry stipulate mortar type based on weather conditions and the position of the masonry on the building.

Where to Find the Rules

Building codes devote a lot of space to regulating materials and assemblies selection and the methods of their installation. To comply with code requirements regarding weather resistance, one would have to check all the chapters pertaining to the many materials found in the project, in addition to Chapter 14 of the model codes on exterior wall coverings, and read through the standards incorporated by reference. This amount of complexity starts to explain the value of good specification writers.

For purposes of illustration, let's look at the case of brick, regulated by model codes' Chapter 21, "Masonry." One of the most important factors in masonry weather resistance is the choice of mortar materials and tooling method. Mortar comes in several grades with differing balances between stiffness and softness. The stiffer ones resist cold better, but the soft ones resist water better. With tooling, some shapes, such as concave, vee, and weatherstruck, are more effective barriers to water than others, such as raked joints. But this kind of information is nowhere to be found in Chapter 21, or anywhere else in the building code. There is, however, a reference [BOCA93 2105.7] to ASTM C270. If one looks in Chapter 35, "Referenced Standards," C270 is identified as "Specification for Mortar for Unit Masonry," 1992 edition. The mailing address of ASTM is also listed there, for those who want to order a copy or have questions.

Requirements

For an introduction to the rules pertaining to wind resistance, see Chapter 5 (pp. 86–87).

Optimizing Rules Related to Weather Resistance of Exterior Materials

This depends to a large degree on the extent to which requirements are written in performance or prescriptive terms. If the wall is required to resist water when driven by winds exerting 25 psi of pressure, one has more options than when it is required to use ⅜-in concave joints of Type SW mortar.

Fire Resistance

Background

See Chapter 2 (p. 63) for a definition of fire-resistance. Once required fire-resistance ratings of exterior walls are determined, project teams must start looking at assemblies that meet those ratings. Since unrated walls will not be accepted, the design team must find a listed assembly that meets its design goals. This is, of course, unless the project has a sufficiently high budget to justify having its assemblies tested by an independent laboratory acceptable to the building commissioner or fire marshal in the project's jurisdiction. In addition to the cost of building the assembly at the testing agency, the tests themselves can have substantial costs. Still, numbers are relative and it might be worth asking for large "signature" commercial projects.

Requirements

At this point in the project, the project team must have decided on a construction classification to which to build. By itself, this says little. Classification must be translated into hourly ratings for each part of the building. To do this, look up the fire-resistance matrix. In model codes published since 1993, this is found in Chapter 6. This table lists hourly ratings for each component of the building, based on its construction classification.

Once hourly ratings for the exterior wall are found, one must turn to one of the books of rated assembly designs, such as the Underwriters Laboratories *Fire Resistance Directory*. Such books contain construction details, in plan and section, along with partial specifications, for all assemblies that have been tested so far. As new assemblies are tested, they are added to the books. The details are grouped by type (wall, column, floor/ceiling, etc.) and by primary materials (composite steel, concrete, wood, etc.). Each lists the hourly ratings for which it has been successfully tested and lists the structural conditions (restrained or unrestrained) under which the rating is valid.

Optimizing Exterior Wall Fire-Resistance Rules

As the degree of fire resistance of an exterior wall depends on the degree of safety intended by a construction classification, the best way to get some juggle space is to reduce the needed construction classification through strategies already discussed.

Aside from this, one can at least hold fire-resistance ratings to minimum levels by keeping exterior walls at some distance from adjacent buildings. (Refer to pp. 108–111 for further elaboration.)

Energy Performance

Background

Mechanical engineers hesitate to start designing heating and cooling systems until some decisions have been made about the exterior wall. The reality is that the thermal properties of building shells determine, as much as if not more than the severity of local weather conditions, the task required of mechanical equipment. Since America was first shaken from its energy-unconscious slumber by the OPEC oil embargo of the early 1970s, building regulators have taken energy dependence to heart. To a large degree, it has been these rules that have prompted the construction of new, energy-efficient houses that feature such traditionally wasteful design elements as volume ceilings and vast expanses of glass.

So where and how do rules go about defining energy performance? Model building codes have Chapter 13, "Energy Conservation," to address the issue. As of now, these chapters are fairly brief (BOCA93's is only two pages), but they are supplemented by model mechanical codes as well as model energy codes such as CABO's *Model Energy Code.*

Such guidelines address not only the quantity of insulation expected in our walls but also such issues as the need for fresh air, with the potential problems related to overinsulation; solar gain including infrared, visible, and ultraviolet radiation; and the efficiency of equipment such as furnaces and hot water heaters.

Requirements

Factors pertaining to roofs are explored at the end of this chapter. Those pertaining to infrastructure are explored in Chapter 7 (pp. 191–192).

ASHRAE 90

The primary source for energy rules is ASHRAE 90. While it contains useful rules for energy performance, it addresses the individual pieces rather than taking a holistic approach. The energy codes written by the model code publishers, explored in Chapter 2 (p. 39) don't generally add much. Their basic function is that of adopting ASHRAE 90 by reference.

ASHRAE 90's prescriptions for performance are organized by climatic setting. They establish maximum levels of enclosure leakage based on the average number of degree days recorded in the project area, setting tougher standards in colder climates. The rules tend to be oriented toward value and try to encourage design solutions that will be cost-effective.

A core technique of ASHRAE 90 is envelope analysis. This is an analysis that focuses on a building's overall performance rather than the results achieved by its individual components. It is a bit like the formula that carmakers use to meet fuel economy targets, allowing the production of a gas guzzler so long as its consumption is offset by a manufacturer's more efficient cars. Under ASHRAE 90, large areas of single-layer glass can be acceptable if offset by compensating increases in insulative value at solid portions of the walls and roof.

Ultimately, calculations are generally required by code officials. Although simple spreadsheets do the job, each jurisdiction has its own slightly different forms.[29]

ENERGY STAR

The Environmental Protection Agency has its Energy Star Program. Since this program affects more than just enclosure, exploration of it is included in Chapter 7 (p. 192).

Optimizing Energy Performance Rules for Exterior Walls

ASHRAE's holistic approach to energy performance is complemented by the concept of whole-building energy budgeting. Energy budgeting looks at the efficiency of a project based on the consumption of comparable projects. The idea is to provide a baseline for evaluating the energy budgets of existing and proposed buildings. Energy budgeting can help justify (or refute) solutions proposed for alterations and new work.[30]

Wall Materials

The task of choosing materials for exterior walls involves some interesting distinctions. Regardless of how they look on the outside, most contemporary exterior walls are made of layered construction. Done this way, each layer has a particular role to fulfill, so materials can afford to specialize. No single material or layer is depended on for all the characteristics needed in an exterior wall.

The material property that perhaps most easily comes to mind is water resistance, seeing as how building shells are supposed to keep us dry. Despite this, water resistance is not generally critical for any particular material, since resisting water is usually the result of the combined actions of several materials. Many components of an exterior wall assembly contribute a portion of the total resistance, including facing materials (what one sees on the elevation), sheathings, flashings, sealants, and felts or wraps. Further, some of the weather resistance is provided by the way these materials are arranged, how they are positioned, and the size of the area they are expected to handle.

Resistance to Fire

Background

Although it seems awfully mundane, there is a strong argument to be made for fire resistance as the most important characteristic of any materials considered for use in an exterior wall. The primary issue to be established is whether the materials appropriate to the design intent are compatible with the chosen fire-resistant assemblies. (See p. 63 for a discussion of the fire-resistance concept.)

Requirements

When tested assemblies are published in such books as the UL *Fire Resistance Directory*, component materials are described, and names of products that have been tested in the particular assembly design are listed with names of their manufacturers. This takes all the guesswork out of determining which products have the requisite characteristics. (See p. 247 for an example of a UL design.)

Optimizing Rules Related to Fire Resistance of Exterior Materials

If the materials listed in tested assemblies are not those you had in mind, and those that are listed are unacceptable to the project team, the only choice is to pay for a new test. This is costly, not only in monetary outlay, but it may also delay your design work while you await approvals.

There is, however, another option, although it may not work in every situation. Tested assemblies are intended to fulfill functional needs, and other than economy, there is no reason to leave them exposed. Build the assembly strictly in accordance with the tested assembly design, and then cover it with exposed materials consistent with your design intent.

Resistance to Aging

Background

Durability is a key factor in achieving low life-cycle costs, and therefore value. Unfortunately, exterior wall materials must resist a tremendous amount of wear and other aging factors caused by sources such as:

- ultraviolet rays, which make soft materials such as sealants and gaskets brittle
- acid rain, which erodes stone
- gravel ballast from lower roofs picked up by the wind and blown into higher walls
- hail, which causes dents
- on lower levels, vandalism and vehicles with bad aim
- misinformed maintenance, such as sandblasting as a cleaning technique

Further, it is harder to service worn exterior walls than roofs or interiors since, except on very small buildings, scaffolding or movable platforms are needed for access.

For historic properties or properties in historic districts, maintenance of exterior walls is critical. This is especially true for rehabilitation projects where virtually everything but the original facade is "upgraded" (developer jargon for "replaced"). (See pp. 256–257.)

Requirements

There are rules bent on saving those owners with short-term focuses from making long-term mistakes. For instance, the *Secretary of the Interior's Standards* prohibit use of sandblasting as a technique for cleaning masonry. Local preservation codes often contain provisions prohibiting "destruction by neglect." Such provisions require performance of ongoing maintenance.

WINDOWS

Context

Fenestration, the pattern and placement of windows and doors on a facade, is often not only the most visible component of an elevation but also the one of which the most demands are made. Windows and doors must be capable of the most complex of juggling acts, controlling one set of conditions when open,

- passing visible light and air, but
- impeding rain, birds, and insects

and another set of conditions when closed,

- passing nothing by default, but
- impeding wind, smoke and fire, and baseballs, and some degree of sound and thermal energy, yet
- allowing a choice of passage or impedance for visible light (spandrel and reflective glass), infrared radiation (low emissivity glass), distinct outlines (pattern, or obscure, glass), and many other factors.

Ratio of Wall to Window

See also Chapter 4 (pp. 109–111).

Transparency of Exterior Wall

As already noted, the percentage of a facade devoted to openings strongly limits the distance a building must be held from existing structures, including itself in the case of an L- or U-shaped building where one facade is brought into close proximity to another. Unless a building is kept a substantial distance from any of its property lines, fenestration must comply with listed maximum percentages. When design teams feel that their design intentions are being overly constrained, they might reconsider their options for positioning.

Protecting Openings from Fire

As with transparency, this is also a positioning-related issue. Openings protectives are expensive, may require periodic maintenance, and are partially expressed on the interior. Still, when wide setbacks are undesirable or impossible, protectives may allow a greater range of options.

Hurricane Considerations

For hurricane zones, where basic wind speeds of 90 to 100 miles per hour or more are anticipated, rules may require that the locations and sizes of openings in exterior walls be carefully controlled [IBC 1609.2 *Building, Open* and *Partially Enclosed*]. The point of this is to prevent buildings from imploding, and openings (once the glazing is blown out) can reduce or increase pressures. Two options are provided:

- "Open buildings": Where every exterior wall is at least 80 percent open. Nothing more need be done, since this prevents wind pressures from building to problem levels.
- "Partially enclosed buildings": Where the total area of openings in the leeward exterior walls and roof is less than 20 percent. Make sure that a majority (at least 53 percent) of the openings in the building's enclosure (walls and roof) are concentrated in the windward wall. This makes the building act a bit like a giant pressure equalization chamber during a storm. In other words, wind can enter, but since it can't "rush out the back door," it just slows down and occupies the space, causing minimal damage.

Size of Windows

Light and Ventilation

BACKGROUND

Where natural sources are code-required, or where a project team decides to exceed code-mandated levels, the team must look to the exterior elevations. Although rooms may get natural light and ventilation from light wells and air shafts, exterior walls usually provide more opportunity. Natural light and ventilation can be built into roofs by means of skylights and scuttles, or built into exterior walls by means of windows and louvers. Since only the story immediately below a roof can benefit from skylights, walls become much more critical providers regardless of whether they are oriented to property lines or to courtyards, light wells, and air shafts.

REQUIREMENTS

For light and ventilation requirements based on the characteristics of the spaces being served, see Chapter 8 (p. 196).

OPTIMIZING VENTILATION AND LIGHT RULES

Artificial light and ventilation are allowed in place of their natural counterparts in virtually all situations. This leaves egress (see next section) as the only remaining code-based reason for locating occupiable rooms on exterior walls.

Escape through Windows (Low-Rise Situations)

Background

All building codes require that people fleeing a fire, regardless of the size of the building they are in, be given at least two completely independent choices of escape route. In smaller buildings, it doesn't always make sense, in terms of daily need, to provide two stairs. Even in larger buildings, the corridors that lead to the alternate escape routes may not be the most direct way out. Firefighters bring ladders to a fire scene and are trained for rescuing people through windows, so long as they are within reach of the ladders. Further, people are most vulnerable when asleep, since they are less likely to become aware of a fire until some of the limited time for escape has been lost. For all these reasons, building codes require that critical rooms near the ground (generally bedrooms within the first four floors) have windows large enough and accessible enough for emergency escape [IBC 1010.5].

Requirements

Requirements generally regulate three parameters: window size, window location relative to the floor, and hardware type. Sizes must be large enough to accommodate virtually anyone (5.7 SF minimum area, with a width of at least 20 in and height of at least 24 in, they must be within climbing distance of the floor (sills no higher than 44 in), and if security screens are in place, they must be operable from the interior without special equipment [IBC 1010.5.3, CABO92 R-210.2]. Of course, the idea of using windows for egress implies either that a ladder, with or without a firefighter, will be there, or that leaping to the ground unaided is better than trying to use internal means of egress.

Optimizing Escape-through-Window Rules

Once again, an exception to these requirements is made for sprinklered buildings. In addition, exceptions are made when rooms have access to corridors served by multiple exits, and even in some cases when the window of escape opens onto an interior atrium.

Optimizing Fenestration Rules

Once again, distance to adjacent structures affects amount of fenestration and fire resistance required of windows. The farther the wall is from adjacent structures, the more flexibility is allowed.

For codes that relax ventilation requirements where windows are located in more than one wall to provide cross-ventilation, design spaces so as to take advantage of the code.

Remember that the fully operable sashes found on casement, awning, and hopper windows allow not only maximum ventilation but unobstructed egress. Pivot windows allow maximum ventilation but obstructed egress. The half-operable sashes found on double-hung, single-hung, and sliding windows provide only half as much ventilating area as illuminating area, even though they do provide unobstructed egress through the ventilating area. Regardless of design, fixed windows and fixed portions of half-operable windows do not count toward egress.

Balconies and Railings

Context

Balconies and railings can do a lot for a building in terms of adding visual interest, if not even a sculptural quality. They add visual and functional depth to facades, and the patterns that they and the shadows they cast create against exterior wall planes enrich the texture of those walls.

As with other issues, taking care of aesthetics just isn't enough. Balconies affect massing, egress, and access to light and air. Guardrails work to prevent people and objects from falling over the edge.

Balcony Design

General Issues

Balconies are only one type of occupiable outdoor space. Other types include rooftop terraces, treated as rooms in terms of egress, and fire escapes, which are no longer permitted except where they already exist [BOCA 1025].

The term *balcony* is used to describe two very different types of exterior spaces: outdoor corridors and outdoor program spaces. Which one they are depends on their placement.

- Outdoor program spaces are dead-end spaces, entered from indoor rooms or hallways. Outdoor corridors are part of a building's system of stairs and hallways, and are used by people on their way to indoor rooms.
- Outdoor corridors are used by occupants escaping a fire. People using outdoor program spaces must first go back into the building on their way out. When egress isn't possible, and within the first four stories, outdoor program spaces may provide opportunities for rescue by firefighters using ladders.

- Outdoor corridors are designed to comply with requirements for exit access (see p. 224). Outdoor program spaces have no unique design requirements, although they must meet general requirements such as those for design load and accumulation of snow and ice [IBC 1004.6].

Balconies bring up interesting regulatory issues, some related to design, others to use.

- They may extend beyond property lines as noted in Chapter 4 (pp. 105–107).
- Their floor area may or may not count toward occupant load, as noted in Chapter 7 (pp. 175–176). Their floor area may or may not count toward Floor Area Ratio calculations, depending on the provisions of the relevant zoning ordinance. Most zoning ordinances count only interior balconies toward FAR.
- If allowed, the enclosure must comply with building codes [BOCA 1014.12].
- Since they are likely to pose potential fire hazards when used for cooking, state and local codes, including fire-prevention codes, may prohibit such behavior even if rental agreements don't.

Enclosure

Enclosing balconies affects the amount of natural light and air available to rooms across the exterior wall from them. Codes consider interior rooms to be "adjoining spaces" of enclosed balconies, so windows in the enclosures can count toward the light and ventilation needs of both spaces. The key issue is whether windows in enclosures are able to serve both spaces. The answer depends on the openness of any partitions separating indoor spaces from enclosed balconies. So long as they are sufficiently open (25 SF minimum) to allow a free exchange of light and air, the enclosure is allowed [BOCA96 1206.2, CABO R203.2]. In other words, enclosures must not limit adjacent rooms' access to light and air.

Railing (Guardrail) Design

Background

Just to keep matters clear, rules regarding guardrails are quite different from those for handrails. Handrails are for holding onto, not for containment, so are designed for ergonomic requirements, not structural ones. Guardrails are required where the floor on one side of the rail is significantly lower than on the other side. Before 1984, handrails without guardrails were considered sufficient to keep people from falling off centerwell egress stairs. Since then, both handrails and guardrails have been required.

But aside from these exceptions, even the basic, primary task of a guardrail isn't easily accomplished. What does it take to hold back a bear-sized baseball fan returning a little too quickly to his seat in the first mezzanine row, afraid of missing some action after a trip to the bathroom? This is why posts are rarely more than a few feet apart. The glass guardrails you've seen at shopping malls extend into steel sleeves embedded deeply into concrete floor structures and are tempered to increase their load-carrying capacity.

Requirements

WHERE?

Guardrails are generally required wherever "the first step is a big one." By that definition, thirty inches is universally considered to be "big." Regulatory bodies will not accept the argument that only adults will be using the space. Babies and the elderly are of particular concern, but fully able adults can also trip or get clumsy. Typical applications include:

- along any side of a stair or its landings that isn't against a wall
- along the edges of porches, decks, balconies, and mezzanines

HOW TALL?

Rules all seem to agree on a height of 42 in. Note that this is 8 to 10 in higher than the required handrail height (see pp. 263–264) [IBC 1003.13.1].

HOW STRONG?

Rules commonly use a two-formula requirement: one part for single-point impact, and another for continuous load [BOCA96 1606.4, ASCE7 Section 4.4]. Railings are not required to be able to resist loads calculated under both formulas simultaneously. Because of these requirements, the screens in a screened porch are not acceptable as protection against a 30 in drop; guardrails are needed as well [IBC 1003.13.3].

- The first formula takes care of situations where a person crashes into a railing. Rules commonly call for railings to resist a point load of 200 PSF applied at any point in any direction.

- The second formula handles the kinds of crowds one might find on a mezzanine at a concert or in the stands at a ball game. The load is generated by crowds concentrated at the edge of a floor being pushed by those who are farther from the rail. A common requirement calls for resisting a uniform horizontal load of 50 PSF at the required 42-in height and a simultaneous uniform vertical load of 100 PSF at the top of the railing (which usually, but not always, is at the required height).

How Solid?

The space between a guardrail's top bar and the floor must be infilled to prevent people, pets, and objects from going through. The spacing of openings in these infill areas is carefully controlled, but not their configuration. (See p. 261.)

Optimizing Guardrail Rules

Infills

Rules don't prohibit guardrails from bending while resisting their loads. This allows for twisted cable infills so long as the degree of flexing isn't sufficient to increase the spacing between adjacent cables. There are no rules regarding the orientation of infill members, so horizontal, vertical, diagonal, and curved pickets, spindles, or balusters are all allowed. Still, vertical orientations are recommended as being harder to climb [LSC94 A-5-2.2.4.5(f)]. Also, just because most contemporary guardrails are either solid or made of parallel rods, there is no prohibition against bars that change width. Remember, the issue being controlled is the size of the openings—not their placement and not the size of the solid material between the openings. A wonderful solution to this rule was demonstrated, slightly ahead of his time, by the baroque Italian architect Francesco Borromini, at the cloister of San Carlino alle Quattro Fontane in Rome.

Bypassing the Maximum Thirty-Inch Drop Rule

A few allowances are made for waiving railings. Guardrails are not required at the business edges of stages or loading docks since they would interfere with normal use. They are nonetheless required at any adjacent stairs [IBC 1003.13].

Other Requirements

Railings can be as low as 26 in when directly in front of fixed seats on the balcony of a theater or an auditorium [IBC 1011.13.1].

Ballustrade detail, San Carlino, Rome. Each baluster is the same shape, but every other one is turned upside down. This way, the spaces between balusters are of constant width.

Exterior Circulation

Context

Fire escapes may no longer adorn the facades of new buildings, but stairs and outdoor corridors continue to influence the design of our buildings' public (and not-so-public) faces. In one example familiar to many, Louis Kahn used stair towers as a major design element in his design for the Salk Institute in La Jolla, California. When stairs and corridors are brought outside primary building enclosures, they take on increased roles with regard to aesthetics and safety.

Smokeproof Enclosures

Smokeproof towers are stairs that are reached only by crossing an outdoor space. Through this technique, smoke leaves the exit access when the occupants do but escapes to the sky while occupants continue into the exit [IBC 909.20 and 1008.3.6, LSC94 5-2.3].

Smokeproof enclosures can be on an outside wall and use a small balcony as the transition from exit access to exit. Alternatively, they can be completely within the building and gain access to the outdoors via open mechanical ventilation shafts.

Outside Stairs

While smokeproof towers are enclosed, outside stairs are open to the air. They must nonetheless be designed so as not to frighten occupants who have acrophobia. Some rules call for "visual protection." The Life Safety Code requires screening at least four feet high on any stair more than three stories high [LSC94 5-2.2.5.2].

When outside stairs are used, special rules must be followed governing the design of the exterior wall adjacent to the stair, including its fire resistance and the nearness of windows to the stair [LSC94 5-2.2.6.3].

Outdoor Corridors

See Chapter 9 (p. 224).

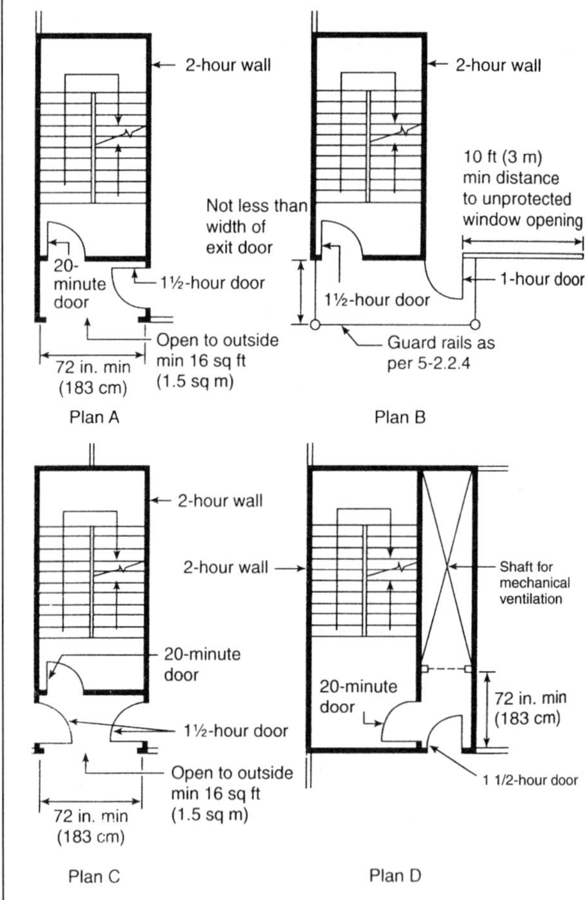

Figure 5-43. Four variations of smokeproof enclosures conforming to code criteria. Plan A utilizes an open air vestibule. Plan B shows entrance to the smokeproof enclosure by way of an outside balcony. Plan C could provide a stair enclosure entrance common to two buildings or two building areas. In Plan D, smoke and gases entering the vestibule would be exhausted by mechanical ventilation. In each case, a double entrance to the stair enclosure with at least one side open or vented is characteristic of this type of construction. Pressur-ization of the stair enclosure in the event of fire provides an attractive alternative and is a means of eliminating the entrance vestibule.

LSCH Figure 5-43: Four variations of smokeproof enclosure. ©NFPA.

WALL AND WINDOW DETAILS

Context

Why This Issue Is Important

GENERAL

A strong and appropriate vision for the basic design of a facade, while necessary, does not provide enough information to build from. For example, deciding that an office building will be clad in a green glass and stainless steel fixed and hopper curtain wall with particular overall dimensions is a good start, but must be followed by other decisions regarding mullion widths, colors and efficiency of sealants, and the shapes, types, and finishes of operating hardware on the windows.

Why Regulatory Bodies Care

To some degree, one can't know if the basic design intent for an exterior wall is feasible until the details are worked out. The proof is in the details, and small things can lead to big problems.

Additionally, details are critical to some rules, such as accessibility and historic preservation. While accessibility only applies to exterior wall design in terms of doors, historic preservation considerations affect most details, albeit only for projects of a historic nature.

Where to Find the Requirements

Make sure you've obtained copies of all relevant documents. Of course, the usual legislated requirements pertain, such as zoning and building codes. In addition, if the project is in a residential, commercial, or industrial development, get copies of any mandated development guidelines. Finally, if it's historic, look into preservation guidelines.

DESIGN

For considerations related to the aesthetic style, color, and shape of exterior walls, see Chapter 10 (pp. 265–268).

INSTALLATION

Background

Requiring that materials of a particular quality be used or that installations meet certain performance requirements may be insufficient to assure overall project quality. For this reason, rules use standards to specify minimum levels of workmanship.

Many owners try to have work rebuilt when they conclude that it is of poor quality. What they are really suggesting is that the work is below minimum acceptable industry standards of workmanship. Owners and builders can easily get into claims and counterclaims over this issue, but the only way to settle disputes is to refer to standards described or referenced in building codes.

One might be able to avoid this sort of problem from the outset by being sufficiently familiar with regulation–referenced standards to point them out to builders at job-site meetings before work is completed. Even earlier, one can incorporate references to these standards into the project specifications on which bids are based. Rules aren't the only documents allowed to incorporate standards by reference.

Requirements and Optimization

Clearly, requirements for workmanship vary tremendously from trade to trade. For exterior walls, including curtain walls and window units, standards do not directly say much about installation or workmanship. However, they do two things:

- They explain methods for testing assemblies to determine quality.
- They set categories of quality related to performance. In other words, if a particular test can yield scores ranging from 0 to 100, the standard might classify a test score of 0 to 25 as Class I, a score of 25 to 50 as Class II, and so on. These classifications can then be used by codes to regulate the types of materials or assemblies of materials that will be accepted.

In this way, standards describe industry-accepted levels of performance rather than workmanship. Performance is related to specific test methods. It is left to design and construction teams to decide how the required levels of performance are achieved. As noted for performance standards in general, this approach allows for a wide range of design solutions without requiring rule optimization.

Accessories

Building Signage

Signs, whether attached to or projecting from buildings, or on their roofs, have a mixed reputation. They are considered either eyesores, as has often been considered the case with billboards, or vibrant energizers that can bring life to public spaces, as with the animated signs at Times Square and Las Vegas or the colorful banners hanging from the streetlights on many Main Streets. Some of this is related to the size, color, luminosity, graphic content, and animation of the signage, and some is related simply to the presence or size of the signage. Signs can be designed to be discreet, and therefore more likely to be approved, or can be designed for maximum impact. Issues to think about include:

- The extent of a sign's encroachment into the public way.
- The height and length of a sign.
- A sign's structure. Codes regulate the use of guy wires or tie-back chains and the structural design of support pylons.
- The "presence" of a sign. Signs attached to buildings are frequently allowed more liberties than freestanding "pole signs." Internally illuminated signs may be subject to more control than externally illuminated signs. Animated signs using flashing neon and rotating message boards are generally subject to tighter controls than fixed-graphics signs.
- Unless pornographic, the message conveyed by signs is generally unregulated, since freedom of speech is protected by the Constitution.

While signs are regulated by most zoning ordinances, they are also addressed by model codes. There is also a separate model code, the Uniform Sign Code, published by ICBO.

Mandated Elements

Some details need to be included somewhere in or on a building just because rules require them. Investors and users don't usually think about any value provided by fire department indicators and aircraft warning lights, but since they help firefighters put out fires sooner and prevent planes from crashing into tall structures near airports, they support basic project goals.

Fire Department Equipment

Background

Equipment that is specifically required to be located on facades includes:

- Hose connections used by fire departments to link standpipe and sprinkler systems to water supplied from fire hydrants via pumper trucks.
- Key boxes, when requested by the code official, containing keys for firefighters to use to gain access to the building. These may be required anywhere access may be difficult, inside or outside a building.
- In some codes, rotating signal lights and/or bells to signal the presence of a fire and locate the entrance leading to the Fire Command Center. (See p. 187.)

These are strictly building and fire-prevention code issues.

Requirements

Hose connections must be located where the fire department wants them, between 18 and 42 in above the ground, with a sign [IBC 912].

Optimizing Fire Department Equipment Rules

Hose connections: The primary exceptions are allowed for:

- Very small systems that get their water from the same domestic water supply that brings water to the sinks and toilets. This might be the case when only an incidental use space must be sprinklered.
- Almost as small systems with less than 20 heads (generally under 3,500 SF).

And obviously, no connections are needed for buildings too minor to have standpipe or sprinkler systems. When you can't avoid them, consider locating hose connections in a recess.

Rotating signal lights and/or bells, when required: Consider mounting the bell horizontally under a soffit.

Aircraft Warning Lights

Background

The lights help airplanes identify the edges of tall buildings that are within minimal distances of established flight paths. They are not terribly intrusive and can actually be tied in with the general roofline design.

Requirements

Requirements are set by the FAA, the Federal Aviation Administration, and are strictly related to the position of the building. Requirements are found in the FAA's Advisory Circular AC 70/7460-1, "Obstruction Marking and Lighting." This document is available free from the Department of Transportation (Distribution Unit, TAD 484.3, Washington, D.C. 20590).

Window-Washing Equipment

Background

Occasionally, one hears of a project where the team forgot to provide for window-washing equipment. This can be an expensive after-construction addition. Little is required if windows can be washed from the ground or from individual harnesses. For larger buildings, where swing stages or motorized catwalks are appropriate, buildings must be detailed to provide anchorages and lifting equipment. Anchorages generally involve modifying a curtain wall to accept proprietary guide devices. Lifting equipment for facades generally takes the form of rooftop davits, while geared rails are used under large glazed ceilings.

Requirements

OSHA includes a review of alternative methods for accommodating swing stages. (See p. 48 for one of the acceptable systems.)

Roofs

Context

Roofs are as important to enclosing interior space as are walls. When pitched, they are visible to passersby, sometimes making the primary visual statement of a design. Flat and low-slope roofs can become important design elements when serving as the base of an outdoor patio or terrace.

Chapter 5 reviewed factors influencing the shape of the roof, including materials selection, historic value, structural loads, dormers and other roof structures, projections, and drainage. This chapter addresses the set of considerations that follow a determination of shape: the performance and stability of the roof as a barrier separating indoors and outdoors, and its role as topmost plenum of the building. Issues of concern for conceptual design include:

- Thermal function, with its impact on total energy consumption and consequent need for passive cooling strategies.
- Accommodation of infrastructure.
- Protection of the roof structure from fires originating inside and outside the building. This is of particular concern in fire districts.

Requirements

Thermal Performance

See the description earlier in this chapter under "Wall Assemblies."

Infrastructural Capacity

Ceilings at roofs are more commonly left exposed than at intermediate floors. Rules require that some juggling be performed where exposed ceilings with upright sprinklers are used rather than suspended ceilings with pendant sprinklers (see p. 216). Because the upright sprinklers spray their water upward, such adjacent elements as lighting fixtures and ducts are likely to interfere with the proper distribution of water. These competing infrastructural elements must be spaced at distances conducive to noninterference.

Skylights

Model codes don't give skylights a break on snow load, figuratively speaking. Resisting snow can be a problem for the type of glass roof one might find atop a large atrium. While codes give a break to greenhouses (defined by code as places where plants are grown for retail sale or research), they do so only when the space below is continuously heated and the glass has an insulative value of under R-2, since snow can be expected to melt before it accumulates. The same logic might suggest some strategies for reducing load at other types of glass roof.

Fire Protection from the Outside

The use of roofs may be limited by local ordinances for properties included in fire districts. Wood decks may be prohibited, but wood water tanks and PVC cooling tower components might be acceptable since they are normally full of water.

Fire Protection from the Inside

The amount of protection needed for structural components depends on the height of the roof structure above the floor. (See pp. 244–245.)

FOUNDATIONS

Context

A foundation is part of the enclosure system of a building. Although it is not typically visible from the exterior, it surrounds below-grade space, protecting it from the elements and from structural forces that conspire to fill the basement with dirt. So while the project team isn't generally concerned with the foundation's aesthetics, it is concerned with its function.

The footings on which the foundations rest are entirely structural in function and generally have little influence over the conceptual design of a project. They will therefore not be explored here.

Foundation considerations with an impact on conceptual design include:

- Appropriateness of foundation system for complexity of footprint.
- Implication for construction schedule and economy related to depth of excavation.
- Interaction between foundation design and excavation shoring technique.
- Available space between foundation and property line related to soil cohesion and sheeting method used. Low–cohesion soils with laid-back or benched excavations require much more room than slurry walls with tie-backs.

Requirements

Foundations must be able to resist lateral loads mandated by codes, including those imposed by soil loads and earthquakes (see p. 115). Flood-resistance requirements are addressed in IBC 1612 and Appendix 31-1:100.1, and the SBCCI Standard for Flood Plain Management. As for energy efficiency, see the description earlier in this chapter (p. 158).

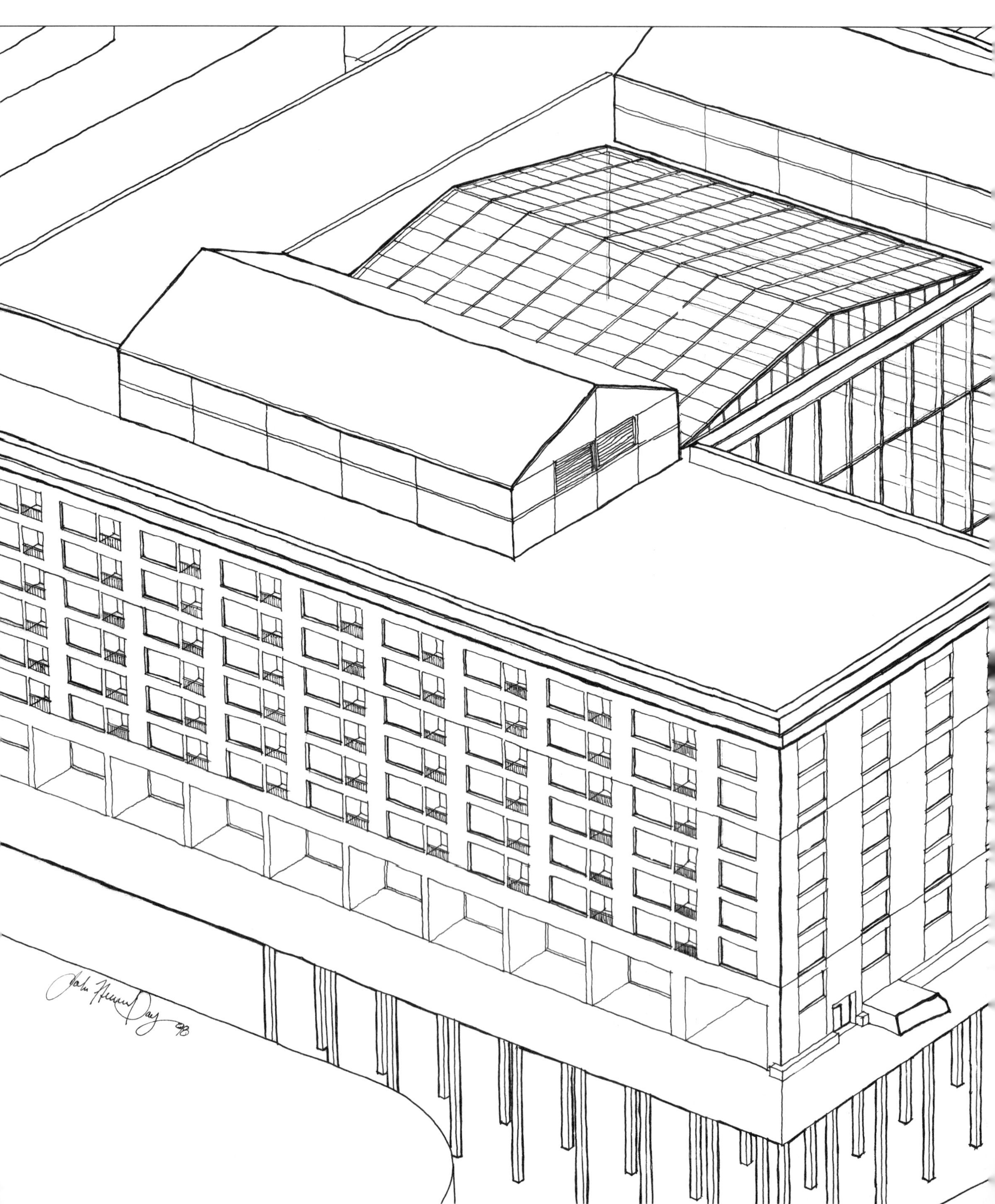

PART IV

BUILDING DESIGN

Now that we have explored the thinking behind rules in general, have an understanding of site selection and design, and of urban design issues, we can turn to design issues that primarily affect only your project team's building. Rules influence possibilities for the design of the base building and everything it contains, including rooms, corridors, bathrooms, stairs, and equipment. This section discusses the following design issues:

- **Designing overall layout**—the ways spaces are organized into a whole

  How can "flow" between spaces be achieved through the design, fulfilling program and aesthetic intent without compromising security and safety?

  How can the design best accommodate the movement of people within it, fulfilling program and aesthetic intent without losing functionality during fire-induced conditions?

- **Designing individual spaces**

  How should typical spaces be designed, such as offices, apartments, classrooms, and shops, fulfilling program and aesthetic intent while providing at least minimal physiological requirements such as light and air?

  How should atypical, or "special," spaces be designed, such as atriums, projection rooms, and courtyards, given the special, potentially hazardous, conditions they carry?

- **Designing surfaces and components**

  What shape should railings have and what texture should floors have, fulfilling functional and aesthetic requirements while being hygienic, structurally secure, and skid resistant?

  How high should countertops be? What kinds of designs are responsive to ergonomic considerations? Even with this one, we could use some guidance with respect to providing for the needs of all users, regardless of physical condition.

- **Designing assemblies**

  What combinations of materials, in what order, with what kinds of connections will fulfill the detailed requirements of building industry rules?

7

FUNDAMENTALS

Designing Overall Configuration

CONTEXT

Now, with some tentative ideas regarding permitted use, site positioning, and massing, design teams can turn their attention to fulfilling project specifics. They're ready to start developing the internal configuration of the project, in plan and section. Before dealing with spaces, the partitions between them, and the elements within them, they need to think through the overall layout.

To this end, rules can help identify answers to the following questions:

- What specific levels of performance is the design expected to achieve? What level of general design excellence is addressed by rules?
- How can "flow" between spaces be achieved through the design, fulfilling program and aesthetic intent without compromising security and safety?
- How can the design best accommodate the movement of people within it, fulfilling program and aesthetic intent without losing functionality during fire-induced conditions?
- How can architectural, structural, and infrastructural components fit harmoniously into a single building?

This section looks first at building performance as a whole. It then turns to fundamental approaches to egress (see p. 62–63 for a definition of egress). It also deals with regulatory preferences for partitioned spaces, or "fire areas," within which disasters can be more easily contained. Last, we will look briefly at those special spaces that tend to trigger additional regulatory provisions.

Nothing in this section is meant to suggest that consideration of overall configuration should or can be ignored until site design and urban design issues are decided, as all design issues are interrelated.

BUILDING PERFORMANCE

Since the profession of architecture first emerged as a distinct discipline, architects have contended that design excellence could not be quantified, that there could be no adequate way to measure such intangibles as ease of circulation and building identity. In 1995, ASTM published a series of twenty-one standards for "whole building functionality and serviceability," coordinated under ASTM documents E1334 and E1679, intended to do just that. These standards suggest that design excellence can be quantified, and potentially could radically alter the process of design delivery.

The twenty-one standards contain lists of design goal categories that apply to almost any project, with nine quality levels described for each category. In each case, Level 9 descriptions describe outstanding designs, and Level 0 descriptions identify designs that fail to consider the issue at hand in any significant way. Among the standards are:

- E1664 Layout and Building Factors
- E1667 Image to the Public and Occupants
- E1669 Location, Access, and Wayfinding
- E1692 Change and Churn by Occupants
- E1701 Manageability

Although these standards are not intended for use as regulations, they can be seen either as restrictive nuisances or as tools developed by experts based on historic precedent, and available to interested project

Occupant Requirement Scale		Facility Rating Scale
9 ☐ Operations require office spaces with high quality image, including distinctive character, excellent appearance, and a well coordinated interior. The spaces must appear generously sized throughout.		9 ☐ ○ **Appearance:** The appearance is excellent within occupied areas, with a distinctive variation in visual character, e.g. lighting levels, colour, ceiling height, shape and size of spaces, materials, views, and very distinctive differences between public and occupied zones, or, the building interior appearance is unified and coordinated, without being bland. ○ **Sense of spaciousness:** Office spaces appear generous, e.g. in large spaces with over 50 occupants, ceiling heights average 2.9 m or more.
	8 ☐	
7 ☐ Operations require office spaces with higher than normal image, including some distinctive features, good overall appearance, and consistent interior. The spaces must appear moderately sized throughout.		7 ☐ ○ **Appearance:** The appearance is good, e.g. the building presents a similar appearance throughout the office areas with some substantial differences in planned lighting levels and character between public zones, reception areas and office areas, or, the building interior is consistent and regular, without being bland. ○ **Sense of spaciousness:** Office spaces appear moderately spacious. The ceiling height is commensurate with the size of spaces, e.g. in a range of 2.7 m–2.85 m in large open office area.
	6 ☐	
5 ☐ An average appearance is appropriate. The office spaces must appear adequately sized throughout.		5 ☐ ○ **Appearance:** The appearance is average. e.g. the building presents a similar appearance throughout the office areas with some minor differences in visual character between public zones, reception areas and office areas. ○ Sense of spaciousness: Office spaces appear adequately sized. The ceiling height is adequate but not generous, considering the size of spaces, e.g. in a range of 2.6 m 2.75 m in large open office area.
	4 ☐	
3 ☐ Visual character is not a priority. Can generally tolerate below average and cramped appearance in the office spaces.		3 ☐ ○ **Appearance:** The appearance is poor, e.g. very uniform appearance throughout office areas, with minimal or no variation in planned lighting levels and character except in reception/lobby areas, or, the building interior appearance is very complex and untidy or badly done in office spaces. ○ **Sense of spaciousness:** Office spaces look cramped. The ceiling height in a large open office, e.g. with over 50 people, is relatively low, (2.6 m).
	2 ☐	
1 ☐ No requirement at this level.		1 ☐ ○ **Appearance:** The appearance is bad, e.g. excessively uniform appearance throughout occupied zones, with minimal or no variation in planned lighting levels and character throughout reception and office areas, or, the building interior appearance is extremely complex, disjointed and untidy or badly done. ○ **Sense of spaciousness:** Office spaces look very cramped. The ceding height in a large open office, e.g. with over 50 people, is relatively low, (2.6 m).

☐ Exceptionally important. ☐ Important. ☐ Minor importance.	
☐ Mandatory minimum level (threshold) =	☐ NA or NR

One of the evaluation sheets from ASTM E1667, for making decisions related to "Appearance and Spaciousness of Office Spaces." ©ATSM.

teams. As such, they fit well within the category of publications addressed by this book. Although they are relatively new documents, project teams might find them to be useful tools in deciding the intended purposes of proposed buildings, and then in fulfilling those intended purposes.

The standards are meant to be used jointly by investment teams and design teams at least twice during a project's development:

- During predesign, to decide design priorities. Here, the investment team reads descriptions pertaining to categories of concern to them and decides which levels of excellence they expect their project to attain for each category.
- Upon completion of design work, and perhaps again during postoccupancy evaluation, to determine the degree to which the project as designed or built suc-

ceeds in meeting priorities set during predesign. Here, the design team uses another set of descriptions as an evaluation checklist. If a design is intended to meet Level 6 parameters for "Image to the Public," but rates only as a Level 4, it might be worth redesigning to some degree before releasing the contract documents for bid.

As these standards find their way into use, it will be interesting to see the impact they have on design excellence and on project team relations.

EGRESS CONCEPTS

No matter what else is true of a building's design, it must facilitate the movement of exiting people during a crisis. Although this is a feature that one hopes will never be needed, all too often it is. It therefore remains a top priority "just in case." Fundamental design concepts have often been based on notions of "circulation," the way people move through spaces; egress simply adds a heightened awareness to the idea. For each project designed, an attitude about egress must be established.

CONTEXT

Why This Issue Is Important

During a crisis, smoke may reduce visibility to zero. Noise will make communication more difficult and can create confusion. On top of that, panicking people do not act rationally or intelligently. How can people be evacuated safely from a building under such conditions?

A lot depends on early detection and suppression. Sprinkler systems have proved tremendously successful in this arena. But while the fire is being beaten back, the basic design of the spaces in a building and of the partitions enclosing them must encourage safe egress.

What Kills? Smoke and Panic.

Warning and egress systems must respond to these two factors. History has demonstrated that people usually do not panic during fires so long as they can see where they are going and there is adequate signage or other information helping them to feel that the situation is in control. Fire isn't as much of an issue since it doesn't usually reach people until they're already dead. From the investment team's perspective, people are much more precious than property, since, to be frank, financial settlements related to loss of life are much more expensive than those arising from loss of property.

What Causes Property Damage? Fire.

In some cases, the water used in fire fighting also causes damage, but that will be explored later. A critical point is that property can be replaced more easily than people. Besides, its value is easier to quantify. Loss from property damage can be managed by balancing project costs against the cost of insurance. Premiums rise as the likelihood of damage increases. The likelihood of damage rises when effective design measures, such as more fire-resistant assemblies or the inclusion of sprinklering, are ignored. It is an equation, one reason for having insurance. When premiums get too high, the numbers support investment in fire-resistive construction and fire suppression.

The Underlying Science

Burning Issues

Fire is a chemical reaction whereby organic molecules get agitated, break down into smaller components, and recombine in the presence of oxygen to form ash and carbon dioxide gas. The ash is mostly carbon.

Fire is a potent force, even in its more restrained form, known as aerobic oxidation or metabolism. Aerobic oxidation is indispensable to life for its ability to release energy for those creatures that can't produce their own through photosynthesis. Ironically, it can be deadly when the release gets out of control. As we'll see in a few paragraphs, this explains a lot of its dangers.

Common fires require all three of the following ingredients:

- **Fuel,** meaning something that burns. It must be organic (have carbon as its primary ingredient, with hydrogen and oxygen not far behind).
- **Heat,** the energy needed to persuade fuels to transform from solids or liquids into gases ("volatize" them), as only gases are combustible. Even logs and gasoline won't ignite until sufficiently hot for some surface molecules to volatize.
- **Oxygen,** the troublemaker that runs off with some of the broken-down pieces as soon as the situation heats up sufficiently for them to go gaseous.

If any of these elements is absent, combustion does not occur. Once done, the recombined materials are very stable and cannot be reassembled into their original organic molecules and oxygen.

Sources and Solutions

Fuel. Fuel is sitting around everywhere, but we can try to reduce it by relying more on inorganic materials. This is what happens when mineral fiber insulation is used in place of cellulose insulation. Another approach is to add binders to hold material particles together on a molecular level, reducing their ability to volatize.

Heat. Since extreme heat is inherently dangerous to living things and essential to combustion, it is best eliminated from an environment. Using water's evaporative cooling properties is an excellent way to remove heat. This is what sprinkler systems do, and it is the reason behind gypsum board's fire-resistive properties. Gypsum contains water and releases it slowly when heated, in a process called calcination.

Oxygen. Isn't this a good thing? Yes, but during a fire, burning fuels compete with living creatures for the limited amount of oxygen available. Too often the burning fuels win. This requires some explanation.

The Oxygen Wars

One might think of smoke as preused air, air whose oxygen atoms have been chained to carbon atoms through the process of burning, also called oxidation. What remains is a mixture of particulate carbon ash and carbon dioxide gas. Smoke is dangerous because it contains insufficient free oxygen to keep the fires inside living creatures burning.

If you remember high school science, plants produce oxygen, and other living things consume it. Plants excrete it as waste since they don't need it (we politely but less accurately say that they produce it) since they get their energy directly from the sun.

Other creatures consume it. They need it because they can't get energy from the sun but must nonetheless keep their bodies' industrial processes working (arms lifting, hearts pumping). This is what goes on in the mitochondria of animal cells. Dieters really do literally "burn off calories." Oxygen allows us to keep our own furnaces stoked and burning. Without it, we can't keep our own fires lit, and we go out like a candle in a jar. Our processes cease and we die.

Conclusions

To summarize, the task of minimizing or eliminating the damage caused by fires, a task that codes attempt to address through their requirements, can be accomplished through any of the following five strategies. Combinations of several are always preferred.

- Control the fire:
  1. Reduce the quantity of available fuels, either by using fewer organic materials or by erecting barriers or arranging "fuel packages" to keep fuel sources beyond the reach of fires.
  2. Lower the heat, generally through evaporative cooling.
  3. Starve the fire of oxygen while maintaining a constant supply to people, a rather difficult task.
- Get people out of the fire's way:
  4. Put barriers between the fire and the people.
  5. Provide safe and efficient means of egress.

Each of these strategies will be examined. Chapter 9 takes a detailed look at means of egress. Chapter 10 discusses ways to reduce the quantity of available fuels through careful selection of materials. This section examines barriers, methods of lowering the heat, and ways to starve fires of oxygen.

Why Regulatory Bodies Care

Public Advocacy

They care because their constituents care. In this case, their constituents include insurance carriers as well as individual citizens. As noted previously, modern American building codes originated at the request of nineteenth-century property insurers confronting widespread fire losses. Since then most governments have seen the value of regulating life safety.

Even though our culture considers loss of life to be the ultimate disaster, to prevent it we must prevent property damage. As property burns, it tends to spread, causing more loss of life if not brought under control. Property protection can lead to reduced loss of life.

The Value of Research

People tend to behave very predictably, especially en masse. Individual behavioral differences seem to disappear not only when people panic but even when they don't, simplifying the task of design. We have only to find out what the common behaviors are. The easiest way to do this is to investigate fire scenes. By applying lessons learned in fires to new projects, design strategies can be developed to reduce loss of life. This is the basis on which the model codes organizations have developed their requirements.

The Nature of the Problem

Determining use group.

Before any discussion of egress can occur, project teams must determine which of several sets of rules apply to their projects. There are several sets of rules because different program types involve different degrees of hazard and are occupied by people with different degrees of maturity, awareness, and mobility. These differences are a reflection of the use of the project. The answers to most questions of egress are therefore rooted in the project's use group. (See p. 61 for more on determining occupancy and use group.)

The Scale of the Problem

Calculating the number of occupants.

Background

In many ways, this is where it really starts. The more people in a space, the more facilities dedicated to exiting are required. Many people have noticed signs near the doors that state the "Maximum Occupancy" of the room, or are familiar with ushers barring entry to an event because of "fire hazard." These examples illustrate the need to keep populations of spaces within the ability of available exits to bring them to safety.

As a term, *occupant load* refers to population count, not structural weight. The quantity and design of doors, stairs, hallways, and other exit amenities are based on this determination.

Requirements

Calculating Load Per Room

There are two possible ways to determine the number of people on which to base exit design. They might be thought of as a basic count and a maximum count. The basic count uses code-supplied formulas; the maximum count lets the project team go beyond the basic count so long as it designs to handle the increased risk.

Basic Count

Use occupant-per-unit ratios, listed on tables found in the codes. These are based on the number of people that could physically fit in the space given the program for which the space was designed. The model codes each use a single table to cover all possibilities [IBC 1003.3.2]. The Life Safety Code has a different table or formula for each use group, and for new projects as well as existing ones, for a total of twenty-one project-specific calculations [LSC 8-1.7 to 29-1.7]. It also contains a summary table [LSC A-5-3.1.2].

The IBC complicates matters by defining occupant load relative to "occupiable spaces." It uses a common definition of occupiable space as any room where work, play, or learning activities occur. That would appear to leave a lot of spaces out. The Life Safety Code ignores the entire issue of occupiable space by requiring simply that load be calculated using its tables. Either way, it boils down to this: If the use is listed in the table, it is counted; if not, it isn't. Since the table doesn't list bathrooms, their area need not be counted.

Why is this? Some spaces attract occupants, and others borrow them. Users may go to a library to use the reading room and stacks, but not to use the bathrooms or corridors. One can usually assume that bathroom occupants are temporarily borrowed from the primary spaces listed in the tables. Corridors are usually populated by people who are simply in transit between destination rooms. In these two examples, counting the program spaces listed in the tables automatically accounts for those who may be in bathrooms and hallways when a fire breaks out.

The unit referred to in the ratio may be amount of floor area (in offices, for example), number of chairs (in fixed-seat courtrooms), or even number of lanes (in bowling alleys). In some cases, additional parameters are also listed [BOCA93 1008.1.2]. Even where units are not listed by number of chairs, fixed seating determines occupant load [IBC 1003.3.2, BOCA93 1008.1.6].

For example, if you want to determine occupant count for a restaurant, first find restaurants in the matrix, listed as "Assembly Use with mixed tables and chairs." Next to this heading you will find the required design load of 15 SF per person, regardless of whether you ever expect to handle that number of people in this facility. Code officials are saying that, in their experience, it is possible for a restaurant to pack 100 patrons in a 1,500 SF dining room if it really squeezed the tables in.

Finally, although roofs, balconies, and patios are not enclosed, they are counted for purposes of occupant load to the extent that any anticipated uses they may have are listed in the tables [IBC 1003.3.9].

Maximum Count

Decide how many people the project team wants to be using the building at any one time, regardless of how many it could physically hold, and then design for that number.

This is something of a reversal of the basic-count option. If the project team decides how many people the managers of the building will allow in, regardless of how many the formulas predict, and this number exceeds the formula, codes require exiting to be designed for the programmed occupant load [IBC 1003.3.1, BOCA93 1008.1.1 and 1008.1.4, LSC 5-3.1.3]. The Life Safety Code is generally comfortable leaving this open-ended, limiting only assembly occupancies to 1 person per 5 SF. The model codes, however, apply this 5 SF maximum to all no matter what the program suggests [IBC 1003.3.4, BOCA93 1008.1.5]. At or above this density, the laws of physics and culture, which generally outrank statutory laws, conspire to prevent egress.

Other Sources of Occupant Load

In addition to the count in a given room, one must also provide sufficient exiting capacity to handle anyone from adjacent spaces who might need to come through the room to get to an exit [IBC 1003.3.3, BOCA 93 1008.1.3]. This includes people fleeing a mezzanine or penthouse area above the space in question [IBC 1003.3.7, LSC 5-3.1.6, BOCA93 1008.2].

People coming down stairs from upper levels need not be added to the occupant count for lower levels so long as they can be expected to stay in the stairs until they exit the building [IBC 1003.3.6, LSC 5-3.1.4].

Optimizing Occupant Count Rules

Count Units Carefully

This is more critical with area units than other types. Some formulas are listed as net area; others are gross. Where net area is the code-listed basis for the count, make sure to count that way. You can assume that nobody will be occupying a concrete column or storage closet.

If based on number of fixed seats, space left vacant to wheelchairs would count as a fixed seat.

Count Occupant Load Separately for Each Program Area

Do this even if areas are not separated by walls. If one space includes areas for seating and other areas for standing, such as restaurants with waiting areas, the waiting area may need to be counted in addition to the seating area, on the thinking that people may end up standing in the foyer when all tables are filled. As the *Life Safety Code Handbook* points out:

> *[The LSC] uses the term* use *rather than* occupancy *for an important reason, because the use of an area may be different from its occupancy classification. For example . . . a classroom in a university . . . although classified as a business occupancy, would have its occupant load based on educational (if of traditional classroom style) or assembly (if of lecture style) use.*[31]

Only Count Spaces That Are Independently Occupiable

As noted above, ancillary spaces not listed on tables can safely be ignored in counting occupant load. As a strategy, this doesn't require much planning, since it happens automatically as a result of the way the tables are written.

Once Code Requirements Are Determined

Projects need not be so tightly packed as codes allow. Project teams may choose to make the dining experience a bit more gracious and reduce the density down to 18, 21, or even 25 SF per person. Whatever is decided, planning for egress must still be based on the formulas. A day or night may come, perhaps a New Year's Eve party, when as many people as can physically fit fill the space. Tables will be pushed together, and people will forget about overloading until the alarms go off. Since even the best crowd-control procedures can fail on occasion, codes steadfastly enforce the maximum crowd numbers.

From a design perspective, these figures can also be used in reverse during initial project programming. For example, if a space is being designed to hold 100 diners, these tables can be used to calculate that the program should allocate at least 1,500 SF to this function, and in all likelihood, more. This shouldn't replace proper programming research such as visiting existing restaurants to get a feel for a density appropriate to a particular project, but it can provide a quick "ballpark" confirmation.

Conceptual Approaches to Exiting

Once you have a good idea of the number of people for whom adequate exiting must be provided, you can begin considering the design of the exits. A properly designed egress strategy allows people to get safely from program spaces to safe places.

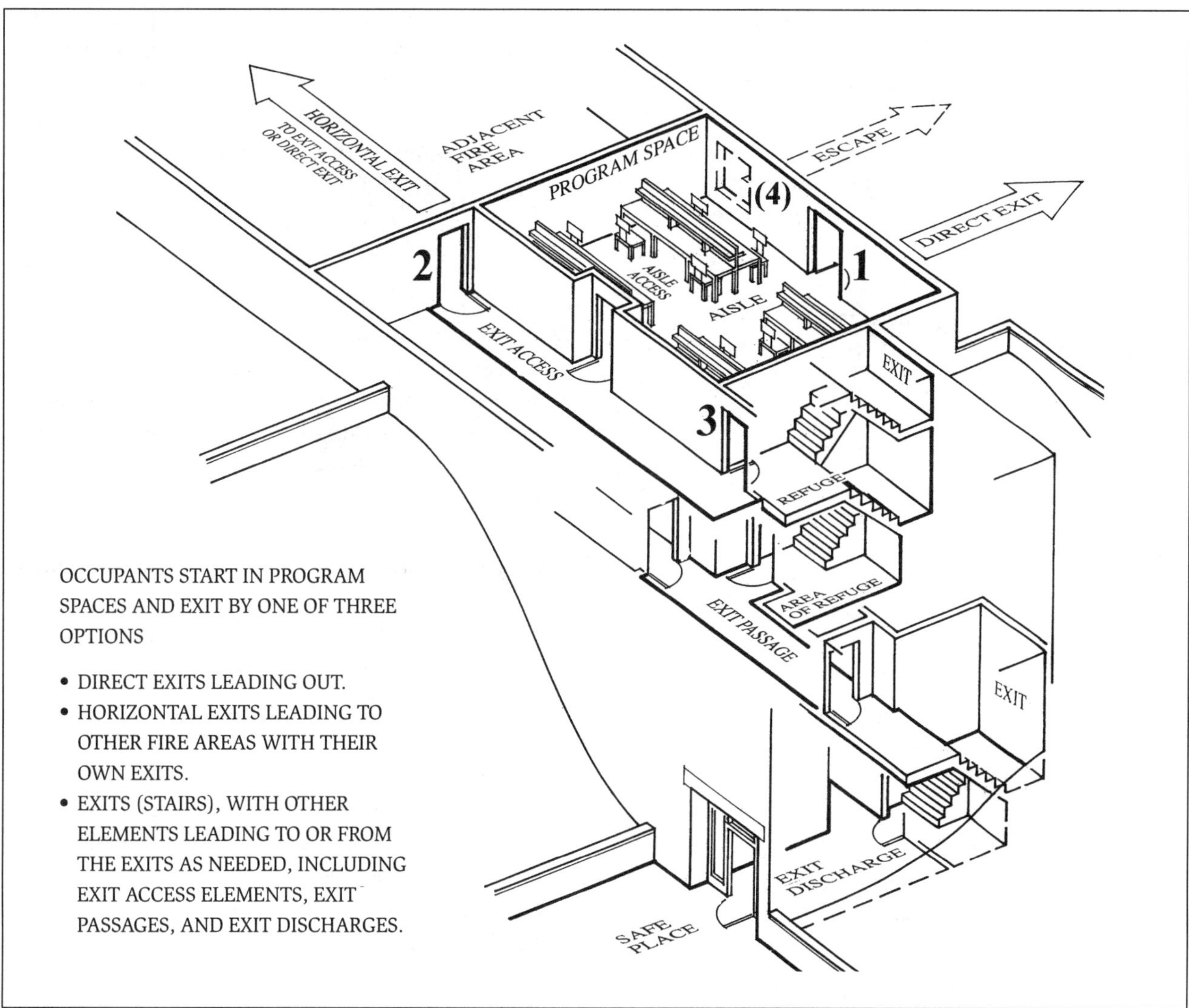

Spaces constituting the exit path.

The Ends of the Journey

Program Spaces

Although not named as such in exiting rules, program spaces are clearly inferred. These places are the reason people are using the building. Usually, program spaces are also the places where fires start. Even program spaces that are far from the origin of the fire are not places people would want to stay during a crisis, simply because they generally need to be, and are, designed for their intended program purposes rather than for protection or egress.

When program spaces are so large or complex that routes out may be blocked, aisles must be designed into the spaces. Where it might be difficult even to reach aisles, aisle accessways must be designed. Neither of these is enclosed—they are simply kept clear of obstructions. Once occupants have reached the doors leading out from program spaces, the considerations described below can take over.

Safe Places

Safe places are the end of the exiting process. They are not named specifically in egress rules simply because they are beyond them. It is important, however, to be clear as to what kinds of spaces qualify as safe places, since the egress strategy must insure that they can be reached.

The most easily defined type of safe space is the outdoors. It is safe because smoke and heat can freely dissipate, and the movement of people is relatively unrestricted, even on urban sites. Codes often require that a "public way," such as a street or alley, be reached as the most suitable termination of an egress strategy.

A less obvious type of safe space is an adjacent fire area. A fire area is also a building, and may appear to the occupants like just another part of the building that is in crisis, except that it won't be full of smoke and flames, although it may contain its share of firemen and occupants catching their breaths. It differs from the building in crisis in that it is separated by a fire wall and horizontal exits. Fire walls and horizontal exits are designed to extremely high fire-resistant construction standards. Because of this, people located in an adjacent fire area are statutorily considered to be as safe as if they were located in an unconnected building or across the street.

Basic Egress Decisions

Once the beginnings and ends of the egress strategy are defined, one can consider the means of getting from one to the other. There are three basic decisions to be made when developing a safe egress strategy.

- **The complexity of the egress strategy:** Direct exits or indirect exits (exit paths)
- **The type of exits used:** Horizontal exits, vertical exits, escapes, or none (because safe places are immediately adjacent to program spaces)
- **The degree of protection provided:** Enclosing egress paths, as required by fire-resistance tables, or eliminating the need for enclosure through the concept of timed exiting (described below)

The magnitude of the spatial differences implied by these differing approaches argues convincingly for their early consideration.

The Complexity of the Egress Strategy

Background

Direct exits, appropriate only for rooms that are adjacent to exterior walls, are possible when rooms can be exited directly to the safety of the outdoors. This strategy is particularly useful in warm climates. When these conditions are not present, indirect exiting can be provided through use of an exit path. Essentially, indirect exits are direct exits with exit access elements and exit elements inserted between program spaces and safe places. They are appropriate for exiting from any room, since they make use of a series of separate, protected spaces to guard people from smoke and fire on their way to grade-level exterior doors.

Direct Exiting

Background

In a direct exit design, corridors are not needed to escape a building in crisis. Rooms which use direct exiting are separated from the outdoors only by exterior walls containing doors. This works only for floors at ground level, but sloped sites can have several such floors. There are two reasons for designing with direct exits:

- Corridors use interior space, and therefore budget, that could otherwise be used for program areas. Of course, corridors are also useful for getting around a building when it's cold or raining outside, and they make handy locations for distributing infrastructure such as ducts and plumbing. Still, it's nice to have the choice.
- Some single-story buildings with direct exiting are exempted from area limitations. This can amount to a big advantage. This exemption is not, however, granted for buildings of the lowest construction classification.

Requirements

Some areas in buildings that qualify for unlimited floor area must use direct exiting. In particular, this includes classrooms in schools and participant sports areas in facilities containing them.

Maximum travel distance requirements tend to limit the size of "unlimited" area projects. Although rooms other than classrooms and sports areas may exit via corridors, such supplemental rooms do not usually constitute a majority of the programmed areas. Such buildings, limited as they are regarding depth from exterior wall (containing exit) to furthest interior wall, may tend toward linear rather than equidimensional plan shapes.

Optimizing Direct Exiting

Designs using direct exiting can also include internal corridors, but they would be intended for circulatory convenience rather than for egress.

Indirect Exiting (The Exit Path)

Background

The exit path is a mainstay of safe design for the overwhelming majority of buildings. It has been developed and proven through more than a century of use. The concept is simply that a sequential series of spaces can be used to safely bring occupants from program spaces to safe places. The basic spatial components of an exit path are illustrated and compared in matrix form below, and discussed in detail in Chapter 9. They can include exit access elements, exits (including areas of refuge and exit passages), and exit discharges. Exits are primary elements, with exit access elements leading to them from program spaces, and exit discharges leading from them to safe places. More complete descriptions follow.

Exit Access Elements

These exit elements lead from program spaces to exits. They most commonly take the form of corridors, and are designed to hold back fire and smoke, and communicate exit instructions to the occupants by means of signage and, in some cases, public address systems. The Life Safety Code considers everything leading up to an exit, including program spaces, to be part of the exit access.[32]

Exit Discharges

Issues unrelated to egress may suggest keeping some distance between vertical exits and exterior walls at the ground floor. Discharges link exits at grade with safe places. Foyers can work as discharges, within limits. Codes allow for a portion of the exit population (50 percent) to exit via a single protected foyer. Of course, in such situations foyers must be designed to the same protective standards as vertical exits.

The Type of Exits Used

Exits allow fleeing people some respite from danger by providing a measure of control over it.

- Horizontal exits are rated doors in compartmented designs that separate and connect fire areas in crisis from and to those that are not in crisis. Refer to the next section on subdividing space for further discussion.

To Get Out of a Room

Use	Name	Type	Required Fire Protection
When the path out of room is sometimes blocked	***Aisle***	Path to Exit	None, but always clear of obstructions
When the path to aisle is sometimes blocked	***Aisle Access***	Path to Exit	None, but readily cleared of obstructions
When a haven is immediately adjacent to the room	***Direct Exit***	Exit	Nothing to enclose
When codes require a supplemental path	***Escape***	Exit	Depends on form of escape (none if windows)

To Get the Rest of the Way Out of a Building (if needed)

Use	Name	Type	Required Fire Protection
When the exit is some distance from the room	***Exit Access***	Path to Exit	Moderate
When haven is on a different level	***Vertical Exit***	Exit	Very High
When the closest haven is an adjacent "building"	***Horizontal Exit***	Exit	Very High
When the best design requires putting vertical exits in different places on different floors	***Passage***	Path within Exit	Very High
When the exit is some distance from a haven, and the intervening space can be dedicated to exiting	***Passage***	Path from Exit	Very High
When the exit is some distance from a haven, and the intervening space can't be dedicated to exiting	***Discharge***	Path from Exit	Very High

Exit path components.

- Vertical exits eventually lead beyond all buildings, and along the way are isolated from danger by enclosing walls and controlled air quality. They may be stairs or ramps connecting upper-level exit access elements to safe places, with or without areas of refuge, passages and discharges.
- Escapes are a form of exit that supplement (but do not replace) vertical or horizontal exits in higher risk uses such as sleeping rooms and high hazard facilities. Outdoor fire escapes were once common, although they are now illegal for new work due to the tendency of their attachments to tear out from heat-induced expansion.[33]

Related elements include areas of refuge and exit passages. Areas of refuge are smoke-protected spaces connected to exits where people can await rescue or plan a course of action (see pp. 224–225). Exit passages allow the position of vertical exits to vary from floor to floor in response to the building's program. Passages must maintain the continuity and protections provided by the exits they connect.

Timed Exiting

Background

For most projects, all elements of the egress path must be enclosed with fire-rated assemblies to protect people as they exit. See Chapter 10 for an extensive exploration of enclosure assemblies. Some projects, however, can take advantage of an alternative that allows for unenclosed circulation spaces.

Timed exiting strategies provide the key to unenclosed circulation, but work only for spaces with high ceilings. When fires first break out, all smoke rises to the ceiling, leaving air nearer the floor smoke-free. Over time, smoke accumulates at the ceiling, slowly thickening, and eventually reaching occupied levels. This takes time, during which exiting people need no protection from smoke. If this time exceeds the fire ratings required of exit enclosures, exits need not be enclosed. Timed exiting calculations are particularly useful for stadiums and arenas [BOCA96 1013] and other very large civic spaces.

Requirements

When designing an exit path using the concept of timed exiting, one compares the time needed to exit against the time needed for smoke to descend to occupied areas. This is a fairly complex calculation, taking into account the following factors:

- the quantities of flammable materials available ("fuel contributed")
- the quantities of smoke generated
- the speed with which the fuels burn
- the volume of the space, and
- the available ventilation.

Not only are specific requirements missing from codes, many do not even mention the concept. An experienced life safety consultant can be of tremendous value.

Using Timed Exiting Strategies

The methods and formulas of timed exiting are described in *Fire Protection Engineering*, second edition, Philip J. DiNenno, P.E., editor, published under the authority of the Society of Fire Protection Engineers by NFPA in 1995, Section 3, Chapters 13 ("Movement of People") and 14 ("Emergency Movement").

Requirements and Implications of Egress Elements

See "Emergency: Occupants" in Chapter 9.

SUBDIVIDING SPACE

Assessing the maximum allowable against the minimum usable.

Context

Why This Issue Is Important

Architecture has always been about making space. Clearly architects don't make space in the literal sense; the space is already there. What making space really means is enclosing and surrounding space, putting walls around it to define it. In a risk-free world where natural law was suspended, walls could be built and positioned based solely on the desire to define space.

However one feels about it, design must respond to risk and natural law. Despite the desire to define space for programmatic or aesthetic goals only, there is a real need to subdivide, or compartmentalize, program areas when a program doesn't fit or when it involves several different uses.

When It Doesn't Fit

When a program can't all fit on a single floor, somebody has to make the difficult decision of how to split it. Since this issue doesn't directly affect safety, welfare, or value, rules don't have much to say, and decisions can be reached by consensus of the project team.

There are two reasons for a program's not fitting:

- The portion of the site available for construction is too small for a single-story project once setbacks, easements, areas with unsuitable soil conditions, and other restraints are taken out of consideration. The way a project team chooses to handle site limitations is not subject to regulation. For further exploration, see Chapter 4.
- The project doesn't justify the costs of building to the "ultimate" construction classification (noncombustible protected), and the more feasible lower classifications impose area limitations. Since this reason is completely linked to building code provisions, the rest of this section explores them.

Rules for allowable floor area set maximums on the amount of space a "building area" may contain on a given floor.[34] Since a "building area" is not necessarily what we generally think of as a building, it is critical to understand the code's definition of this term. This provides an excellent illustration of the cost of making assumptions.

Building codes don't always use the standard definitions one might find in a dictionary. We commonly think of a "building" as the stuff enclosed by the facades and roof. Codes define a "building" as any structure intended to be used or occupied [IBC 201, UBC91 403]. That doesn't seem significantly different from our common understanding. On the other hand, "building area" is defined as the space enclosed by exterior walls and/or "building separation walls."[35] The primary impact of this distinction is that the amount of area a project team is allowed to design is based on "building area" areas, not "building" areas [IBC 502.1, BOCA93 202.2].

In terms of optimization, this means that a building's exterior walls may contain one or more "building areas," each with its own allotment of allowable area. From this distinction arise the design possibilities unique to compartmentation.[36] Some significant design opportunities are possible when one realizes that allowable area limitations can be overcome simply by putting several "building areas" into a single exterior or skin. Without questioning, and then confirming, the critical definition of "building" or "building area," project teams might miss this opportunity.

When a project program must be subdivided to fit the allowable "building areas" within the "building," the method of subdivision may matter, particularly when several uses are involved.

When a Program Involves Several Different Uses

Some projects are clearly intended to house several uses. For example, convention hotels frequently combine sleeping quarters (residential), restaurants and conference facilities (assembly), retail shops (mercantile), service-oriented businesses such as travel agencies and banks (business), and parking (low-hazard storage). Since almost all building code rules depend on which use group applies, how does one know which rules to use? This is particularly true when the different uses vary with respect to their hazard level, degree of occupant mobility, or expected number of occupants.

Four ways of handling mixed use are available:

- Ignoring them. This approach is allowed only when one use is overwhelmingly dominant and the others are clearly incidental.
- Making all use areas comply with the rules for the most restricted use.
- Putting them in separate "building areas" as defined above.
- Putting the uses into separate "fire areas" as described below.

The first three sound if not familiar, at least understandable. The fourth option, a "fire area," is a portion of a building controlled by one tenant or use, and therefore requiring enclosure by fire-resistive floor and wall assemblies. Many projects contain only one fire area, since they contain only one tenant or use. When there is more than one, the separating walls and floors are referred to differently in different codes. (See pp. 239–244 for an extensive review of these assemblies.)

The Difference between Building Areas and Fire Areas

Fire areas must fit within building areas, but a single building area can contain more than one fire area. Although they may be arranged side by side like books on a shelf, building areas may not be stacked; they must each be enclosed by exterior walls or foundation-to-roof "building separation" walls. Still, one of the fundamental issues that a design team must resolve early in the design process involves both types of areas: the compartmentation of a design into building areas and fire areas.

Resulting Code Optimization Issues

As noted earlier (see pp. 133–137), the size of a building area is limited so as to maintain a balance among three issues:

- The number of occupants
- The potential for fire based on use group
- The fire resistance of the construction

With safer construction, building areas may be larger. With low-hazard uses, building areas may be larger. Larger areas mean that fewer partitions are required, bringing greater design freedoms. Managing building area and fire area strategies is critical to optimizing overall layout.

Why Regulatory Bodies Care

By compartmentalizing buildings and fire areas, one creates the potential for barriers to access and egress. If left unenclosed, they have the potential to become conduits for fire and smoke.

Codes are concerned with classifying and grouping the diverse parts of a project program by use group. Often, not all of them fall under one category. By dealing with this, codes can ensure, to some degree, that the needs of particular occupancies are appropriately addressed.

Within a single use group, codes require that some portions be subdivided so that only areas that are easily managed are kept continuous.

Where to Find the Requirements

FINDING OUT WHAT CODES SAY

The provisions for mixed use group occupancies are a good place to start [IBC 313]. In addition to this, project teams also need to understand strategies necessary to providing adequate separation between the groups. Make use of the tables, found in every code:

- The fire area separation matrix, which lists minimum fire-resistance ratings required to separate any of the possible use group adjacencies [IBC 313.1.2, UBC91 Table 5-B].
- The table of minimum fire-resistance ratings required of building elements [IBC 601, BOCA93 602, UBC91 Table 17-A].

Find out the maximum amount of space codes will permit within a single building area. This will depend on the construction classifications to which the project team is willing to build. The more fire resistant the classification, the larger the allowable building area.

Analyze the project program to see what works in terms of subdivision into modules. Smaller modules allow the project team greater choice among alternative construction classifications.

FINDING OUT THE DEGREE OF FIT BETWEEN PROGRAM AND ALLOWABLE SIZE

Generally, the process starts by developing a set of quantitative and qualitative descriptions of the spatial requirements for each room in a project. This is called a program.

For Individual Floor Plates

Quantitative and qualitative program needs are often converted into adjacency diagrams to help analyze the physical relationships between programmatic elements. Diagrams may be developed to explore physical adjacency, visual adjacency, acoustic adjacency, or some other relationship. These diagrams are commonly called bubble diagrams after the circles that are usually used to represent program elements. On large projects, such diagrams can be big and complex. The complexity can be handled graphically using levels of specificity, the graphic equivalent of outlining.

To Organize Multiple Floors

Project adjacency diagrams can be judiciously broken into subdiagrams or modules containing program elements that make sense as a group. If levels of specificity were used in the adjacency diagrams, this step will be partially done. Project teams may find that certain sizes reappear with some frequency among the modules. This exercise can help tremendously in deciding on minimum planning sizes.

Project teams sometimes prepare a special type of adjacency diagram, called a stacking diagram, showing the relationship of the individual modules to one another. While the stacking diagram is intended to show the most elongated response (minimum module size, maximum stacking depth), it can easily be adjusted by assembling modules into larger groups.

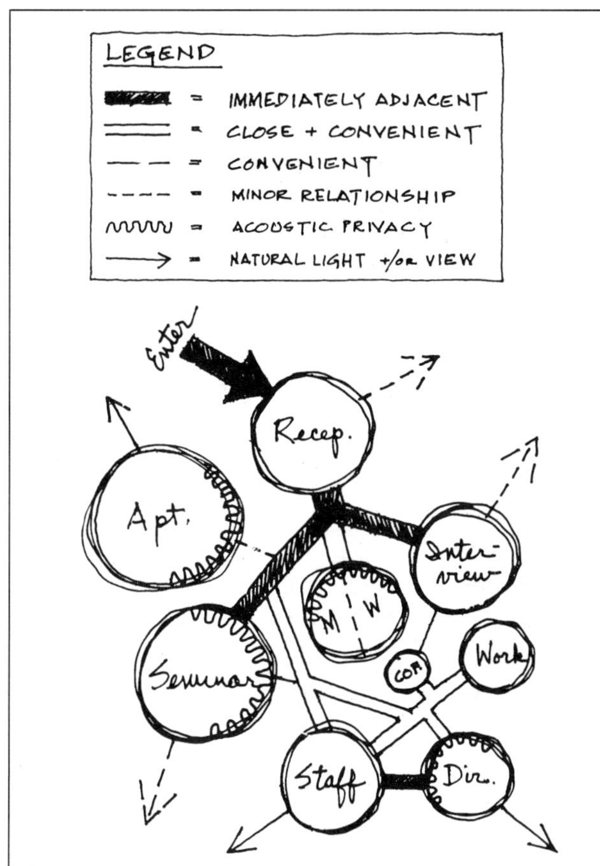

Adjacency ("bubble") diagram. ©Mark Karlen and Kate Ruggeri.[37]

The resulting diagrams might be expressed as vertical sections or exploded isometrics of the modules. Either way, it is critical that such diagrams be recognized only as planning tools and not as design responses to the program.

Single and Nonseparated Use Groups

Background

Single Use

Single use occurs when every significant space in a project is actually occupied by a single "use group." This is much more restrictive than requiring that individual spaces be suitable for a given single use.

Incidental Uses

(Which May Be Present in Single- or Multiple-Use Projects)

Most projects contain at least a few minor spaces that don't reasonably fall under any of the dominant use groups of the project. Common examples of such spaces include boiler rooms (industrial) and storage rooms (low hazard). Such spaces are called incidental uses and are not considered to be distinct use groups. Codes list specific incidental uses, in some cases indicating maximum sizes for which the designation applies [IBC 302.1.1, 313.1.1.1].

Nonseparated Uses

This situation occurs when all use groups are located within a single building area, enclosed by a single set of building separation walls [IBC 313.1.1].

Requirements

Single

When all the program areas in a given project fall under a single use group, no further restrictions apply.

Incidental

The only design features codes seek for dealing with the special needs of incidental uses are fire-rated partitions, ceiling, and floors to separate them from the rest of the project and, in some cases, sprinklering for protection [BOCA93 302.1.1].

Nonseparated

If the project team opts to keep all uses together in a single fire area, all the uses must be designed to comply with the rules for the most restrictive one. This includes that use's limit on allowable area and height, within which all the other uses must also fit. It might be wise to check a few different provisions against each use before concluding which use is most restricted, but determination of the "winner" of this dubious distinction usually isn't difficult.

Optimizing Compartmentation of Single or Nonseparated Use

To Determine Fit between Program and Allowable Size

- For individual floor plates

  Use bubble diagrams (adjacency diagrams) to help you decide how many square feet should go on a given floor.
- To organize multiple floors

  Adjacency diagrams (of whole programs) must be extrapolated into stacking diagrams (floor by floor).

Single

Needs no strategies.

Incidental

Designations are quite explicit. Considering how very small the areas in question are, and how individualized some of the situations might be, code officials are likely to work with project teams on this one. The one real difficulty is likely to arise where the incidental space requires sprinklering when there is no intention to sprinkler the building in general. Taking advantage of incidental use status is still preferable to the alternatives. Besides, for the most part, incidental areas really are dangerous. The cost of required separations and protections is money well spent.

Nonseparated

While the worst to be said of stiff fire-resistance requirements is that they add to the cost and put a damper on open, flowing spaces, much graver consequences are attached to the allowable size limit. Once the design reaches the maximum size, money and clever design won't make any difference.

But if the total program is small enough to fit within the size limits of the most restricted use, nonseparated use is clearly the optimal approach, since no use group separations will be necessary. Project teams still need some separations to fulfill other requirements.

MIXED (SEPARATED) USE GROUPS

Background

With the mixed use group approach, each use group is housed in a distinct fire area separated by fire barrier assemblies, exterior walls, and distance [BOCA93 313.1.2 and 313.1.3].

Requirements

Each fire area may be the full size allowed by Table 503. Further, building codes put no limit on the number of use groups that can be combined this way. One more thing—there's no requirement that only different uses can be separated into different fire areas or building areas.

If you're thinking about the possibility of huge, site-inappropriate buildings, remember that zoning codes will remain in place.

Optimizing Compartmentation of Mixed Use Groups

Put areas of the same use group together so as to minimize the length of required fire barrier walls.

A very viable way to get around the area and height allowances of building codes is to separate a building program, even if it contains only a single use group, into multiple building areas.

OPTIMIZING COMPARTMENTATION RULES

When It Doesn't Fit

Go to Higher Construction Classifications

As construction is designed for greater fire resistance, size is allowed to increase. Clearly, costs per square foot increase as well, but they may be justified due to the reduced need for expensive fire barrier assemblies, and the operating cost factors associated with the day-to-day inconvenience of small fire areas.

Include Sprinklers

Use an automatic fire suppression system. This brings very large bonuses, often far outweighing the costs of the system when compared with the costs of smaller building areas. BOCA93 506.3 and 504.2 allow one additional story, 20 additional feet, and a 100 percent increase in area (200 percent if fewer than three stories).

Divide Programs into Multiple Building Areas (Compartmentation)

Division into building areas is an option available to any project and allows for avoiding Table 503 limitations through use of building separation walls with horizontal exits.

Increase Street Frontage

See Chapter 5 (p. 137).

Pay Special Attention to Exceptions Allowed for Specific Types of Projects

Codes provide numerous specialized exceptions to allowable size rules based on special uses. Examples include heights allowed for auditoriums due to their special acoustic, sight-line, and stagehouse rigging needs; heights allowed for high-hazard manufacturing that needs unusually tall spaces to accommodate equipment; and areas allowed for schools with direct exits. Seek out such tailored rules.

Examine Zoning and Site Issues

If zoning and site issues are a problem, you may need to pursue a variance, rezoning, or another site. Optimized building code factors do not displace or supersede problematic zoning issues.

When It Needs to Be Separated

When you can't beat 'em, join 'em. Provide the separations required to achieve subdivision strategies, then provide openings to penetrate those separations. (See pp. 248–255.)

Related Issues

For information on fire separation within fire areas, see Chapter 9 (pp. 217–218). Even within a given use group, certain assemblies must be fire rated and will therefore separate space.

REINFORCING CULTURAL IDENTITY

Context

Cultural identity is a design issue with broad implications. The International Style, a twentieth-century architectural movement with its roots in Europe, theorized that vernacular architecture and historic references were inappropriate considerations for the design of new buildings. Its proponents felt that there were no valid precedents for the types of buildings then needed, since the conditions they faced were viewed as unprecedented.

As a result, for several decades designs were built, throughout the world, that embodied few conscious references to the historical cultures of the societies that built them. They did, however, very consciously embody their contemporary cultures.

We have therefore returned to the notion that it is appropriate, and likely inevitable, for all buildings to be reflections of their cultures. But cultures are constantly changing. This requires, if we are to retain some collective memory of our histories, that notable examples of each historical period be preserved. Identifying the most appropriate approaches for doing so is an issue of concern to project teams.

APPROACH

Background

With historic properties, there is value to be gained from maintaining the built record of a property's evolution. Like marking the height of a child on a door jamb, preserving elements of a building's growth and change helps us, as a culture, remember where we've been.

When additions or renovations are done so faithfully to existing properties that new cannot be distinguished from old, some of the lessons contained in the historic property are lost. Rehabilitation guidelines therefore require that new work be clearly distinguishable from existing, while also requiring new work to be sensitive to the design of the existing. This is a fine line to tread.

Requirements

The Secretary of the Interior's Standards for Rehabilitation (1996) directly addresses these ideas through the following mandates (numbers listed are the citation references):

- 3. Each property shall be recognized as a physical record of its time, place, and use. Changes that create a false sense of historical development, such as adding conjectural features or architectural elements from other buildings, shall not be undertaken.
- 4. Most properties change over time; those changes that have acquired historic significance in their own right shall be retained and preserved.
- 9. New additions, exterior alterations, or related new construction shall not destroy historic materials that characterize the property. The new work shall be differentiated from the old and shall be compatible with the massing, size, scale, and architectural features to protect the historic integrity of the property and its environment.

These imply that there is an intrinsic value in retaining manifestations of cultural identity that justifies the effort invested. Project teams should not be too quick to conclude that retaining our links to the past is too much trouble.[38]

Optimizing Rules Related to Reinforcing Cultural Identity

Due to the growing numbers of existing buildings, the shrinking numbers of unimproved sites, the greater budgetary efficiencies required for contemporary building ventures, and the increased focus on the environmental and economic advantages of recycling, broad reuse of the available building stock is inevitable. The recognition of cultural identity as a consideration for doing so can provide great value at a relatively low level of additional investment.

ALTERNATIVE DESIGN APPROACHES

Some types of projects are special, requiring design approaches that differ fundamentally from more common projects. Teams must be aware of rules pertaining to such approaches from the outset. While building codes refer to projects of "special use and occupancy," project teams might think of them more broadly, as alternative design approaches. Some such rules may even suggest design ideas for projects that aren't required to comply.

The next few sections will briefly examine the specialized rules for high-rise projects, enclosed malls, modular buildings, and underground projects. Atriums are also considered a special type by codes, but inasmuch as they are usually spaces within building rather than building types, they are addressed in Chapter 8 (pp. 202–204).

There are other special project types that don't impact most building projects sufficiently to address here. Project types in this category include:

- Parking garages, with their load of gasoline-tanked vehicles and noxious fumes.
- Movie projection rooms using the dangerous combination of nitrocellulose or acetate film and electric arc lamps.
- Stages and other theater spaces, with all their associated and sometimes flammable and structurally questionable temporary constructions, scenery, costuming, special effects, and electrical systems, all in close proximity to large groups of assembled people.
- Amusement parks, due to the fact that their occupants are "intentionally confounded" or distracted, making it more difficult to discern means of egress.
- Airport traffic control towers, due to the critical nature of their function.
- Buildings that handle, use, or store hazardous materials, for obvious reasons.
- Swimming and diving pools.

In addition, there are some types of projects that rules don't address at all, but that require special thinking. One example is that of convention centers and arenas. While some rules scattered throughout the codes address issues of concern to such project types, they are not gathered under a subcode as is true of the examples listed above.

HIGH-RISE DESIGN STRATEGIES

[BOCA93 403, BOCA87 Section 602, UBC91 1807]

Context

Some projects have to, other projects choose to. Some projects are required to be built according to the provisions for high-rise buildings—those that contain occupiable space higher than can be reached by the ladders of a fire truck (generally those with floors more than 75 ft above the street). For lower projects, the project teams may decide to follow high-rise design requirements for the benefits they bring, such as design flexibility. On the other hand, they can increase project costs substantially and, in some cases, reduce design flexibility. Considering the reach of these design strategies, it is a good idea to decide early whether the project will follow these rules.

According to BOCA93, all buildings designed under high-rise requirements, whether high-rises or not, must include:

- **Sprinklers** throughout, except in the case of open parking structures and electrical or telephone substations.

  An even tougher alternative is offered whereby projects may be built to lower construction classifications with lower fire-resistance ratings at vertical chases in exchange for having more manageable sprinkler systems, providing secondary sources of water, and sprinklering the vertical chases.
- **Smoke detectors** in all mechanical and electrical equipment rooms, in addition to the generally required detectors in occupied rooms.
- **Annunciator panels,** preferably graphic, to indicate locations of fire.
- **Voice signaling (alarm) systems** to alert occupants to specific problems and give them exiting instructions through distinctive audio tones and spoken instructions.
- **Telephone or intercom systems** and a **fire command station,** to allow firefighters dispersed throughout the building to communicate with their battalion

chiefs, directing operations from the command station.

- **At least one specially sized and equipped elevator** that can be commandeered by authorized users (firefighters, rescue squads, and paramedics) reaching every floor.
- **Emergency generators,** installed in fire-rated enclosures, to supply power, lighting, and exit signs.
- **Security overrides** allowing locks in exits to be disabled from the command station. Locking fire stair doors from the stair side prevents people from using stairs as a way to gain unsupervised entry to secured areas. During emergencies, however, such locking would limit the movement of firefighters through the building.

Yes, it's a long and fairly expensive list, but it can save on other costs and yield long-term benefits in the form of reduced property and liability insurance premiums for the investment team.

Some of the items on the high-rise list are needed only for high-rises, so let's review them here.

Operations

FIRE COMMAND STATIONS

Background

A fire command station is a room into which information is relayed regarding location of fire and status of various systems such as sprinklers, elevators, and air handlers; and from which fire-fighting operations can be directed and occupants can be addressed [IBC 913.2, 403.7].

Requirements

In general, fire command stations contain electronic and telecommunications equipment, located on walls or in consoles, and designed by electrical and fire protection engineers. This room must be surrounded by minimally fire-resistant walls and generally has no windows, so has little impact on a building's typical users. Frequently users aren't aware of its existence.

Although these functions do not require a large room (IBC requires 96 SF minimum), its location must be approved by the local fire department. The fire department will undoubtedly want it on the ground floor in a location that is visible and easily accessible, a combination that is undoubtedly at odds with other goals of the investment and design teams.

Fire command centers contain annunciator panels. These are boards that light up to indicate where fires have been detected, where sprinkler systems have been activated, the location and status of all elevators, and other information vital to planning the counterattack, including the location where the fire was first detected. Panels may be nongraphic, containing lists of zones accompanied by indicator lights, or graphic, representing the floor plan and/or section of the building with lights actually showing the location of the indicated area.

Local fire departments often require some prominent signage indicating the location of the room. This ranges from a small sign outside the door to large red lettering on the door facing the lobby and a rotating beacon on the exterior building facade. Fire drills give local firefighters a chance to familiarize themselves with the layout of buildings in their district, including the locations of their fire command centers.

Optimizing Fire Command Station Rules

The position of the command center and how the position is communicated (through signage and/or flashing lights) are most often issues of concern for design teams. It may be worth offering an increased frequency of fire drills in exchange for slightly reduced presence of the command station, since drills provide opportunities for local firefighters to get familiar with a building's layout.

Fire Department Communications

TELEPHONE LINKS TO COMMAND STATION

Background

Large buildings have the potential for large fires. Firefighters are often out of voice and sight range of their supervisors for extended periods of time. Further, several fire companies may be working on a fire at once, and their actions must be coordinated. Some kind of two-way communications systems is required [BOCA93 403.6].

Requirements

The key component is a dedicated system with connections in all elevators, elevator lobbies, and exit stairs, all of which terminate in the fire command station. The system might use telephones, intercoms, or two-way public address systems. Alternatively, nondedicated fire department radios may be considered acceptable.

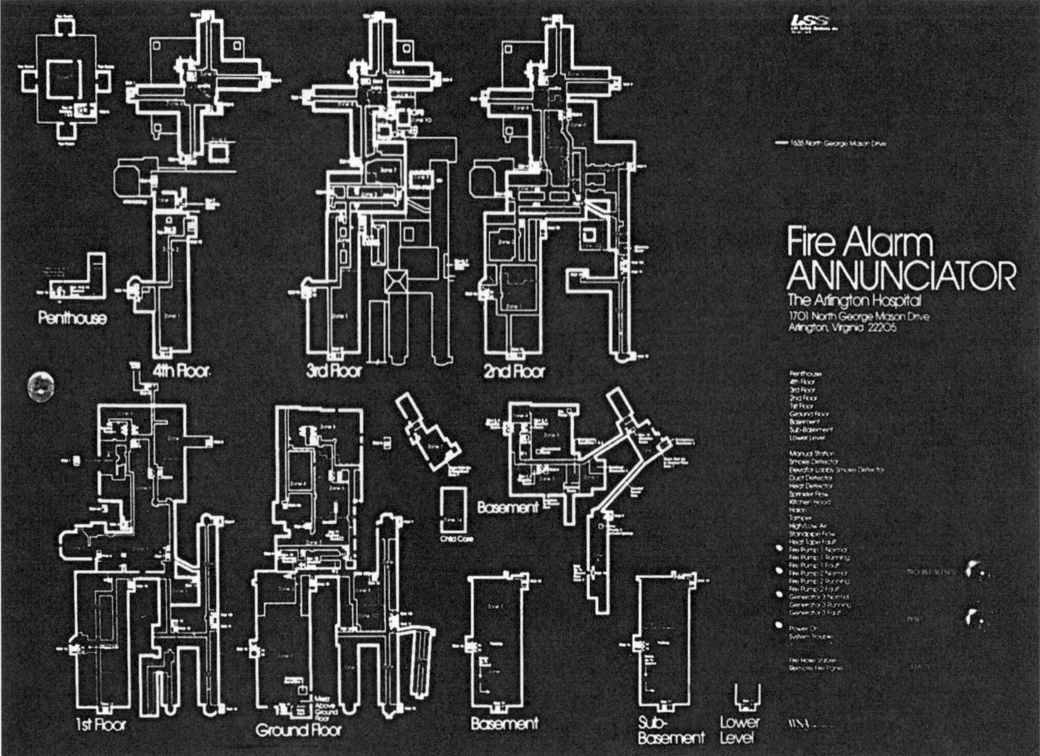

A graphic annunciator panel for a multistory hospital. Photo by the author.

Firefighters can carry telephone handsets with them and plug them into the dedicated systems whenever they need to. When not being used, the jacks can be ignored since they cannot be used for normal phone calls. In addition, they are small, so are relatively easy to integrate into an interior design.

Intercom and public address systems have the drawback and advantage of being available to unauthorized personnel, and the advantage of hands-free operation. They also involve more equipment, which adds cost and may be subject to vandalism or breakdown. The equipment is also large enough to make integration with an interior design more difficult.

Public Communication

Alarm Systems

High-rise buildings generally require more effective communication devices than other types of projects, partly because occupants can be so far from safety and partly because a fairly large proportion of occupants may simply be visiting the building at the time of the fire, and therefore not be very familiar with its layout. Exits are closer and occupant count lower in smaller buildings; with higher hazard and factory buildings, occupants are generally more familiar with the layout. (See pp. 212–215 for specific requirements.)

Elevators

There are no elevator issues unique to high-rise projects. The requirement that each floor be served by at least one elevator large enough to hold an open horizontal ambulance cot applies to any building over four stories high, not just high-rise buildings. (See pp. 209–211 for information about conveyances.)

Project teams that opt, for other reasons, to use the rules for high-rise buildings for projects under four stories high will need to meet this provision, even though other buildings under four stories high are exempt [IBC 403.8].

Fire Suppression Systems

For low-rise projects, the incorporation of sprinklering is an option that carries a more liberal set of rules. For high-rise projects, it is required [BOCA96 403.2]. (See pp. 215–217 for information smoke and fire control.)

Optimizing High-Rise Design Rules

The benefit of designing to high-rise requirements is safer design. Other projects probably don't need so much safety. Low-rise buildings are relatively safe by nature. Therefore, compliance with high-rise rules generally isn't justified for projects that aren't required to comply. The costs of all the added systems simply do not return much value for the investment unless the project is big. Low-rise projects can achieve some of the key benefits of high-rise design simply by sprinklering.

Enclosed Mall Design Strategies

Context

History

Strip shopping centers have been around for a long time. It's just that until they were recessed behind parking lots after World War II, they consisted of commercial buildings, side by side, occupied by retail establishments and fronting on urban streets. Perhaps it is unknown whether the evolution into the mall form was one of building two strips face to face on a large tract to form an internal "street" or of demolishing an existing vehicular street running between existing storefronts to form the first "pedestrian street." Either way, outdoor "malls" flourished until the first one was enclosed, creating a climate-controlled shopping environment incorporating multiple businesses. In each step of this evolution, safety became more of an issue.

Commercial streets were relatively safe. They were easily accessible to fire trucks, front and rear. Exit distances to public ways were never further away than the runs to the front or back doors of the establishments. The freestanding nature of each building slowed the spread of fire from building to building.

Since outdoor malls were designed for pedestrians rather than vehicles, their designers had to deliberately plan for fire department access to the "pedestrian streets." In addition, since multiple stores shared single structures, they were separated only by tenant "demising" partitions. A further difficulty was posed by the fluid nature of store boundaries—as stores expanded or different stores moved in, the demising partitions needed to be adjusted, potentially compromising fire resistance and structural independence.

With the enclosure of the malls, the "pedestrian streets" had the potential to become fire traps, spreading smoke and flame from store to store, limiting access to natural light and air, and dramatically increasing exit distances.

Background

The IBC defines "covered mall building" as a building containing a "mall," and several tenants and occupancies. It defines a mall as "a roofed or covered common pedestrian area" that contains the main entrances to more than one tenant and has three or fewer levels open to one another. "Anchors" are defined as buildings that meet exiting requirements "independent of the mall," so are not part of the definition of a covered mall building.

Requirements

The building codes [IBC 402] require enclosed mall buildings to be designed with:

- Ceiling-height fire partitions between tenant spaces (even when ceilings aren't rated) but none between tenant spaces and malls [IBC402.4]. This particular requirement is rapidly evolving and will most likely be revised before final publication of the IBC.
- Distances to egress measured to mall exits, not tenant entrances [IBC 402.5.4].
- Special formulas for calculating occupant load, based on a multiplier of leasable area [IBC 402.5.1].
- Minimum mall widths (20 ft) intended to handle egress needs [IBC 402.6].
- *No limits on area,* so long as building is not of combustible construction and is at least 60 ft from the closest building (once past the anchor buildings and any parking structures) [IBC 402.7].
- At least half of the required exiting from theaters and nightclubs, but not restaurants, leading directly outside rather than into the mall [IBC 402.5.3].
- Limited amounts of plastic used on the mall facades of tenant stores, when used either as signage or as decorative panels. Such limits are stated in terms of percentage of area (20 percent), size (36 in high if horizontally oriented, 96 in x 36 in if vertically), distance from adjacent tenants (18 in), and design (with edges and backs wrapped in metal) [IBC 402.15].
- Kiosks, when used, of a maximum size (300 SF), a minimum distance (20 ft) from other structures, built without untreated wood products, and protected by sprinklers [IBC 402.11].

Optimizing Enclosed Mall Rules

Some general code requirements work favorably for malls. For instance, the general exemption of fire-rating requirements for roof construction, including beams and trusses, more than twenty feet above a floor, means that most mall roofs can be left exposed, without sprayed fireproofing or gypsum board encasement [IBC Table 601, Note 3A]. Although this provision is available for any project, the rules for malls go one step further. For roofs less than twenty feet above the floor, malls may use Type IIB or IIIB construction, for which roofs and columns require no rating, with unlimited areas.

Malls may benefit from project teams looking at other requirements and strategies as well. It may be appropriate to design egress based on timed exiting, and malls with more than three open stories must also fulfill the requirements for atriums.

Mobile and Modular Unit Strategies

Regardless of nomenclature, the type of construction known alternatively as mobile, modular, or manufactured units includes two types of building:

- Units that may be relocated several times during their usable lives, built with wheels and with enclosing sides all around.
- Units intended to use the highway system only once, to move from the factory where they are built to their final sites. They may be missing walls so that several units can be bolted together to form a larger space.

In both situations, common issues include the magnitude of structural loads incurred while in transport and the need for anchorage to resist uplift.

Due to the fixed size limitations, codes allow the use of smaller components than for similarly sized site-manufactured construction. For example, a special 3/8 in thickness of gypsum board is allowed, a size that is prohibited for site-manufactured units.

Specific requirements are found in CABO Appendix C and BOCA96 420.0. As of the working draft, the issue had been left out of the IBC.

Underground and Windowless Spaces

Context

This category includes portions of above-grade buildings that are below grade, earth-bermed and roofed above-grade buildings, below-grade buildings, and any spaces with no windows. Since the 1970s, these rules have also applied to some energy-efficient projects located primarily below ground.

Roughly defined, underground building provisions apply to parts of buildings whose occupants lack ready access to grade except by ascending stairs or ramps. As noted in Chapter 5 (p. 141), codes contain different provisions for several types of basement. Chapter 5 described above-grade basements as those located under first floors more than 6 ft above grade, and below-grade basements as those under first floors less than 6 ft above grade. Codes further define deep basements as a special type of underground building where occupied spaces are far below grade (more than 30 ft below exit level). For over a century, designers have included deep basements under high-rise buildings.

Below-grade and deep basements have the advantage of not counting as stories in building code allowable height calculations. This will not matter to many high-rise projects, since allowable height has no limits for protected noncombustible construction classifications.

Nonetheless, deep basements are common with high-rise construction. Codes become concerned when projects have two or more occupied basement levels at least thirty feet below exit level, and apply even stricter rules beyond 60 ft.

The key issues are egress, followed by structural load and access to natural light and ventilation. At these depths, egress may have to overcome inadequate ventilation, disorientation, and fire department access problems. During a fire, exiting people sometimes forget that direction to safety is up—it's counterintuitive.

Requirements

Rules may mandate certain construction classifications, compartmentation strategies, and sprinklering [BOCA93 405].

For windowless spaces, codes contain a limited number of references, requiring some kind of smoke ventilation for areas that might trap smoke, and sprinklering if other requirements haven't already mandated it [IBC 408.8 and 903.2.12.1]. SBC has a rule mandating fire department access panels on perimeter walls where there are no windows for access.

Since the Life Safety Code is arranged by use group, it contains an entire section dedicated to the needs of underground and windowless buildings [LSC 30-7]. Those provisions concern only the need for sprinklering, separation of exits at the level of discharge, emergency lighting, smoke evacuation of exits or program spaces, and signage in exits.

PROVIDING FOR SYSTEMS

STRUCTURE

The Impact of Structural Design on Architectural Design

The impact of structural design on building shape was discussed in Chapter 5 (pp. 150–153). Here we will consider its impact on building layout.

There are structural design questions that should be addressed during the conceptual development of a project. These questions generally require close collaboration between architect and structural engineer, with input from the owner if structural needs impose on occupied spaces. The issue goes beyond the size of structural elements to address:

- Their locations relative to architectural elements.
- Their general organization, e.g., in an orthogonal or radial grid or irregularly distributed. Few decisions are as basic as this, or as deserving of early design commitment. Some structural systems are conducive only to orthogonal use; others are much more open-ended.
- The distances between them. Long spans require deeper structural members or one of the more sophisticated systems, such as post-tensioned concrete or cable structures.
- The physical stiffness or flexibility inherent in the system. Bouncy structures crack relatively brittle finishes such as plaster, tile, and brick. Severely bouncy structures lead to perceptual discomfort (fear, or at least, concern) among occupants.
- The adaptability of the structural system to future modification. Deciding a few years after completion to cut a new stair between two floors of a post-tensioned concrete office building can cause project teams a few sleepless nights and wishes that the floors had been built with concrete fill on steel decks.

Selection of Structural System

The decision that will most influence the issues listed above is the selection of the structural system. This decision is influenced, in turn, by:

- Seismic concerns: Braced steel frames and masonry shear walls respond quite differently to seismic loads. Designs using them are therefore granted different allowances [IBC Table 1613.4.1, BOCA96 Table 1610.3.3].
- Fire resistance: Certain structural materials are appropriate only for certain construction classifications. For example, heavy timber can be used only for a heavy timber construction classification, wood stud framing only for combustible classifications.
- Stiffness and deflection. (See pp. 270–271.)

  Quality of materials, including strength: Each of the model building codes and many of the standards set minimum acceptable specifications for compressive and tensile strength, shear strength, strength of connectors, etc. These become the bases for structural design. So long as a design team knows how strong a given material will be, it can specify the amount of material needed to resist the expected loads.

Integrating Structural and Architectural Design

Structure can be integrated into demising elements between spaces, or can be exposed within a space. The former takes more planning; the latter carries opportunities for architectural enrichment. These issues beg for collaborative exploration.

INFRASTRUCTURE

Infrastructure is a term used to refer to all the ducts, pipes, and wires in a building used to make the environment comfortable, handle our bodily functions, power our appliances, and meet our communications needs. Since these issues strongly affect our health, safety, and welfare, minimum standards are set by rules.

Occupants don't usually pay attention to these mechanical, plumbing, and electrical systems (until they're uncomfortable), and the significance of these elements for project feasibility or conceptual design could easily be underestimated. While wires are small and easily incorporated into many designs, ducts are much larger, and the design implications of infrastructural decisions is enormous. Lighting depends completely on the amount of sunlight the design allows in. The design of heating and cooling systems depends completely on how well insulated the building is, how efficiently it uses sunlight, and where mechanical plants are located within the building.

These factors have always been important factors in schematic design, but now they are being joined by the drive toward sustainability for its effect on the financial bottom line. There is a generally increasing industry-wide awareness of the short- and long-term costs of inefficiency and "throwaway" consumption.

While it is beyond the scope of this book to review specific optimizations of infrastructural systems, it is important to take a cursory look at the growing trend toward sustainability.

Energy Efficiency

Context

This section deals with energy efficiency in building systems. (For energy efficiency in enclosure design, see p. 158.)

Energy efficiency is no longer an option. Many codes are incorporating minimum efficiency standards for thermal performance. The IBC allows exceptions only for portions of buildings that "are neither heated nor cooled" or whose maximum energy usage, as designed, is under 3.4 Btuh/SF, which translates to about 7,000 Btu/SF/year. In effect, this forces all buildings to comply, since usages of 50,000 to 80,000 Btu/SF/year are considered reasonably efficient for office buildings designed using advanced energy techniques.

Of critical importance is recognition of operations—the way the building is used—as often the most significant factor in efficiency. The most efficiently designed building won't achieve its aspirations if lights, coffee makers, and copiers are left on at night, faucet washers are allowed to drip, and toilets are flushed twice by each user.

Gray Water Plumbing Systems

Plumbing is starting to be associated with energy efficiency, and project teams are starting to look at "gray water" recycling for meeting needs that don't require fresh water. With gray water systems, the waste from washing fixtures (sinks, tubs, and showers) is filtered, disinfected, dyed, and then reused for flushing toilets and urinals. The Audubon Society headquarters by the Croxton Collaborative, architects, in New York City was an early commercial user of this technology. The 1995 IBC includes rules for optional gray water systems in its Appendix C. By putting them in an appendix, local governments are able to decide whether they want to use the technology or not, since they can adopt or reject the appendix separately from the rest of the code. Cats and dogs may want to start drinking out of water bowls.

Each of the model codes has provisions for energy management, and some of the code organizations have published stand-alone energy codes. The IBC, in Chapter 13, simply adopts one of them by reference, the ICC Model Energy Code.

Federal Rules

The EPA has initiated several programs intended to increase the energy efficiency of buildings. The Energy Star Buildings Program, although voluntary, has enrolled 280 million SF of space as of July 1997. Additional information is available through the Internet (www.epa.gov/appdstar/buildings/). The program includes five stages:

- Green Lights Upgrades, with its own Web site at www.epa.gov/appdstar/green/glb-home.html
- Building Tune-up
- Load Reductions
- Fan System Upgrades
- Heating Plant and Cooling System Upgrades

Applicability to Existing Buildings

None of the energy codes or energy chapters in model codes require energy efficiency of existing buildings that aren't planning to renovate, except in the case of federal buildings. There, reductions in energy usage were mandated by executive order twice—once during the energy-conscious Carter administration, and once in the early 1990s, when the goal of a 20 percent reduction over a ten-year period was set.

Optimizing Energy Efficiency Rules

Efficiency can sometimes balance to the bottom line when upgrades are offset by utility company rebates. For several years now, energy utilities have offered to pay for part of the cost of exchanging older, inefficient mechanical units for new, high-efficiency units in both residential and commercial projects. The advantage to them is that such efficiencies help them continue to meet soaring demand and meet increasingly tough pollution and energy usage standards while they struggle to build additional capacity.

8

ROOMS

Designing Spaces

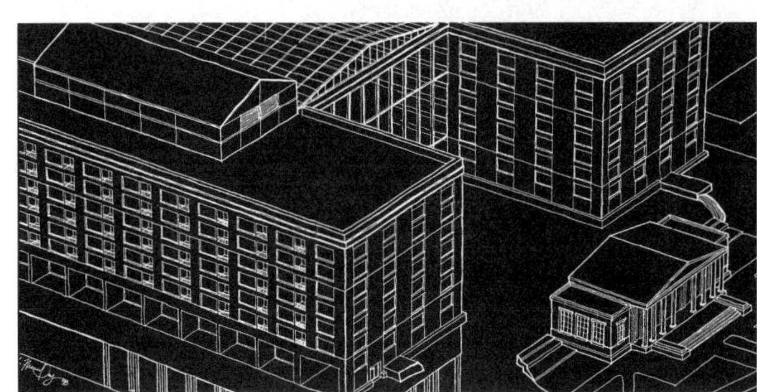

CONTEXT

Why This Issue Is Important

People can't live or work just anywhere, despite evidence to the contrary. Even those who don't care about their surroundings are or would be bothered by truly substandard environments. (Kids—don't try this at home!)

Rooms, or program spaces, are the settings within which a project's action takes place. Whether common or unusual, they must fulfill certain design parameters to provide for human activity. The only regulatory constraints on human activity are issues of permitted use (as discussed at pp. 92–98).

Considerations extend beyond use, however, to proportion, environment, and hygiene. Room proportions must be sufficient for their purposes. Environments must supply breathable air and adequate light and be sufficiently quiet. Concerns for hygiene require that waste be managed and that sources of infection and disease be minimized.

Why Regulatory Bodies Care

The basic unit of built environment is the room. Concerns for health, safety, and welfare must originate there. Some minimum standards can be and are set, despite the importance of allowing for a diversity of individual lifestyle decisions (where they do no harm).

MEETING BASIC NEEDS

What are the minimal spatial requirements for human happiness? The answer is clearly tied to culture, which changes with place, time, and wealth. Building standards that would have been considered beyond luxurious two generations ago are now considered minimal. Building codes written for the middle class share some responsibility for raising the standard of housing beyond the reach of those who find themselves homeless. In Japan, some airport hotels are being built with cubicles barely larger than individual refrigerators on their sides, where weary travelers can sleep, watch TV, or catch up on their letter writing.

Still, conventional wisdom suggests that if no standards are set, the health, safety, and welfare of the poorer segments of society may be put at risk by investment teams who work this market segment. History has shown the deleterious effects of living and working conditions that are cramped, airless, and dark. Standards of living can be a difficult subject involving difficult choices.

In setting standards, codes do not yet have definitive classifications of space. They hint at several dichotomies but break their classifications at the first opportunity. The draft IBC has taken nomenclatures from each of the model codes, so temporarily embodies this chaos. Perhaps when published, the IBC will achieve consistency. Among the listed classifications are:

- Habitable Spaces v. Occupiable Spaces, as in IBC 200: *Definitions* and 1207. Then there's also Intended for Human Occupancy, as in IBC 1202.3 and 1204.1, and Occupied Space, as in IMC 402 (the International Mechanical Code). This last term appears to refer to both Habitable and Occupiable Spaces.
- Interior Spaces v. Uninhabited Spaces, as in IBC 1202.1 and 1202.2, where the former is undefined and the latter refers to attics and under-floor areas. Then there's Unoccupied Spaces, as in IBC 1208, referring again to attic spaces, but also to crawl

Interior — Intended for Human Occupancy

Class	Subclass	Function	Height	Ventilation	Light: Natural	Light: Artificial	Dimensions	Area	Notes
Habitable	Primary	Live, sleep, eat	7'-6"	4% of floor area (nat) or mech per Table 403.3 in the IMC	Net glazed area = 8% of floor area	Averaging 10 fc @ 30" A.F.F.	7'-0" min. in any direction	1 @ 150 SF, others 70 SF	Height reduction allowed for beams, furred ceilings, etc.
	Secondary	Cook	7'-0"				3' clear btwn counters	N/R	
	Supplemental	Live, sleep, eat	Sloped 5-0" to 7'-6"	↓	↓	↓	N/R	N/R	35 SF max. counts toward req'd area
	Useable	Live, sleep, eat	Under 5'-0"				N/R	N/R	Not counted toward required area
Occupiable		Work, play, learn	7'-6"	↓	↓	↓	N/R	N/R	
Subsidiary	Primary	Circulation	7'-6"				N/R unless egress path	N/R	
	Secondary	Bath, closet, laundry	7'-0"	↓	↓	↓	N/R	N/R	
	Leftover	Low closet	Under 7'-6"				N/R	N/R	
	Basement	Any use	6'-8"				N/R	N/R	
	Unfinished	Enclosed parking	7'-0"	Per IBC Section 406			N/R	N/R	
		Mechanical room	N/R	Per Mechanical Code & NEC			N/R	N/R	
Uninhabited		Attic spaces	N/R	1/150 of area	N/R	N/R	N/R	N/R	
		Crawl spaces	N/R	1/150 of area	N/R	N/R	N/R	N/R	

An attempt to clarify intended characteristics of building spaces.

spaces, which may or may not be different than under-floor areas.

In light of all this confusion, the matrix above may shed some light. On the other hand, the rules may not be as classifiable as this chart suggests.

A brief description of some of these spaces may help clarify matters.

1. **Habitable spaces** generally have the most stringent design requirements, since people in them may be asleep (semi-incapacitated) or handling food (where cleanliness is a significant issue). Habitable spaces include spaces where people live, sleep, cook, and eat. Of these, kitchens are singled out for special rules. Subsidiary spaces such as bathrooms, corridors, closets, and mechanical equipment rooms are not considered to be habitable spaces.
2. **Occupiable spaces** are where people work, play, or learn. One might think of them as where people do everything but reside.
3. **Service spaces** are suboccupiable spaces that are nonetheless large enough to move around in. They include
   - primary service spaces such as bathrooms, corridors, and closets;
   - secondary service spaces such as basements, parking structures, and mechanical equipment rooms.
4. **Uninhabited spaces** are spaces that are accessible but not normally used, and generally not climate controlled.

As the above matrix suggests, separate sets of rules are enforced for each of these types of spaces. While there is overlap in the rules, the spaces can't quite be viewed as subsets of each other, since each is bound by at least a few unique requirements.

Proportion

Area, Length, and Width

Requirements

Requirements are noted in the matrix above. Length and width are handled with a single requirement that must be met throughout a space, eliminating the difficulties involved in applying length or width requirements to round or angular rooms.

Implications

The UBC and IBC appear to include an unstated assumption. At the heart of the question is whether the minimum size for a dwelling unit is one occupiable room or two occupiable rooms, of which one is habitable and the other is the bathroom.

- First, they require that at least one room in every dwelling unit be 150 SF minimum [IBC 1207.3]. Since there is no accompanying minimum required number of rooms, one seems justified in concluding that a developer could get approval for a 1,500 SF project with ten dwelling units of 150 SF each.
- Second, they have separate, tougher requirements for efficiency units. These requirements mandate living rooms of 220 SF minimum when used by one or two people, plus 100 SF for each person beyond two [IBC 1207.4, UBC94 310.7]. Using these rules, the smallest ten-unit project would be 2,200 SF.

This seems to suggest that there is an acceptable form of dwelling unit more modest than an efficiency. While the codes appear to assume that the two-room efficiency dwelling unit is the smallest acceptable, they do not explicitly say so.

As a side note, habitable rooms beyond the 150- or 220-SF main rooms, if provided, are required to meet minimal area (70 SF) and dimensional (7'-0") requirements [IBC 1207.1, 1207.3].

Height

Background

Minimum height requirements acknowledge:

- the taller users of our buildings, who otherwise must adopt poor posture to avoid head injury.
- that smoke will rise, allowing sufficient space below it to crawl to safety so long as the space is sufficiently tall.
- the role of height in allowing convective air movement, which helps keep spaces heated or cooled evenly.

Height is measured to the highest ceiling surface from the floor level immediately below it, with some allowance (4" to 6") made for beams, coffering, or "other obstructions" such as light fixtures.

Requirements

Habitable and occupiable spaces share a single requirement (7'-6"), with a lower requirement (7'-0") for service spaces, such as kitchens, bathrooms, laundry rooms, and spaces that are used only in passing, such as corridors. A third requirement (6'-8") is allowed for basements without habitable rooms [IBC 1207.2, CABO92 R-205].

Optimizing Height Rules

For rooms with sloped ceilings, half of the room's minimum required floor area must meet the minimum height requirement. The other half must be at least five feet high. Any floor area provided beyond the minimum requirement may be of any height.

For example, the codes require a minimum room area of 70 SF. A 120 SF attic room could be designed with 35 SF of floor area located beneath a portion of the ceiling that slopes from 10'-0" to 7'-6", another 35 SF sloping from 5'-0" to 7'-6", and a third area of 50 SF under a ceiling portion that slopes from 5'-0" to 1'-0".

Number of Rooms

Background

Chapter 5 addresses the relationship between number of units and size of lot (p. 133). While number of rooms is not the same as number of units, they are related. Design teams must be sure that the number of rooms in a project is compatible with the number of units determined through planning analysis.

But what is considered to be a room? If some portion of a room is surrounded by partitions, is it a separate room? Is the height or opacity of the partitions a factor? These questions are somewhat clarified in the definition of "adjoining spaces" [IBC 1204.2.1], or what CABO refers to as "Alcove Rooms" [CABO92 R203.2]. This definition considers several rooms as one for purposes of lighting (and perhaps ventilation) when the intervening partitions occupy less than half of the intervening plane, and the open portion is at least 25 SF and 10 percent of the larger room's floor area.

Requirements

Interestingly, there does not appear to be any general requirement for minimum number of rooms. In the specific case of efficiency dwelling units, the IBC (but not CABO) requires at least two rooms: a living room and a bathroom. While not requiring a kitchen, they do require a sink, a stove, a refrigerator, and a closet somewhere in the unit. For other types of units, however, codes are mute.

ENVIRONMENT

Light and Ventilation

Background

Imhotep, one of the first architects of Western civilization, responsible for the stepped pyramid at Xozer in Mesopotamia, was deified many centuries later by the Romans as their god of medicine. The Romans must have seen a clear relationship between the design of the built environment and the health of the occupants. (Imhotep must have done a few housing projects as well as that pyramid.)

Modern rules have also recognized the connection between built projects and the health of their occupants. In fact, one of the early forces behind the first modern codes was the desire to do something about squalid living conditions in many big cities. As a keystone in the effort to provide healthy environments, codes require adequate light and ventilation, although they are quick to accept some or all of each requirement from artificial sources.

Although rooms may get natural light and ventilation from light wells and air shafts, exterior walls usually provide most of it. (See the discussion at p. 160.)

Requirements

Codes have minimum requirements for light, air supply, and air exhaust. Under certain conditions, however, they allow the requirement to be met by lightbulbs and fans rather than the sun and a breeze. Those conditions are met when a window isn't needed for emergency escape (required only for the first four floors above grade), and when adequate artificial means are provided [IBC 1204.1, CABO92 R-203].

LIGHTING

Building codes express requirements for light as minimum levels of illumination on walls and work surfaces. While they require that all occupiable spaces have openings in the exterior wall proportionate to their floor area (8 pecent minimum), they permit all of it to be provided artificially by requiring "natural or artificial" sources [BOCA93 1204.1]. Additionally, they require that artificial lighting be capable of providing the full requirement [BOCA93 1207.1].

There are, as yet, no legal mandates for environmentally friendly and cost-free daylighting. Project teams have the option of using natural light or not, so long as the minimum levels of lighting are achieved.

VENTILATION

As with lighting, rules leave the choice of natural or artificial ventilation to the project team. While building codes require that all occupiable spaces have openings in the exterior wall proportionate to their floor area (4 percent minimum), they permit all of it to be provided artificially when they require "natural or artificial" sources [IMC, 401.2, BOCA93 1205.2].

Note that the exterior wall opening required for ventilation is half of that required for lighting. This allows for the use of double-hung and sliding windows, since their two-panel designs eliminate the possibility of being open more than 50 percent of the way.

The old City of Chicago Building Code (Chicago has since adopted BOCA) had a very sophisticated approach. It allowed for varied ventilation requirements depending on the extent to which designs provided for airflow. It defined typical rooms very carefully, as those:

- with windows in only one exterior wall,
- where the farthest point in the room was less than 16 ft from a window, and
- where the window didn't face a wall less than 10 ft away.

In such cases, the required window area was 5 percent of the room's floor area. Where parts of a room were more than 16 ft away, the requirement jumped to 6 percent. Where windows were provided in two or more exterior walls, the requirement was reduced to 4 percent [Chicago87 81.1-14].

Further, Chicago required all sleeping spaces (bedrooms) to have natural ventilation. This explains why one finds operable windows on the residential ninety-second floor of the John Hancock Building.

Optimizing Light and Ventilation Rules

Artificial light and ventilation are allowed in place of their natural counterparts in virtually all situations. This divorces occupied space from exterior walls and roofs.

Air Quality

Background

Imhotep probably didn't worry much about the quality of the air wafting through his projects. Unfortunately, we must. Sick Building Syndrome and Building Related Illness are implicated in many health problems. "Comparative risk studies performed by the EPA . . . have consistently ranked indoor air pollution among the top four environmental risks to the public."[39]

- Sick Building Syndrome (SBS): Refers to situations whereby "occupants experience acute health and comfort effects" while they are in a building, from which they recover fairly quickly on leaving the building. SBS symptoms do not fit known illnesses. Symptoms include skin and respiratory problems, headaches, fatigue, and nausea.
- Building Related Illness (BRI): Refers to recognized illnesses that are directly attributable to airborne contaminants in a building, which need time to incubate, and for which recovery requires more than simply leaving a building. Symptoms are those of recognized illnesses and include fever, chills, coughing, and muscle ache.[40]

Building codes' opinion on air quality is limited to the provisions of the mechanical codes. A more considered approach, although not a mandated one, is taken by the EPA's Indoor Environments Program. As more health care costs are attributed to issues of indoor air quality, codes will most likely take a more aggressive stand to preempt the problems. The following review of research findings and voluntary rules is intended to give readers the kind of advance normally supplied by mandated rules.

Environmental Hazards

Irritants

- Gases thrown off by new plastics, fresh concrete, carpets, and other volatile materials until (if ever) they stabilize. The process of emission is frequently referred to as off-gassing. A visible example of this can be seen in the thin layer of vinyl deposited on the interior surface of windows in new cars that have vinyl dashboards.
- Cleaners, adhesives, solvents, cosmetics, paints, hobby materials, and other chemically active materials used indoors that are sources of Volatile Organic Compounds (VOCS; see also pp. 91–92).

Allergens

- Natural allergens such as pollen
- Chemical allergens

Noxious Substances

- Products of the incomplete combustion of natural gas, wood, or other fuels originating in furnaces, appliances, or fireplaces, such as carbon monoxide. These are strongly affected by building design, and infrastructural design in particular.

Radon

- Found in soils under projects, a known natural carcinogen. It can't be cleaned up since it is integral to the geology of a place. It can only be released by ventilation so that it doesn't accumulate.

Odors

- Related to normal kitchen or bathroom activities
- Related to inadequate maintenance and deteriorating materials

Germs

- Introduced by sneezing and coughing

The EPA Indoor Environments Program was initiated in 1995, although its work started some years earlier as the offices of radon and indoor air. Specific concerns of this effort include radon, Environmental Tobacco Smoke (ETS—there's got to be an acronym for everything), and general pollutants such as particulates, fibers, mists, molds, fungi, bacteria, and gases. Readers who haven't yet reached for a face mask—keep reading.

EPA studies of human exposure to air pollutants indicate that indoor levels of pollutants may be two to five times, and occasionally more than one hundred times, higher than outdoor levels . . . It is estimated that most people spend about 90 percent of their time indoors. Comparative risk studies . . . have consistently ranked indoor air pollution among the top four environmental risks to the public.[41]

In addition to outdoor sources and sources related to building users and activities such as food preparation, cleaning, sick people, and smokers, the EPA has identified the following building components as sources of indoor air pollutants:

Sources of Negative Factors

Components and Furnishings

- Soiled or water- and moisture-damaged materials or furnishings, since they support microbial growth, including molds and mildews
- Dry plumbing traps, since they allow the passage of sewer gases (this occurs only at fixtures that are rarely used)
- Volatile materials or furnishings, since they contribute VOCs or inorganic compounds, or off-gas in general
- Friable materials, since they contribute particles (particularly problematic when the friable material is asbestos)
- Materials that attract or hold dust and pollen, since they maintain particles that irritate, and support a wide variety of mites and parasites

Equipment

- Drip pans, ducts, coils, and humidifiers, since they serve as incubators for microbial growth
- Improperly vented furnaces, stoves, and fireplaces
- Ducts, since they collect and distribute dust and debris
- Office equipment, since it releases VOCs and ozone

Requirements

The BOCA93 Mechanical Code names restrictions only for particulates from combustion and noncombustion sources, smoke, and sulfur dioxide. It actually leaves itself open to many other possibilities by prohibiting "nuisance pollutants," a form of pollution not quite defined [BOCA93 M-1700].

The EPA suggests six basic control strategies for reducing indoor air pollution:

- Managing pollution sources, including removing, substituting, and encapsulating them. (See pp. 256–260).
- Providing localized exhaust at the sources, such as fans and fume hoods.
- Increasing ventilation to dilute bad air. Fumes don't readily dissipate in buildings with high-performance enclosures that tightly control air changes out of concern for energy efficiency. New ASHRAE standards address this issue.
- Controlling exposure by designing to segregate users from sources spatially, and by scheduling to segregate them temporally.
- Filtering the air
- Providing user education and postoccupancy evaluation.

(See also pp. 207–208, 217–218, and 259–260.)

Noise

Background

Rules are primarily concerned with how noise, as opposed to sound, is handled. The public has generally decided that concert hall acoustics don't require regulation. It is, however, quite concerned with noise levels caused by airplanes on takeoff, by vibrating rooftop mechanical equipment, and by interference from noisy neighbors.

Airplanes are controlled through zoning ordinances, and vibrating equipment is handled through mechanical codes. Model building codes regulate the last category. They require acoustic separations between adjacent apartments and between apartments and corridors in multifamily residences. Sound levels inside the spaces do not appear to be a concern of residential regulation.

For institutional projects, such as schools and psychiatric hospitals, the deleterious effects of inappropriate sound levels would seem to justify regulation. This is the type of rule that project teams would best seek in state codes, listed with other regulations related to health, health care, or education.

Model building codes regulate soundproofing for residential occupancies only and are primarily concerned with privacy. Building codes focus on regulating the design of sound barriers such as partitions and floor assemblies, rather than regulating the sound itself. They specify minimum levels for ratings that measure the amount of dampening provided as sound moves through walls and floors to other spaces: Sound Transmission Class (STC) and Impact Isolation Class (IIC). (See pp. 258–259.)

Taking the opposite approach, OSHA regulates noise levels within a space and is concerned with hearing-related injury. Soundproofed partitions don't do any good when the sound is generated in the occupied room. Perhaps because of this, OSHA doesn't address construction at all, but dictates the amount of noise to which workers may be exposed. It specifies sound levels beyond which workers must be provided with protective earplugs or headphones. No rules address the Noise Reduction Class (NRC) of finishes lining a space, perhaps because the noise levels remaining are more important than the amounts absorbed, or because sound absorption affects reverberation time more than it affects sound levels. Besides, high NRC values represent only one solution, since occupants may be able to reduce the volume of sound at its source or can wear ear protection if they so choose.

Requirements

The IBC requires fixed ratings of 45 for both STC and IIC, and only for partitions enclosing tenant spaces [IBC 1206, BOCA93 1214]. This is not difficult so long as spaces are enclosed in partitions.

OSHA simply says that "protection against the effects of noise exposure shall be provided when the sound levels exceed those shown in [the following charts]." It doesn't try to reduce sound levels, it just deals with the problems noise can cause [29 CFR 1910.95(a)]. Besides, the noise levels permitted are fairly generous—an eight-hour day at 90 dB (decibels), 100 dB for up to two hours, and 115 dB for fifteen minutes. It even allows impact noises of 140 dB [29 CFR 1910.95 Table G-16]. Except for some concerts and factories, it's unlikely that noise levels will ever reach these levels. As for solutions, factory workers can use ear protection, and arena managers can keep an eye on sound engineers.

For most common projects, project teams have the option of solving noise problems with absorbent interior finishes rather than by handing out earplugs. More important, sound levels at such projects are unlikely to trigger regulations, so design responses are up to the project teams.

Optimizing Noise Rules

Since the dominant rules address the control of STC and IIC ratings, focus on them. There isn't much getting around the idea that good fences make good neighbors. Project teams can let space flow all they want within tenant spaces, but tenant separations must provide the required acoustic isolation.

Thermal Comfort

Rules specify temperature ranges within which occupied spaces must operate. They require that supplemental heat be provided as needed to maintain required temperatures. Even in rental units, heating system failures during winter months can constitute violations of building code [IBC 1203].

HYGIENE

Plumbing Count

Background

How often do women complain about the lines at public rest rooms? As with elevators, one would hope that plumbing fixture count would be related to wait time. Well, after many years of codes that required equal minimum numbers of fixtures and, consequently, much longer wait times, codes finally woke up to the realization that women need more toilet fixtures than men since tending to some of their bodily functions requires more time.

Requirements

QUANTITY

The number of fixtures required varies by type of fixture, but all codes relate the number of fixtures to the number of occupants. Until recently, plumbing codes calculated the number of occupants using a different formula than the building codes used to calculate occupant count for egress. It stands to reason that the same number of people would be in the building just before a fire as during times of typical bathroom usage. This situation appears to have been reconciled; the International Plumbing Code defers to the building code for occupant load calculation.

A few formulas in the table are worth noting:

- In general, as in past codes, the same number of fixtures are required for men as for women. An exception is made for bodily elimination fixtures (water closets and urinals) in certain assembly uses. Women's rooms are now scheduled for twice the number of these fixtures than are men's rooms in theaters, museums, stadiums, arenas, and churches. Perhaps this will avoid future intermission hassles. If stadiums seem an unlikely place for double the demand from women, bear in mind that women attend sporting events more than in the past, and that stadiums also provide venues for concerts, rallies, and religious gatherings.
- Societal norms are still changing. While the IBC allows half of the required bodily elimination fixtures to be urinals, the figure was as high as 80 percent in the old Chicago Code. Old codes (and likely many local amendments to the model codes) also allowed for continuous urinal troughs to be used in

OCCUPANCY		WATER CLOSETS (Urinals see Section 420.2)		LAVS	BATHTUBS/ SHOWER	DRINKING FOUNTAINS (see Section 411.1)	OTHER
		Male	Female				
ASSEMBLY	Theaters	1:125	1:65	1:200	—	1:1,000	1 srv snk
	Nightclubs	1:40	1:40	1:75	—	1:500	1 srv snk
	Halls, Museums, etc.	1:125	1:65	1:200	—	1:1,000	1 srv snk
	Coliseums, arenas	1:75	1:40	1:150	—	1:1,000	1 srv snk
	Churches	1:150	1:75	1:200	—	1:1,000	1 srv snk
	Stadiums, pools, etc.	1:100	1:50	1:150	—	1:1,000	1 srv snk
Business (see Section 404)		1:25		1:40	—	1:100	1 srv snk
Factory and industrial		1:100		1:100	see Section 412	1:400	1 srv snk
Dwellings		1:unit		1: unit	1:unit	—	1 kitchen snk, 1 clothes washer connection per unit

Excerpt from Table 404.1—"Minimum Number of Plumbing Facilities," 1995 International Plumbing Code. © ICC.

place of individual urinals. In more recent codes, trough urinals are outright prohibited [IPC95 401.2]. Here's a curious allowance, added to the City of Chicago Code back in 1975:

In toilet rooms for females, urinals may be substituted for water closets, not to exceed one-third (⅓) of the required total number.

Perhaps they were thinking of the women's urinals that were test-marketed following the release of Alexander Kira's *The Bathroom*[42] in 1966 but that failed to attract a following. Still, an enlightened provision.

CONTROL
Codes writers must decide the number of and relationship between people who will be sharing any given toilet room. Under what conditions may bathrooms be located communally rather than in individual units? Dormitories and apartments clearly have different agendas, not based entirely on responding to the market.

LOCATION
How far apart, horizontally and vertically? Where dedicated bathrooms are not required for every space, most codes allow bathrooms to skip floors, so long as occupants on every floor have access to bathrooms on an adjacent floor. The handicapped must be able to access these bathroom by elevator or ramp. This means that every third floor must have a bathroom for any particular sex. For example, a women's room on the fourth floor will serve floors three and five, so more bathrooms aren't needed until floors seven and one to serve occupants on floors six and two.

ACCOMMODATION
Accessibility rules require that a minimum number of bathroom fixtures accessories (such as toilet paper dispensers), and bathrooms themselves, be accessible to the disabled. This has led to some interesting cases, including one where the owner of an ice-skating rink was required to include accessible fixtures even in rest rooms that were accessible only from the ice. The owner's sense that only able-bodied, sighted, skaters would be able to get to these bathrooms missed the point—some skaters, whether due to age or infirmity, would benefit from rules that require grab bars and the positioning of accessories to be within easy reach. Still, there would have been no point in enforcing Braille signage or minimum dimensioning for wheelchair-bound users. check rules carefully for the degree to which bathrooms need to be accessible and for the minimum populations that trigger compliance

Optimizing Plumbing Count Rules

One traditional optimization factor has been the use of urinals in place of water closets in men's rooms. They require less floor area and take less time to use than water closets, so are more efficient fixtures. Rather than require their use, however, codes allow the option of substituting a certain percentage (half, per IBC 420.2) of the required water closets in men's rooms with urinals, leaving the total fixture count unchanged.

Food Service

Background

Commercial kitchens, whether serving restaurants and

cafeterias or serving institutional settings such as hospitals and schools, are bound by hygiene rules written into individual local or state codes. They have broad implications for the design and construction process, influencing such decisions as:

- Capacity of exhausts and location of supply registers for mechanical systems.
- Choice of materials and finishes for floors, walls, and ceilings but with special emphasis on the floors. Surface porosity (affecting ease and effectiveness of cleaning procedures) and slip resistance are addressed.
- Interior amenities affecting sanitation, such as sneeze guards.
- Lighting levels and protection against breakage of lamps.
- Specialized equipment such as walk-in refrigeration.
- Disposal facilities.

Hygiene rules generally have a few limited goals in mind:

- Safe food handling. This requires that facilities for washing be provided and that adequate refrigeration be available.
- Safe environment. Since food can't be prepared without exposing it to the environment, provisions must be made to screen out or otherwise eliminate disease-carrying insects such as flies and roaches, to control air quality, and to facilitate cleaning.

Requirements

Without going into detail, it is worth noting the following common requirements:

- Ceiling finishes may be limited to those with easy-to-clean surfaces without fissures. Mylar-faced lay-in panels are a typical response to this requirement.
- Rough materials including concrete block may need to be filled and painted to achieve a smooth finish.
- Carpet will most likely be prohibited throughout food preparation areas and bathrooms used by kitchen staff, other than unusual exceptions.

As for the amount of facilities required, with federally run institutional projects such as prisons, hospitals, and schools, rules often specify minimum amounts of food storage space required as square feet per user. Such rules are found with other rules for the facilities, not with other rules for health. Therefore, look under the Bureau of Prisons rather than under the Food and Drug Administration or U.S. Department of Agriculture.

Optimizing Food Service Rules

Some issues tend to be very tightly constrained; others are grossly simplified and allow everything to be negotiated. Frequently, the best way to handle either situation is to get to know, and then work with, the code officials, since many food service requirements are determined locally. This has recently been getting more difficult, since many health departments now refuse to look at designs until they are at least halfway complete. This is in response to efforts to reduce the size of government.

Optimizing required clearance can yield significant gains. Space must be maintained around many pieces of equipment to allow for cleaning and for hot vent pipes and stacks, resulting in some very large rooms. Generally, clearances are not required when individual pieces of equipment can be rolled away to provide access needed to clean. Many pieces of equipment can be rolled if fitted with casters and flexible connections. Pieces that can't be moved are those that must be positioned directly over drains and those connected to fixed vents.

Medical Facilities

Background

While proper design of food service facilities can go a long way toward preventing disease and reducing its communicability, medical facilities are critical to combating disease. These codes are also developed on a state-by-state basis.

Issues addressed include:

- Handling, including storage and containment of radioactive and contagious materials
- Choice of finishes with respect to cleanliness and static grounding
- Design guidelines for medical gases, such as oxygen and vacuum lines

Requirements

Without going into detail, it is worth noting the following key requirements:

- Maintaining acoustic privacy is a more difficult design issue than usual since insulated ductwork is prohibited. This is because such materials form an ideal breeding ground for contagions and irritants, such as viruses and mildews, due to the warmth, darkness, and easy access to germs that come in with recirculated air.

- As with kitchens, finishes are limited.

DESIGNING SPECIAL SPACES

Some spaces have such unusual characteristics that the rules for common spaces are simply not relevant. Atriums, courts, yards, mezzanines, and penthouses are such spaces. Each has special considerations related to fire spread, egress, and ventilation.

Some codes have sections for other special spaces. The SBC has a section governing the design of nuclear fallout shelters. Its requirements apply to the structures only when they are in use as fallout shelters. It addresses egress, ventilation per occupant, and occupant load, but has no unusual requirements for construction classification or fire resistance [SBC91 Appendix L Section 514].

ATRIUMS

Background

When floor openings connect two or more stories, there is some degree of risk. Openings do not impede the flow of fire or smoke, and could quickly conduct it throughout the area they serve if not designed to prevent it. Openings can be handled in one of three ways:

- They can be wrapped in fire-resistive partitions, as long as that doesn't compromise other design goals.
- They can be left open where doing so is expressly allowed by rules. A complete list of such situations is found in Chapter 10 (p. 248).
- If none of the above situations apply, floor openings can be designed as atriums.[43]

Requirements

There are two main ways to control the risks inherent with atriums [IBC 404, LSC94 6-2.4.6]:

1. Before atriums are affected, slow the spread of fire and smoke to them from occupied spaces that are burning through the use of:
    - Smoke skirts: Panels that extend downward from ceilings at the edges of atriums. They hold smoke at those locations to give it a chance to exhaust through direct exhaust systems before backing up into the atrium. They are frequently made of unframed glass.
    - Direct exhaust systems: Powered, contained, ducted smoke systems that whisk smoke out of buildings before it spreads.
2. Once atriums have been affected, slow the spread of fire and smoke from them into as-yet-unaffected occupied spaces surrounding them.
    - Enclose the multifloor spaces, even if not right at the slab edges.
    - Pull the smoke "plume" away from adjacent occupied spaces and toward the middle of the atrium.
    - Pull smoke out of the building with rooftop scuttles (hatches on fusible links or other actuated releases) and fans.
    - Provide closely spaced sprinkler systems to establish water curtains where needed. (See p. 217.)

Optimizing Atrium Rules

> **CASE STUDY**
>
> John Portman, an architect, developer, and urban designer, was the visionary who saw possibilities where others saw only code requirements. He single-handedly invented the modern atrium building in 1967 when he convinced code officials in Atlanta, Georgia, to let his radical design for the Peachtree Hyatt hotel be built.
>
> Consider the boldness of this proposal. It was the first project in the world with a soaring interior atrium, glass elevator cabs moving, hoistwayless, within the atrium, tiers of surrounding balconies, and major commons spaces at the base of it all. Just as important, it was the first project with the configuration, components, and infrastructure to make it work. Portman soon followed with similar projects in San Francisco, Detroit, and Los Angeles, and the genre that is now repeated in various forms throughout the world was established.

Courts and Yards

Background

Many projects contain outdoor spaces that are needed to bring natural light and ventilation to otherwise "landlocked" interior parts of the buildings. Such spaces fall under the code requirements for courts and yards. These two spaces are slightly different. Both must be open to the sky, except for limited encroachments, and located on the same property as the building they serve. However:

- Courts:
  1. are surrounded on at least three sides by building walls, giving them depth, length, and height;
  2. have floors that can be located at any level;
  3. are intended to be paved.
  4. *may not be occupied* except for cleaning. Where it is intended that occupants use the space, the natural light and ventilation it might otherwise provide cannot be counted toward meeting light and air requirements.
- Yards have no minimum enclosure requirements, so may have depth and height without length. However, their floors must be at grade or lower.

Other definitions and distinctions were and are made in some model codes, and may be found in some of the local amendments to model codes.

- Model codes have distinguished closed courts (fully embedded) from open courts (located along a project's exterior wall). The IBC has dropped this distinction, classifying all courts as open courts.
- In addition to the types noted above, the City of Chicago Code distinguished between:
  1. Rear Courts, which would now be classified as yards insofar as they were located between buildings and rear property lines.
  2. Through Courts, which were passages open to the sky that connected to public ways, open courts, or other through courts at each end.
  3. Recesses, which were shallow, three-sided undulations in the walls of courts.

Requirements

Codes are primarily concerned with proportion. [IBC 201 and 1205, BOCA93 1202, 1212, and 1213. NFPA does not list court requirements except as they apply to other components.]

- Horizontal Dimension: Rules require that courts and yards be wide enough to prevent fire spread. The IBC sets a minimum width of three feet, doubling it to six when windows face each other across a court. For multistory courts, increases in both width 1 ft per floor) and length (2 ft per floor) are required as the court gets taller. This same rule applies to yards, but only with respect to width.
- Vertical Dimension: Courts, but not yards, require a ten-foot minimum court height except where courts are three-sided and the open side is facing a yard or public way.

The amount of fenestration facing a yard or court is not limited so long as the surrounding spaces are part of a single building area. (See pp. 180–181.)

Optimizing Courtyard Rules

What constitutes "occupied"? If a court is used as a passage, does that disqualify it as a source of natural light and ventilation?

What constitutes "open to the sky"? Are glass covers allowed? Can operable clerestory windows compensate for a covered top?

Mezzanines and Penthouses

Background

Mezzanines and penthouses, as defined by codes, are special levels that do not count as stories. (See pp. 141–142.)

Mezzanines are partial floors located in their entirety within individual rooms. They are considered to be continuations of the floors immediately below them. Basic mezzanines may not be separated by partitions from the rooms containing them. Their occupants are allowed to pass through the floors below en route to exits.

Penthouses are unoccupied rooftop mechanical rooms.

Requirements

Mezzanines

Exiting: The occupant loads of mezzanines are added to the occupant loads of the rooms containing them. Exits from those rooms are calculated for the cumulative load. The exit path from the mezzanine to the exit on the floor below must be designed with aisles and aisle accessways if required, and in compliance with common path limits, as if the mezzanine space were located on the floor below.

Enclosure: If partitions are desired to separate mezzanines from the rooms containing them, additional requirements must be met: the mezzanine must be served by at least two means of egress and must have direct access to at least one of them (without first making occupants pass through exit access corridors). Exterior walls of mezzanines are subject to the provisions for other exterior walls in the same project [BOCA93 505, LSC94 6-2.5].

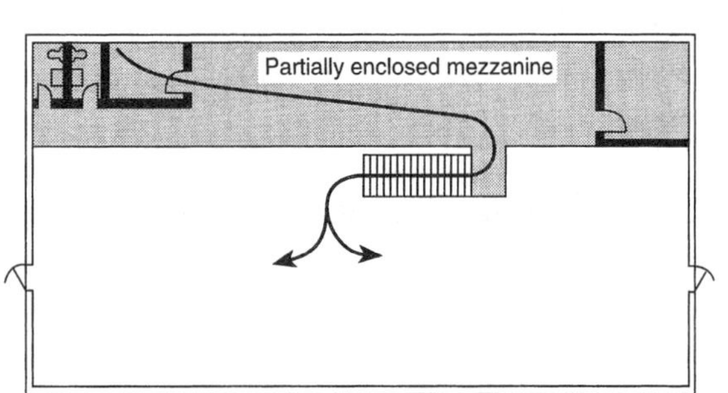

Figure 6-22(a). Open mezzanine. Enclosed portions of mezzanine have occupant load of not more than 10 to satisfy openness requirement. Single exit access allowed if within common path of travel limitations.

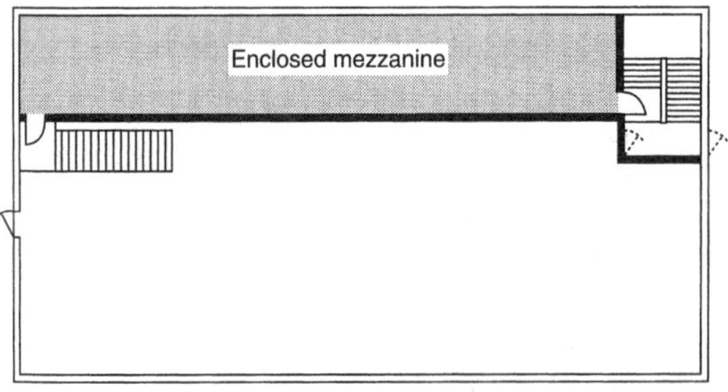

Figure 6-22(b). Enclosed mezzanine. Mezzanine allowed to be enclosed and considered part of floor below if within 1/3 area limitation, minimum two means of egress, and direct access to exit on mezzanine level.

LSC Figures 6-22(a, b). Comparing the requirements of open and enclosed mezzanines. ©NFPA.

Penthouses

Exiting: Exiting from penthouses follows the same rules as exiting from other floors.

Enclosure: Exterior walls of penthouses are allowed one benefit. They are granted a cap on the maximum fire resistance required (generally 1–1½ hours, with noncombustible and, in some codes, protected construction stipulated) regardless of the generally required exterior wall rating, as long as they are recessed some distance (at least 5 ft) relative to the exterior walls of the floor below [IBC 1511, BOCA93 1502, 1510.3, LSC94 6-2.5].

Optimizing Mezzanine Rules

Bear in mind that the occupant load exiting from the floor immediately below the mezzanine will most likely be the highest in the building. Exits may need to be made larger than required for other floors, all the way to the ground, as a consequence.

When designs require that mezzanines be enclosed, at least one exit path from the mezzanine must lead directly into an exit.

Miscellaneous Spaces

Rules identify several types of spaces as warranting special considerations. Such spaces include:

- Stages and platforms, as a consequence of the intense use of flammable materials for costuming and scenery, and in recognition of the hazard potential of fire in a theater [IBC 410].
- Projection rooms, as a consequence of the dangers posed by the materials traditionally found within them: nitrocellulose film and carbon arc lamps. These rules must be given tremendous credit for the safety record movie houses have had. If one were looking for a disaster, one could have hardly found a more likely candidate than the innocuous project room, bringing a highly effective fuel and an incredibly intense heat source together in a single piece of equipment in a single room at the back of a crowded theater [IBC 409].

9

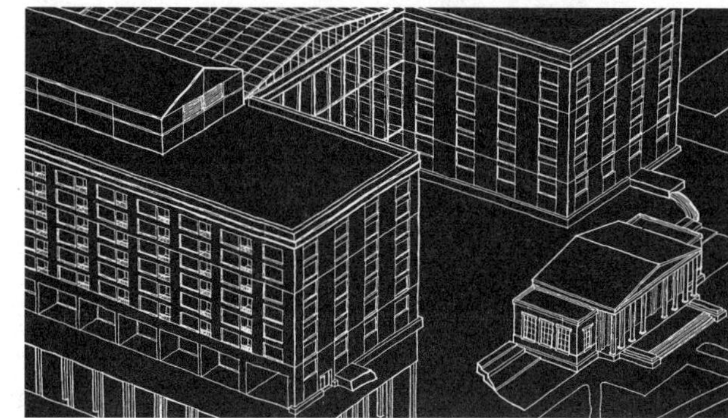

CIRCULATION

Horizontal and Vertical Openings Between Spaces

IT DOES LITTLE GOOD to have well-designed spaces enclosed by carefully considered partitions unless the doors, passages, and other openings that link them are equally well considered.

GENERAL (NONEMERGENCY) CIRCULATION

This topic comprises several components and issues, including:

Purposes of Circulation

- **Universal design:** Those who are not disabled have much to gain from designs that more easily accommodate larger percentages of the population.
- **Medical evacuation:** Occasionally, paramedics must be able to access a building quickly to evacuate a person with a medical emergency.
- **Deliveries:** Although usually unnoticed by occupants, deliveries are a daily function that a building's basic design must facilitate.
- **Collections:** Designing to accomodate trash storage and pick-up requires attention to health considerations and to transfer and transport problems.
- **Infrastructure:** Pipes, ducts, elevators, dumbwaiters, and similar components must be able to pass horizontally and vertically through floors and walls without compromising the structural and fire-resistant characteristics of those assemblies.

Elements of Circulation

- **Optional Stairs:** Stairs beyond those needed to meet the minimum number of exits required by codes.
- **Conveyances:** Elevators, escalators, manlifts, and dumbwaiters can increase the ease and speed with which people and products move through buildings.

PURPOSES

UNIVERSAL DESIGN

Context

Universal design is a new term for the idea that everyone needs accessibility, not only the permanently disabled. Few, if any, rules use the term, however.

Users are starting to expect more, leading to a broadening of the idea of handicap accessibility to embrace that of "universal design" — design that accommodates the widest range of user capacities. The wheelchair-bound are most often associated with accessibility needs, but there are of course other accessibility considerations. For some users, "disabilities" that hinder access will be long-term or permanent. The young and the elderly, for example, may each have trouble with reach, strength, and coordination; the unusually tall or short or heavy may have trouble with reach, passage, or endurance.

Other "disabilities" hindering access may not be addressed by the ADA. They may be short-term, such as broken limbs, or even momentary.

Examples of limited normal limited abilities include:

- Limited muscle control: Shoppers whose hands are occupied with bags cannot easily operate knobs or switches. Parents pushing strollers, shoppers pulling wheeled carts, and travelers pulling wheeled luggage all have trouble with grade changes.
- Limited vision: Parents whose attention is diverted by young children may not see obvious visual cues.
- Limited hearing: Commuters and joggers listening to music on personal stereos may not hear other sounds in the environment.

Although the numbers of people facing such short-term obstacles have not been systematically studied, many people do find themselves in such positions for brief periods several times a week. Thus all users are occasionally in need of environmental amenities that enhance access.

Conceptual approaches to universal design are as critical as detailed solutions (see pp. 260–269 for the latter). Universal design is not something that can be "stuck on" after the schematic design has been approved. To be effective, it must be a formative consideration.

Medical Evacuation

Background

The primary consideration is less simple evacuation than evacuation without unnecessary trauma. The injured or sick must generally be removed in a prone position on an ambulance gurney. The gurney can't be tipped upright to fit in a small elevator.

Requirements

Practically speaking, all elevators used by the public are now required to be of a size that requires a hoistway with interior dimensions of approximately 6'-8" x 8'-6". The numerical transposition makes the dimensions easy to remember. This requirement is cited in several places, if not in so many words:

- Buildings of any height designed under the rules for high rise construction are required to have at least one elevator which can hold an open horizontal ambulance cot of 24 in x 76 in [IBC 403.8].
- Any building over four stories tall has the same requirement [IBC 3002.4].
- Any elevator in any building required to be accessible to the disabled must conform with CABO/ANSI A117.1.

That standard sets a minimum interior dimension of 5 ft to allow a person in a wheelchair to turn completely around to face the doors without having to enter the cab in reverse [IBC 3001.3].

Building codes don't require any particular hoistway size, but the smallest elevator on the market capable of meeting any of these mandates happens to be a 2,500 pound capacity unit. Such an elevator requires the hoistway noted above.

As for use of the elevator once it has been provided, see IBC 3003.

Facilitating Deliveries

Background

Deliveries come in large and small packages. Most will fit through the same spaces that accommodate people.

Unusual Deliveries

The occasional very large delivery can be handled in one of several ways, none of which are addressed directly by codes. Individual panels of curtain walls can be temporarily unglazed to provide a large passage. Some items can be lifted in elevator shafts even when the elevator is too small, by stopping the elevator one floor below the desired floor, forcing open the hoistway doors with a special key, and loading the item onto the roof of the elevator cab.

The occasional extremely large delivery will require more drastic solutions, which are also not addressed by codes. Large pieces of mechanical equipment, such as boilers, air handlers, and cooling towers can either be lifted by crane or helicopter, or lowered through areaways beyond a buildings exterior wall.

Normal Deliveries

The first step in designing for deliveries is to consider the approach of a truck to the building. This is regulated by the on-site loading provisions of zoning codes, which require certain minimum numbers of parking berths. (See p. 125.)

The next step involves provision of loading docks, freight elevators, and passages sufficiently wide to accommodate deliveries of the sizes expected.

Requirements

Other than the loading provisions referred to above, there are generally no requirements related to deliveries. Once the delivery is dropped off, rules take the attitude that markets will force the provision of whatever forms of delivery are needed. An office building without a separate freight elevator may have trouble attracting tenants. If it doesn't, it probably wasn't a public safety or welfare issue.

An exception is sometimes found in hygiene rules that require that biomedical waste or other deliveries (usually being delivered out of a building) be moved through a building following a different path than that used by the general public.

Facilitating Collections

Background

The occupants of buildings produce dry waste—including trash, garbage, recyclables, debris from minor repairs and construction, and old appliances—that must be collected periodically by truck. There are rules for such materials and the related operations, since they can pose health and infestation problems and in extreme cases of neglect may affect community identity.

Sources of problems arising from collections infrastructure include:

- **Incinerators:** They release pollutants and are potential fire hazards.
- **Rubbish and recycling chutes:** They provide vertical passages that link individual floors of a building needing fire separation. They may pose an explosion hazard from dust or lint build-up.
- **Dumpsters** located at the bottoms of such chutes: They provide a breeding place for vermin and microorganisms, are a potential supply of combustible and perhaps hazardous materials. Unfortunately, it is not unusual to see dumpsters located near mechanical system intake grilles, where they can do the most damage.

Requirements

There are few rules for chutes that affect conceptual design or strategic planning, as long as sprinklering is included. Key provisions relate to the fire enclosure and sprinklering of chutes and the "termination rooms" to which they lead [BOCA96 2807, NFPA 82].

Accommodating Infrastructure

Background

Infrastructural elements come in large and small sizes. Small elements such as individual ducts and pipes can be slipped between framing members or reinforcing bars. They are generally thought of as "penetrations," minor interruptions that can fairly easily be closed off against fire by sealing the space between the element and the wall or floor. (See p. 125.)

Larger elements include risers and chases, which are large vertical and horizontal mechanical elements. They provide passage for large ducts or for clusters of smaller ducts, pipes, and conduit. They require three things:

- deliberate structural adjustment of the floor or wall assemblies through which they pass to avoid weakening them,
- enclosure as required for conveyances, and
- dampers used as protectives to prevent smoke or flame from traveling through risers and chases.

Once risers and chases get to places from which their utilities can be distributed, those utilities move out of the risers and chases and into occupied spaces, often by way of plenums and interstitial spaces. Interstitial spaces are low-ceilinged service floors sandwiched between occupied floors, and they are generally found only in buildings with very intense utility needs, such as hospitals and laboratories. Plenums are thinner distribution spaces of two types:

- **Open spaces** [IBC 2805.2, 2805.3]: Horizontal spaces between ceilings and floors, either above a suspended ceiling or below a raised floor. They are open to the degree that ducts and piping can be installed wherever they are needed.
- **Contained spaces** [IBC 2805.4]: Horizontal spaces between adjacent joists within the floor structure

itself or vertical spaces between adjacent studs in a framed partition. They are contained in the sense that the joists or studs are asked to do double duty as structure and to make a "duct."

Codes provide for such spaces to contain piping, including sprinklers, insulation, ducts, both electrical and fiber-optic wiring, and pneumatic tubing. They regulate the use of combustible materials and the "stopping" (filling) of gaps.

As is the case with hoistways, infrastructural shafts for one utility can create significant barriers to the distribution of other utilities. For example, plumbing risers, if uncoordinated during design work, can block the distribution of ductwork.

Requirements

Dampers

Dampers generally do not interact much with other components of a project, so their design is not critical to a project's schedule. They are expensive, however, and any design approach that avoids the need for dampers makes budget available for more visible features. Where dampers can't be avoided, they will need some space, so attention must be taken not to pack ducts too tightly [LSC94 6-2.4.2]. (See p. 254.)

Enclosure of Risers and Chases

Understanding the implications of enclosure requirements is absolutely critical to the schematic design of high-rise building cores. Cores usually contain supply and return air ducts, elevator shafts, and toilet facilities. They may also contain intake and exhaust air risers and electrical and communications closets. Since the basic design approach is to wrap all of these components in slab-to-slab fire-resistant enclosures, the designer faces a challenge in delivering each utility from its enclosure to its destination without passing through the other enclosures. It often helps to make enclosures as compact as possible and to make the larger elements work before locating smaller ones in the remaining space. One might choose to avoid enclosure entirely when permitted. For this, see p. 202.)

Plenums

Rules classify plenums as noncombustible and combustible. For noncombustible plenums, they address the quality level of materials used, referring to NFPA 80 (The National Electric Code) for electrical wiring, UL 1887 for plastic piping, ASTM E84 for the fire resistance of insulation, and the International Mechanical Code. For combustible plenums, they limit the size allowed between separations, requiring the use of "draftstopping" to manage the size [BOCA96 2805].

Optimizing Infrastructure Rules

Due to enclosure requirements, it is important that infrastructural systems be arranged to avoid conflicts. It is not unusual for design teams to realize, fairly late in a project's design phase, that mechanical risers have been located next to hoistways, plumbing stacks, or even bank vaults. As noted, this makes distribution of infrastructure very difficult.

The future will likely see airflow requirements rise as a way of managing Sick Building Syndrome (SBS). (See pp. 197–198.) Such requirements will translate into reduced core efficiencies as more floor area is dedicated to air shafts and ducts. More space will also be consumed by the additional equipment needed to condition this increased flow of air.

The only alternative to larger risers and ductwork is increased airspeed, but raising the speed causes whistling. Whistling can be controlled to some degree, but generally through the use of duct liners. Unfortunately, liners provide breeding places for the microbes that lead to SBS, thus possibly aggravating the problem that increased ventilation requirements are designed to treat.

Dampers can sometimes be avoided by comparing the protection provided by dampers relative to duct walls. Dampers are commonly made from 20 or 22 gage sheet metal. Where ducts are also made from such heavy stock, as might be the case with large ducts to prevent sagging, the thick duct walls resist fire passage as well as any dampers that might be placed within them. In such cases, dampers might be seen as redundant by code enforcers.

ELEMENTS

OPTIONAL STAIRS

Stairs can be included in buildings even when not needed to fulfill minimum exit count requirements. In such cases they are not subject to the design rules for required exits.

If nonrequired stairs are designed to less stringent requirements than required stairs, it would seem that they would be less safe during fires, and therefore should not be used in emergency situations. This raises the question of identification: when exiting, occupants are unlikely to examine an available stair to determine if it is intended for use during a fire. Also, one of the primary rules for required vertical exits is that they be enclosed. If optional stairs remain unenclosed, they could transmit smoke and fire from floor to floor.

These apparent problems are handled by the rules for openings in floors and by the rules for atriums. Optional stairs can be designed as would any freestanding element in a space, such as an office cubicle. During a fire, an optional stair would remain usable as long as the space around it was habitable. Actually, the lack of enclosure on optional stairs makes it easier for occupants to know when the spaces surrounding stairs are becoming dangerous.

CONVEYANCES

Background

Conveyances are moving elements: elevators, dumb waiters, and handicapped lifts; escalators and moving walks; and pneumatic delivery systems. Rules address the machinery itself as well as the way it is used.

An important safety issue conveyances present is that of penetration. Unless they are carefully designed, penetrations can provide easy passage for smoke and flame. Of course, this is not an issue for handicapped lifts and most movable walks, which generally do not penetrate floor slabs.

Penetration is usually handled through the use of rated and gated hoistways. The combination of walls and doors resists the passage of smoke and flame. For pneumatic tube systems, the same idea applies, although a tube takes the place of the hoistway. This is not, however, the only solution. Escalators, sloped moving walks, and glass elevator cabs either can't or aren't enclosed in gated shafts. For them, the basic design approach is similar to that used for atriums; it involves perimeter sprinklering and smoke barriers.

ELEVATORS

Background

The Problem With Elevators During Emergencies

Elevators are critically important during a fire for allowing firefighters to get around a burning building quickly and efficiently, hauling hoses and other equipment to where they are needed, and reaching trapped occupants, including the handicapped, to bring them to safety. Those signs one sees posted next to elevator call buttons that say "Do not use in case of fire" are not there because elevators are unsafe—quite the contrary. The Life Safety Code explains that:

In high rise buildings, towers, or in deep underground spaces where travel over considerable vertical distance on stairs may cause persons incapable of such physical effort to collapse before they reach the street exit, stairways may be used for initial escape from the immediate area of danger, and elevators may be used to complete the travel to the street [LSC94 A-7-4.1].

Deep underground spaces may be more strenuous with less height, because the direction of egress is up. Descending stairs is tough on the skeletal system, but ascending stairs is tough on the cardiovascular system.

The history of elevator use in emergencies is checkered. On the negative side, elevators and their hoistways were often death traps. Smoke and flames would spread quickly through elevator shafts from floor to floor. Touch-sensitive thermally activated call buttons also caused several accidents by stopping elevators at burning floors instead of bypassing them on their way to the ground.

On the positive side, back when it was common for elevators to be manually controlled by trained operators, emergency use was somewhat safer. In fact, before 1956, elevators were given minimal credit in calculating quantity of vertical exit needed. Each 2500 pound elevator was considered the rough equivalent of 7" of stair width.

Now, with almost universal automatic control, operation in atypical situations is less predictable. On the other hand, current requirements for highly rated shaft enclosures and doors and the discontinuation of heat-sensitive controls have eliminated some of the worst problems of the past. Unfortunately, others have taken their place.

The following reasons justify prohibiting the use of elevators for general evacuation (this list adapted from one in the Life Safety Code Handbook[44]:

- While waiting for an elevator to arrive, people may panic or be exposed to smoke or fire.
- Once pressed, a button can't be released. If someone presses the wrong button in haste, in error, or by leaning against the control panel, the elevator will stop, regardless of conditions at that floor.
- Elevators don't work until their doors are closed. People crowding their way into an elevator can prevent the doors from closing. Similarly, any piece of debris that falls across the door threshold when the doors are open can stop the elevator at that floor.
- Unless positively pressurized relative to the fire's location (where air pressure can easily rise by wind blowing in through broken windows or by the expansive character of hot gases), an elevator hoistway can distribute fire and smoke through a building. Moving elevator cabs act like pistons to exacerbate the distribution.
- The elevators may be unavailable, since firefighters have special access to override the call buttons.
- They have very limited capacity, so require more time to evacuate a large number of occupants.
- Without power, no elevator can move. Motors in both electric and hydraulic elevators cease working; electrics are further halted by guide rail brakes applied automatically during power outages. Power can be lost as a result of damage to elevator cables or to general building power distribution cables or equipment.

The exception under which they are still accepted as required means of egress is in relation to areas of refuge and horizontal exits [BOCA96 1006.6]. (See pp. 179, 224–225.) Although they are accepted for use during emergencies, they still may not be counted toward the total number or width of exits required. Design teams still need to find space in their designs for stairs.

The Usefulness of Elevators for General Circulation

Even though elevators can't replace fire stairs, they are absolutely necessary for many projects, and receive plenty of attention from codes regarding their general use. Codes are responsible for the implementation of many of the advances in elevator technology over the years. Despite all those adventure movies, rules assure that elevators can't fall to the bottom of the shaft if their cables are cut, and that people can't be crushed when hiding in an elevator pit or on top of the cab's ceiling.

The number of elevators is normally dictated by market considerations (who will buy a building if access to upper levels is inconvenient?). In fact, The Life Safety Code says

It may reasonably be assumed that in all buildings of sufficient height to indicate the need for elevators, elevators will be provided for normal use; and for this reason, no requirements for mandatory installation of elevators are included in [this code].[15]

Still, codes get involved with other aspects of their selection and use.

Requirements

Elevators and their machinery are addressed by ASME/ANSI A17.1 and A17.3. Requirements are almost completely different for electric elevators and hydraulic elevators. A project team is well advised to work with an elevator consultant who is well versed in elevator codes and standards. For those who are starting from scratch:

- **Electric elevators** can cover any height, and are fast, but require a lot of maintenance. They are pulled by cables, guided by rails (which they grab if the cable is cut), and counterbalanced by weights sharing the hoistway. The cables run over sheaves positioned directly over the hoistways in machine rooms and are pulled by motors and controlled by computers, both of which are also located in the machine rooms.
- **Hydraulic elevators** can traverse only up to six stories and are slow, but reliable and inexpensive. They are lifted and lowered on either one-piece or telescoping pistons and so need no counterweights or cables. The pistons are raised and lowered by oil controlled by pumps. The pumps are connected to the pistons by hoses, so the machine rooms housing them and the controlling computers can be located anywhere within hose reach of the pit.

Overrides are spaces between the tops of cabs when at top floors and hoistway ceilings, and are required for all elevators, just in case someone or something is on top of

A portable handicapped lift used for train passengers at stations without raised platforms. Photo by K. Wyllie.

the cab as it rises through the hoistway. Pits are required to be sufficiently deep for a maintenance worker to remain in the pit, if necessary, when the cab is in its lowest position. For electric elevators, pits contain shock absorbers designed to catch falling cabs in the incredibly unlikely event that their cables should break and their rail brakes should fail. These aren't necessary for hydraulic elevators since the oil in their pistons, which must slowly drain out, impedes a fast descent.

Overrides and pits become a particular design issue in those rare high rise projects sufficiently tall to use "skylobbies." In such buildings, express elevators make nonstop connections from the ground floor lobby to the skylobbies, where occupants board local elevators that stop at the floors between the skylobbies. Since local elevator banks don't connect with the ground floor, they may occupy the same plan space as other local elevator banks, so long as they aren't vertically adjacent. If they were adjacent, the overrides from a lower bank would need to occupy the same space as the pit from an upper bank. When there are at least three tiers of local elevators, hoistways can alternate between two sets of plan positions, thus allowing for some plan efficiencies.

Handicapped Lifts

Handicapped lifts are platforms enclosed only by railings, operated by scissors mechanisms, and powered by hydraulic pistons that raise the lift by pushing the scissors apart and lower it by allowing gravity to compress it back. They are an inexpensive and sometimes portable way to lift people or goods short vertical distances. They are generally used to connect ground levels with entrance landings or main floors with mezzanines.

Lifts rarely blend in with the architecture of a project, and therefore don't quite fulfill the ADA's intention of maintaining handicapped users' dignity. On the other hand, they get the job done in situations where pleas of hardship make more integrated treatments, such as elevators, unfeasible.

Escalators

Even though escalators could be used as stairs during a power failure, they do not meet the required design characteristics for stairs, such as rise-to-run ratio, handrail design, and length of run between landings [BOCA96 1006.6].

The issue of fire spread at floor openings is only an issue where openings require enclosure. (See p. 248.) In such cases, it can be handled with automatic shutters or closely spaced sprinkler systems (water curtains) in a sprinklered building, or by fixed enclosures in unsprinklered buildings [BOCA96 3011]. (See also see pp. 215–218.)

Optimizing Conveyance Rules

For detailed guidance, consult ASME/ANSI A17.1, the Safety Code for Elevators and Escalators and ASME/ANSI A17.3 Safety Code for Existing Elevators and Escalators. These standards describe provisions for manual control of elevators by emergency crews such as fire fighters and paramedics.

EMERGENCY: OCCUPANTS

One aspect of circulation provides a way to get out of buildings during emergencies, by routes known technically as "exit paths" or "means of egress." This section discusses the specific requirements codes impose on the exit path, and discusses strategies for optimizing each. (As such, it follows on the discussion at pp. 173–180.)

The design concepts explored below apply to several components of the exit path.

THE SYSTEM

CAPACITY

An exit path must be

- wide enough to accommodate the numbers of people who would be using it during a crisis, and
- close enough to be reached or short enough to be negotiated before the dangers overtake the occupants.

Height can also affect egress conditions: tall spaces are safer longer than low ones, since smoke rises. Codes assume standard heights—those mandated by general requirements for occupied spaces—but allow for less restrictive designs when arguments based on timed exiting are presented. (For timed exiting, see p. 180.)

Allowable length is slightly different for each element of the exit path, so each will be discussed separately.

Required Width

Background

Required width, the width of the path before deducting encroachments, is based on the speed with which people can move. For example, stairs and doors in the exit path can safely accommodate fewer people than corridors of the same width, since people slow down when going down stairs and opening doors. These concerns apply equally to all elements of the exit path. Rules are given for clear width, the actual amount of space available between two enclosing surfaces. This is not the same as nominal width. For example, a 36 in wide door provides substantially less clear width than 36 in, once the width of frame stops and the thickness of the door in its open position are deducted.

Requirements

FORMULAS

Codes use multipliers to set minimum widths. Multipliers are the number of inches or millimeters of width needed per occupant. One determines the required minimum width by multiplying the number of occupants by this multiplier. In the model building codes, fast-moving elements, including corridors and aisles, have one multiplier (0.2 in/occupant), slower elements, including stairs and doors, have another (0.3 in/occupant). In the Life Safety Code, a few more alternatives are given, to more accurately accommodate the needs of different occupants.

In addition, multipliers vary depending on the hazard level of the use group, or occupancy, with higher hazard occupancies requiring more width per person.

Certain minimums are set even when the formula wouldn't require it. After all, someone whose office door leads to an exit access corridor serving a total occupant load of 50 would not be able to squeeze through a 10-in wide corridor (50 occupants x 0.2 in/occupant).

Some older codes used a more complicated system. It required certain numbers of "units" of egress width for certain numbers of occupants. Twenty-two inches was set as a single unit of exit width, based on the idea that this was the minimum width of exit path that would accommodate almost any adult, even when wearing winter garments or carrying bags. The codes also set the minimum width at two units, or 44 in. This method of calculation was a type of step function, where additional occupants didn't change the required width until the specified break points were reached.

Codes using this type of formula allowed half-units of 12 in to be used for occupant load beyond the 44-in base. This was supposedly on the logic that people will pack fairly tightly going down a two-person wide path, especially with a railing on each side as one would find at a stair, but that beyond two people, occupants don't march so neatly. Stairs or corridors 56 in wide were found to handle more occupants than 44-in paths, if fewer than 66-in paths. The more recent move toward incremental formulas beyond the two-person width acknowledges that wider exit widths can handle incremental increases in occupant load.

Aisles, Corridors, and Exits

In general, and as noted, exit paths are required to maintain a total width of at least 44 in. For residential stairs, the width can be reduced to 36 in. In both cases, handrails are allowed to encroach on the minimum dimension. For stairs serving areas of refuge (discussed later in this chapter), the minimum width is 48 in, but does not include handrails, making the effective minimum width about 55 in [LSC94 5-2.12.2.3].

Doors

Minimum widths for doors in required exit paths is 32 in. For this reason, 36 in doors are generally provided, since they can achieve the required clear opening without special hardware. A maximum width of 48 in is set because wider doors open too slowly as a result of their long swing radiuses. An alternate standard width of 41½ in (until recently 44 in) is required for doors used for moving beds, as in hospitals [IBC 1004.2.1].

Encroachments

Background

No path is wider than its narrowest dimension, so encroachments must be very carefully considered. Encroachments are elements that intrude on required width, such as:

- Doors swinging into the exit path
- Door knobs and handrails
- Water fountains
- Standpipes in exit stairs
- Furniture, whether fixed or movable

Codes could simply require that the required minimum width be measured at the narrowest spot. This position is considered to be unnecessarily strict, since occupants can get past minor encroachments without slowing down as long as they shift their body positions slightly.

Requirements

NFPA measures width as a net clear dimension, making general exceptions only for minimal projections below waist height, such as wall-mounted handrails. This height is due to the fact that people are wider at the shoulders than the waist, partly due to build, and partly due to the swaying motion typical while walking [LSC94 5-3.2].

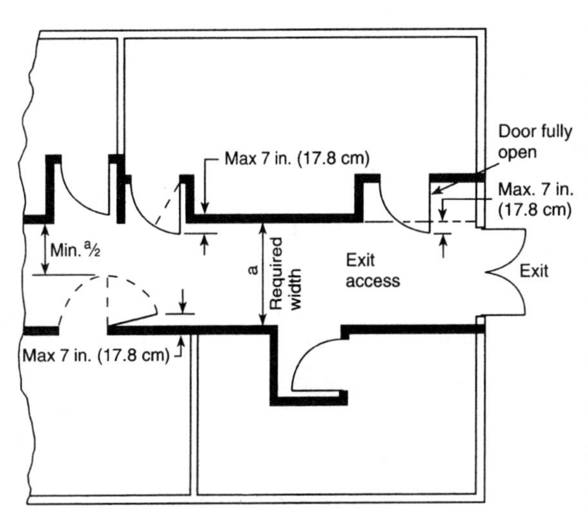

Figure 5-13. Door swing into a corridor. Doors that swing within a recessed pocket of the corridor so as not to protrude into the required corridor width provide the best arrangement for clear passage through an exit access corridor. Doors that swing 180 degrees to come to rest against a wall and do not extend into more than 7 in. (17.8 cm) of required corridor width provide and acceptable arrangement. A door swinging 90 degrees to come to rest in the path of travel is considered as not encroaching excessively on the exit access corridor width if not more than 7 in. (17.8 cm) of the required width of the corridor remains obstructed. Any door swinging into the corridor must leave unobstructed during its entire swing at least one half of the required corridor width.

LSCH Fig 5-13. Door recess designs. © NFPA.

Optimizing Encroachment Rules

At Stairs and Corridors

Set The Encroaching Element Into A Recess

This is particularly effective for an encroaching door swing, but just as easily applies to other encroaching elements. Where designs call for thick walls anyway, such as might be needed to hold the plumbing at a bathroom, small recesses can be made simply by holding door jambs as far to one side of the wall as possible. For thinner walls, wall recesses can provide opportunities for closets or built-ins.

Use Flush Hardware

Use of recessed pulls or push plates, where permitted, may avoid the encroachment caused by knobs.

Recessed fountains. Photo by the author.

Using Balanced (Pivot) Hardware

Traditional hinges cause doors to swing from one edge of the door, resulting in the maximum encroachment possible for a door of a given size. With balanced hardware, where doors rotate about an axis that is between the door's two vertical edges, the encroachment can be reduced. Unfortunately, such hardware also reduces the clear width of the opening, resulting in no net benefit.

Water Fountains

Accessibility codes require that fountains project sufficiently far from walls to allow wheelchair-bound people to get their legs under the fountain. Whether someone is drinking or not, this extension causes problems. Recessing fountains into wall niches is the only viable way to avoid encroachment.

Locate Standpipes In Corners Of Stairs

Standpipes can be, and in most designs are, tucked into the corners of exit stairs. Width is measured along the direction of travel, so corners are usually excess space. By putting standpipes there, this potential source of encroachment is eliminated.

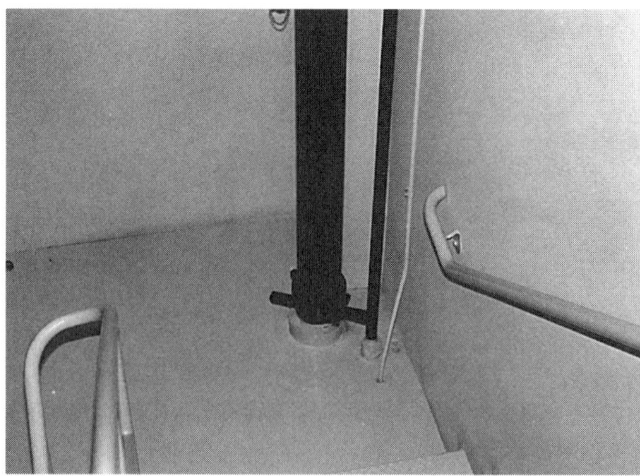

Corner standpipe positions. Photo by K. Wyllie.

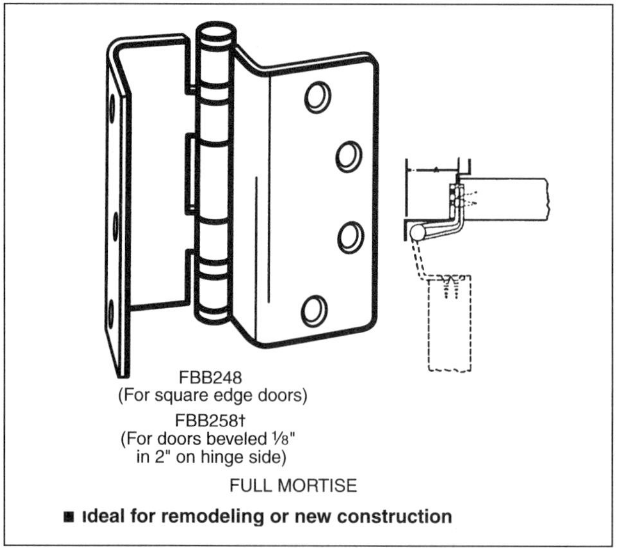

Swing-Clear Hinges. ©Stanley Corporation.

Furniture And Fixtures

Inspectors complain loud and long about this maintenance issue. Required exit paths clogged with all kinds of trash cans, vending machines, lockers, and other obstacles pose real problems, and have been the direct cause of many fire-related deaths.

The solution is to plan for such elements—"a place for everything and everything in its place." Don't leave interior design and operational issues of this type until base building design is completed. By then it's too late. Also, make sure that building staffs know that anything missed then must go somewhere else now.

Make The Exit Element Wider Than It Needs To Be

A projecting elements is not an encroachment if the required width is met by measuring to the projecting element. The beauty of this solution is that no special design tricks need to be used. If additional width is needed to fulfill program requirements, the entire code issue becomes moot. If not, the additional space it needs does impact the project budget.

At Doors

When rules require doors to be 32 in wide, they refer to clear openings. A 32 in door with butt hinges provides a clear opening of about 29 in, once ⅝ in is subtracted on the strike side for the stop, and 2¼ in is subtracted on the hinge side for the door thickness and the hinge's knuckles. With center or offset pivot hinges, the clear dimension is even smaller.

In very tight situations, where larger openings can't be used, or where an existing door must be retrofitted for accessibility, a 30 in door can be made to provide a 32 in opening by using "swing-clear" hinges instead of standard butt or pivot hinges. This type of hinge moves the door itself completely out of the opening. Their disadvantage is aesthetic—they leave part of each hinge exposed when the door is closed.

Smoke And Fire Control

In order to maintain the conditions necessary to life safety, paths for egress must be able to serve as barriers to such environmental hazards as fire, smoke, and heat, and must be able to quickly control any that originate in or enter the path. The elements of the egress path differ in the degree to which they are required to accomplish this, with more demanding requirements made of elements that are closer to the end of the egress path. Nonetheless, all elements respond to some degree to each of the following issues:

General Suppression: Sprinklers, Etc.

Background

Codes generally refer to sprinklers as automatic fire suppression systems. Such systems are commonly regarded as cheap insurance; their effectiveness has been proven repeatedly. They are widely held to be the single most significant factor in raising building safety.

Agent

There are several materials capable of putting out a fire, even if we ignore the protagonist's preferred agent in Gulliver's Travels. Each is appropriate for a particular situation. The choice of suppression materials is primarily a question of program and budget, not regulation. Once the agent is selected, however, the design of the overall system becomes highly regulated by codes.

The purpose of a suppression system is to detect fires and deliver suppressive materials to them. All systems consist of heads that detect the fire and disperse the agent, and pipes that transport the agent to the heads. The types of systems are:

- **Wet:** In this system, all pipes are always filled with water. The water puts the fire out by removing its heat by evaporative cooling, until it is below combustion temperature. It is simple and works for most fires. In some settings, however, the agent is as much an enemy as the fire. Buildings housing valuable documents or electrical equipment are not good candidates for wet agents. Valuable items that get wet can be restored by freeze drying, but the process is expensive, time consuming, and cannot handle large items. Buildings where systems must pass through areas subject to freezing are similarly unfit candidates because of the potential for bursting pipes.
- **Dry:** These systems are identical to wet systems, except that compressed air is forced into the pipes ahead of the water. This means that so long as none of the heads open, all the pipes contain only air. Once a head opens, the air is pushed out by the water, and water starts to flow. This is the preferred system where pipes must pass through areas that are exposed to freezing temperatures, such as basements and outdoor soffits.
- **Halon substitutes** (Halons are chlorofluorocarbons, or CFCs, and so are being phased out due to their impact on the planet's ozone layer): These agents inhibit the volatilization of fuels. In other words, they let the fuels get hot, but prevent them from bursting into flame, regardless of how much oxygen is in the room. They does not displace oxygen, are breathable, unlike carbon dioxide, and act almost instantaneously. They are, however, very expensive, the agent more so than the hardware. The Halon substitutes, as hydrochlorofluorocarbons (HCFCs), are not quite as effective as the Halons, and their behavior is not yet fully known.
- **Water Mist** (finer than that produced by a humidifier): Removes heat, does much less damage than water, and can spread into crevices like Halon. While they hold great potential, these systems have been in use in the United States less than five years and are

fairly experimental. The electromechanical complexity of such systems is directly related to the size of the water particles they produce. Finer mists work more effectively, but are much more expensive to produce. When the these systems are developed to the point that their benefits are reasonably justified by their costs, their use will increase.

Position

Halon substitutes and mists are distributed by special delivery systems, including special pipe sizes, special ways of storing agents, and special nozzles.

With water systems, either wet or dry, water is distributed through pipes and dispersed by heads. The orifice through which the water flows is temporarily blocked by a heat-sensitive glass capsule functioning as a stopper. When the temperature is sufficient, the glass shatters, letting the water out of the pipe. Immediately in front of the nozzle is a deflector that turns the stream into a spray.

Capsules are available in various shatter-temperatures. Since hot air rises, a "volume" space might normally have very hot air just below the ceiling. In order to keep the heads from activating when there isn't a fire, it is necessary to specify capsules rated for the appropriate activation temperature.

Postings and warning lights outside a room suppressed with Halon 1301. Photo by the author.

1. **Ceiling:** Ceiling-mounted suppression systems use heads spaced about 15 ft apart throughout a ceiling and 7 ft from the walls. So-called extended coverage heads work with a 20-ft spacing. They don't work if too high (over 55 ft) because by the time the heat is sufficient to activate the heads, it is also hot enough to evaporate the sprinkler water before it reaches the fuel. Even then, sprinklers are worth including because they keep roof structures wet, postponing their failure. For lower ceilings, they protect structures and control fires very effectively. There are two types of systems:
   - **Pendant:** These are used for most suspended ceiling applications. The deflector is below the nozzle, and the entire head assembly is under the distribution piping. This arrangement conceals the piping above the ceiling, exposing only the head. Water shoots downward. With this system, there are three types of heads:
     - Surface: Heads extend about 3 in below the ceiling.
     - Recessed: Heads are only semi-exposed, extending about 1 in below the ceiling
     - Concealed: Heads are recessed, and covered by metal disks. Upon activation, the head descends to a fully exposed position. Extended coverage heads are not yet available in this form.
   - Upright: These work very effectively where there is no suspended ceiling. The head extends upward from the piping, capped by the deflector. Water shoots upward, hits the deflector, then falls down. With this system, heads are always exposed. Extended coverage heads are not yet available in this form.
2. **Wall:** Where ceiling-mounted systems are inapplicable, heads can be wall-mounted. So called "side wall sprinklers" spray water horizontally past their deflectors. Spacing for common side wall heads is dependent on many factors, including layout and hazard classification, an assessment of quantity and arrangement of combustible materials established by NFPA 13. For extended coverage heads, coverage is a constant 12 ft by 24 ft area.

Distribution Methodologies

In practice, design teams usually establish only the layout of sprinkler heads, leaving the layout and sizing of supply piping to fire suppression system contractors.

Closely spaced sprinklers used at an escalator opening. Photo by the author.

Requirements

The model building codes address suppression in Chapter 9. The Life Safety Code addresses it in Chapter 7-7, and of course, there are other NFPA publications specifically addressing suppression, such as NFPA 13. OSHA has its requirements in sections 1910.158 through 1910.163, with detection covered in section 1910.164. Interestingly, it considers water systems separately from other extinguishing agents.

Optimizing Suppression Rules

Once the decision to sprinkler is made, there are two questions:

1. If on first analysis, water does not appear to be a good candidate for the suppression agent, are there any design solutions that could make it feasible?
   - Identify a routing for piping that precludes unheated areas.
   - Modify the design to cluster all of the areas sensitive to water damage. Provide a dry, mist, or Halon-t type system for this area only.
2. How can systems be designed to be effective without being visually distracting?
   - Use recessed or concealed heads.
   - Use fewer heads by specifying extended coverage types.
   - Expose heads as part of an "expressed" ceiling.
   - Integrate head locations with the design of the ceiling—make them part of a pattern.
   - For historic properties with ornate ceilings, use side-mounted heads where the rooms are sufficiently small. Where walls are too far apart, side-through heads can still reduce the number of ceiling-mounts to a minimum.
   - Ultimately, consider the visual disruption caused by sprinkler heads in relation to the widespread disruption caused by fires that suppression systems might have controlled.

Closely-Spaced Sprinklers: Water Curtains

Closely-spaced sprinkler heads (six feet on center) form temporary fire separations. They use water to contain fire without the imposition of solid barriers. Common applications include:

- Escalator penetrations
- Atriums

Deluge Systems

In some cases, very intense sprinkler installations are used when hazard level is particularly high. They dump large volumes of water, and all of the heads in the system are activated when a fire is detected, not just those heads over the fire, as with typical systems. The McCormick Place Convention Center Annex, built in air rights over railroad tracks, is exposed to the risks inherent in the fuel-loaded diesel locomotives which pass below it. In response, it was built with a "vulnerable belly" soffit equipped with sprinkler heads capable of dropping eight gallons per square foot per minute if an overheating train is detected.

Exhaust: Pulling Air Out

Background

The first thing that needs be said about controlling smoke with exhaust is that it isn't necessary for most projects. Exhaust systems are primarily required for atriums, since most other commercial situations have, and are adequately handled with, sprinklers.

For those large projects for which exhaust is considered, rules require that in the event of fire any equipment circulating air through several spaces be turned off as a first step in maintaining air quality. This includes heating or cooling systems that use ducted or plenum air handling and fans that pull air from more than one space. Any attempt to contain smoke or fire would fail if smoke could ride air handling systems to new locations within a project, and if flames were constantly fed more oxygen by incoming air. For many projects, codes require that power to HVAC systems be tied to the fire detection systems, so that power is cut as soon as an alarm is triggered. Corridors that are served by a building's HVAC system must be designed with cutoffs just like primary spaces.

Once air is no longer being circulated by air-handling systems, one can work toward controlling it. Techniques for keeping smoke from moving from a burning area into a space that is still safe include:

- Limiting travel distance. Although perhaps a "trick answer," this is actually the preferred technique for most projects. When distances to exits are short, design teams need not worry about preventing smoke from moving into safe areas. Occupants avoid inhaling smoke simply by holding their breaths.

  Pulling it away before it has the opportunity to enter. Smoke can be pulled away by forcing it outdoors on its way to the safe space. Once outdoors, it will rise and dissipate in the air rather than re-enter the building. A window can do this, as can a short outdoor corridor, or roof scuttles (hatches) on fusible links [LSC94 5-2.3].
- Working with high ceilings, and possibly with a reduced level of compartmentation than would normally be required, to create a very large holding area for smoke. (See the case study of Potomac Mills Mall at p. 243.)

Requirements

In General

- LSC94 5-2.3.7 and 5-2.3.8
- BOCA93 1015.5 and 1015.6
- Mechanical Codes
- ASHRAE standards
- SFPE (Society of Fire Protection Engineers) Handbook
- NFPA Standards (particularly Standards 92A and 90B)

Pressurization: Pushing Air In

Background

Some spaces, such as exits, must remain smoke-free during a fire, regardless of whatever else happens. Chief among these are exits in high-rise buildings [BOCA93 1015.2]. An effective way to keep smoke from moving from burning areas into unaffected spaces is to keep the unaffected spaces so full of air that there isn't any room for additional air, including air that is smoke-laden. This is called pressurization.

Anyone who has struggled to open a door against the press of winter winds knows that air flow moves only in one direction at a time. Wind generally blows snow into a room whenever the door is opened. Objects already within the room, no matter how small or light, do not generally get blown outside.

This principle can be applied very effectively to controlling the movement of smoke. Mainly, it requires a fan to continually force air from an uncontaminated source (the great outdoors) into the pressurized space. Any other provisions are just so much icing on the cake. For instance, if the pressurized space isn't perfectly airtight, it generally won't matter so much, since air will flow outward from any gaps rather than inward.

Requirements

[NFPA 92A, LSC94 5-2.3.9, BOCA93 1015.7]

Crowd Control

Communication

Background

People need to be directed during a fire. Even when calm, they may not know where to go, and time is a factor in the onset of panic. Alarm systems are activated automatically by fire detectors or sprinkler systems. They can also be manually activated. For projects with fire command stations, they can also be controlled from these. At the least, alarms get occupants' attention and signs help direct. While people may not consciously notice or recognize directions due to duress, they may follow them anyway.

Communications efforts are intended to control fear to prevent panic, and they come in two varieties:

Exit Signs

Exit signs are intended to be as easily visible as possible by the greatest number of occupants. Rules require them to be located near the ceiling at the end of every length of exit access.

There is an ongoing debate regarding placement of the signs. Their familiar location at the ceiling is being challenged by those who claim that such locations are quickly obscured by smoke and that people tend to stay close to the floor while exiting because the air there is more breathable and cooler. Advocates counter that most people exit before corridors fill with smoke, that they are most likely to see the signs when mounted at head level, and that low signs might be obscured by furniture or other people. This issue has not yet been resolved, and until it is, signs are generally mounted high.

ALARM SYSTEMS

Where alarms (sometimes called "signals") are required, there are three varieties to consider: Audible, visible, and voice. Audible alarms have the advantage of being able to penetrate opaque objects such as walls and doors. Visible alarms can work with the deaf and in loud environments. Voice alarms can actually direct people.

Requirements

Design teams concerned about excellence, and with sufficient budget to achieve it, should design communications systems as a team, carefully determining locations for alarms and signage to coordinate with the locations of all the lighting fixtures, sprinkler heads, mechanical diffusers, and miscellaneous pieces of equipment competing for space on a project's walls and ceilings.

EXIT SIGNS

Rules can be very selective when it comes to exit signs, requiring specific sizes of signs and lettering, in addition to specifying locations where they are required.

ALARM SYSTEMS

Alarms must be perceptible from all elements of the exit path, program spaces over a minimum size (1,000 SF), and dwelling units or guest rooms [BOCA93 917.9].

- **Audible.** Rules don't require much except loudness and proximity—the alarms must achieve specified decibel levels within a certain distance of each space. It is up to the project team to decide how to achieve that. NFPA72 allows ceiling-mounted and recessed alarms, but requires that wall-mounted alarms be located at least 7'-6" ft above floors and at least 6 in below ceilings [IBC 907.1.2, NFPA72 6-3].
- **Visible.** These must be provided in public and common areas, in private rooms where designated specifically, though not necessarily exclusively, for handicapped use, and where the normal sound levels are so high that most people wouldn't hear alarms unless they exceeded OSHA rules for allowable noise. Visible alarms use intense flashing strobe lights; sophisticated electronics are used to synchronize the flashes so as not to cause seizures in occupants suffering from epilepsy [IBC 907.1.5.1, NFPA72 6-4].

  The NFPA notes that:

  Visible signaling is a very complex topic. For this reason, the Code presents prescriptive requirements rather than performance requirements, such as those for audible signaling.[45]

- **Voice** (Textual). This type of alarm talks people through crises. It is required in elevators, exit path components, dwelling units and hotel rooms, and tenant spaces over a minimum size. Location requirements are as for visible alarms [IBC 907.1.6, NFPA72 6-7].

SPACES

We now move from general design issues to design considerations for individual exit path components. (For a conceptual discussion of each of the components described below, see pp. 176–180.) The requirements discussed below deal only with design issues related to egress.

PROGRAM SPACES

To recap, these are the spaces where primary user activities take place. Even they are required to meet a few minimal egress goals. (For program space issues other than egress, such as meeting health and hygiene needs, see pp. 196–202.)

Maximum Distance

Codes influence room size and shape by restricting the distance allowed from the furthest point in a room to any doors which lead out. The greater this distance, the more time it takes for occupants to get to a separated

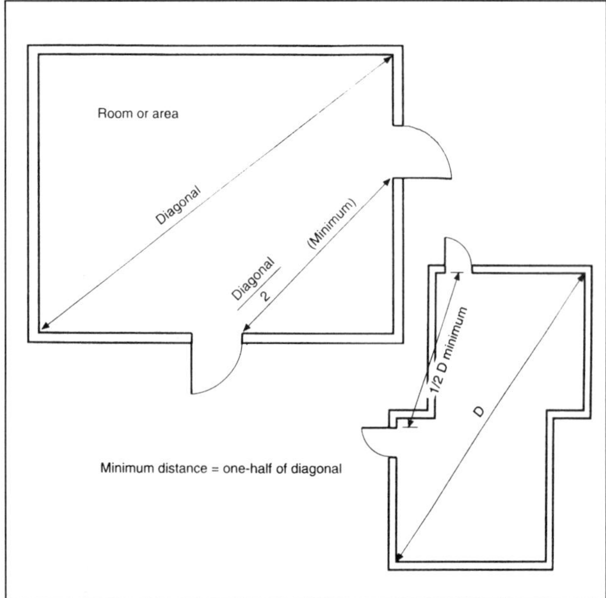

LSC94 Figures A-5-5.1.4 (a, e)—Illustrating the method of calculating distance between exits from a single room. ©NFPA.

(fire rated) exit access or direct exit. A commonly used measure is 75 ft, since occupants can reasonably cross this distance while holding their breaths. This assumes a couple seconds for occupants to get their bearings, and then ten to fifteen more to traverse the distance. Greater distances are allowed when a choice of exits is provided [LSCH App 5-11.1].

Location of Doors

When more than one door leads from a single program space into an exit access corridor, the distance between the doors that are furthest apart is governed by one simple rule: They should be keep as far apart as practicable, but at least half the longest dimension (usually measured on the diagonal) of the room being served.

Maintaining A Clear Path: Aisle Accessways And Aisles

Experience has shown that even large spaces can pose obstructions to egress. Cafeterias can be filled with tables and chairs, auditoriums and stadiums can trap occupants in their seats, and even offices can be cluttered with file cabinets and boxes of paperwork. Under normal conditions, there is no rush and obstacles are barely noticed, only to make their presence felt when people can least afford to deal with them.

When codes permit relatively large individual program spaces, especially in assembly occupancies, they require that unobstructed paths, called aisles, be maintained through the spaces. In addition, aisle accessways are required where occupants might have trouble getting to aisles due to the presence of other occupants, or of restaurant or library chairs which are not tucked neatly under tables or desks, or of theater or stadium seats not folded against backrests. While both aisles and aisle accessways lead to enclosed fire-resistant exit elements, aisles are permanently unobstructed, while aisle accessways, which lead to aisles, may be obstructed or unobstructed depending on the actions of other occupants.

A related egress concept is catchment area. When, for example, a program space has four exits, for design purposes one would think of the space being divided into four catchment areas, one corresponding to each exit. The occupants of each catchment area would be expected to egress through a particular exit, assuming it wasn't the source of the fire. People aren't absolutely predictable and can't be expected to exit through the specific door that the designers had planned for them. Still, the general distribution of occupants moving from the seats and aisle accessways in their catchment areas into the aisles serving those catchment areas, and then into exit access elements and exits appears to work.

Finally, all aisle accessways and aisles must be designed for traffic flow in either direction. This means that from one end to the other, they must be wide enough to meet the minimum width requirements. One can't assume that people will be able to move in the direction of their nearest aisle. This does not mean that all aisle accessways and aisles must have parallel sides, but it does mean that they must include a parallel-sided space as wide as the minimum requirement [LSCH94 p. 225].

In general, although they are not enclosed, these egress elements must conform to the width, length, dead-end, and common path of travel requirements of exit access elements described in the next section. Of course, occupants within aisle accessways and aisles are not protected by rated walls, but maximum travel distance rules ensure that occupants can get to exit access elements in time.

Specific design issues are listed in BOCA93 1011, 1012, and LSC94 Sections 8-2.5.6 through 8-2.5.11. The NFPA sections deal with aisle accessways and aisles serving seating without tables, seating with tables, and in grandstands, bleachers, and folding seating.

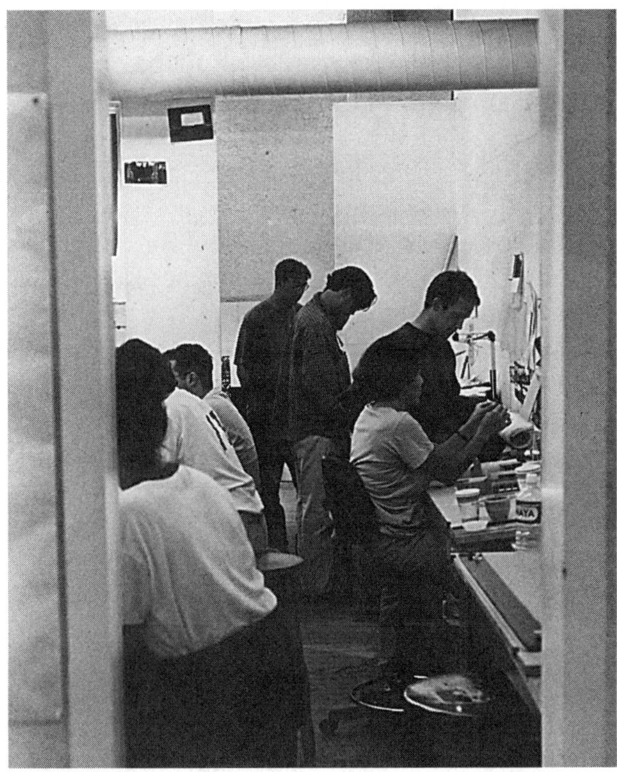

An aisle accessway in a drafting studio. Although crowded when in use, a path is made when occupants exit and stools are pushed aside. Photo by the author.

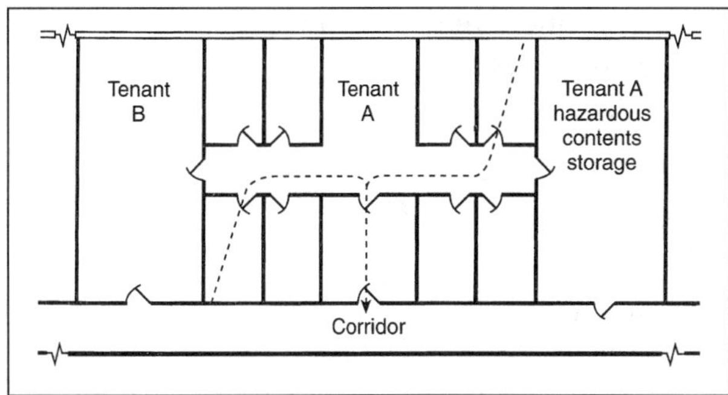

LSCH Figure 5-65 — Exit access through adjoining rooms and spaces. ©NFPA.

Exiting Through Adjacent Spaces

While it isn't necessary to design projects with every room touching an exit access element or direct exit, there are limits to layering rooms. As a general concept, rooms should be supplemental in nature, and their occupants must exit through an adjoining space must have control over that space. It's not unlike leaving keys with parking lot attendants when cars are parked more than one deep from the lane [LSC94 5-5.1.7 and 5-5.2.1].

One interesting distinction is that the Life Safety Code prohibits exiting through such spaces as kitchens, closets, and bathrooms. BOCA permits all three so long as they are not the only exits [BOCA96 1006.2.1].

Optimizing Program Space Rules

With fixed-seat spaces such as auditoriums, so-called continental seating provides an alternative design to traditional row-and-aisle seating [LSC94 8-2.5.7]. With traditional seating, the number of seats in a row is restricted because the tight spacing of rows can make egress difficult. Under the rules of continental seating, longer rows are allowed, but the spacing between rows must be wider. Further, with traditional seating, doors are provided at the ends of aisles, while with continental seating they are provided at the ends of the rows.

Designs which qualify under the provisions of "smoke-protected assembly seating" are allowed additional freedoms [LSC94 8-4.3].

Exit Access Paths

Once again, these are the elements that connect program spaces with exits, usually in the form of corridors between spaces or aisles within them.

Layout

Exit access routes must comply with a few basic design concepts. If they aren't properly arranged around the floor, people will not be able to get to them with adequate speed and safety when necessary.

Distance To Exits

Although exit access routes are somewhat protected, people still need to move through them fairly efficiently, since they will not isolate people from fire indefinitely. Codes limit the total distance an occupant may be required to travel to get to an exit. The following rationale gives a good idea of the performance levels that codes intend to achieve through their prescriptive rules, and may inspire some useful approaches to optimization.

The factors upon which the maximum allowed travel distance are based include:

1. the number, age, and physical condition of building occupants and the rate at which they can be expected to move,

2. the type and number of obstructions (e.g., display cases, seating, heavy machinery) that must be negotiated,
3. the number of people in any room or space and the distance from the farthest point in that room to the door,
4. the amount and nature of combustibles expected in a particular occupancy, and
5. the rapidity with which fire might spread, which is a function of the type of construction, the materials used, the degree of compartmentation, and the presence or absence of automatic fire detection and extinguishing systems [LSCH p. 118].

All codes regulate total length of exit travel, the distance from the most remote in a program space to the door of an exit. Distance is related to use group, with extra distance allowed for sprinklered buildings. The most restrictive distances are for unsprinklered hazardous occupancies (25 ft), the most liberal for factory and sprinklered storage occupancies (400 ft). Length is measured as people walk rather than as crows fly, that is, around partitions rather than by the shortest line connecting remote corners with the doors providing egress [BOCA93 1006.5].

Dead Ends and Common Paths of Travel

Two situations related to poor layout spell danger for an exiting occupant: dead ends and common paths of travel [LSC94 5-5.1.6].

- **Dead ends:** Dead ends occur when spaces in the egress path are configured in such a way that exiting occupants can inadvertently move further from exits rather than closer. When occupants en route to exits mistakenly run into dead ends, they are forced to retrace their steps, losing valuable time.
- **Common paths of travel:** Unlike dead ends, common paths of travel problems occur even when occupants have not lost their way. Occupants sometimes face situations where all possible escape routes pass through one corridor or door before diverging. When fires block the single corridor or door, such common-path-of-travel problems can turn tragic. Common paths occur in the first parts of egress paths, in or near program spaces. Proper planning can minimize the danger. When occupants standing in the farthest corners of rooms or floors need to evacuate, codes address the distance designs can require them to run before providing a choice of exit path.

Common paths of travel and dead ends are quite distinct and must be considered separately during design. The concept of dead ends seems to be better recognized and understood. Ignoring common paths of travel and recognizing only dead ends in hallways that extend beyond fire exits can lead to missed opportunities and inadequate designs.

When dead ends and common paths of travel are allowed, their lengths are usually limited by rules. Dead ends are prohibited entirely by the Life Safety Code except where specifically allowed (but it specifically allows them in most situations). The Life Safety Code's Table A-5-6.1 regulates lengths of common paths of travel. Even if neither is restricted, exit path length is limited by allowable distance to exits.

Distance Between Exits

As with distance between doors leading from program spaces into exit access corridors, the distance between

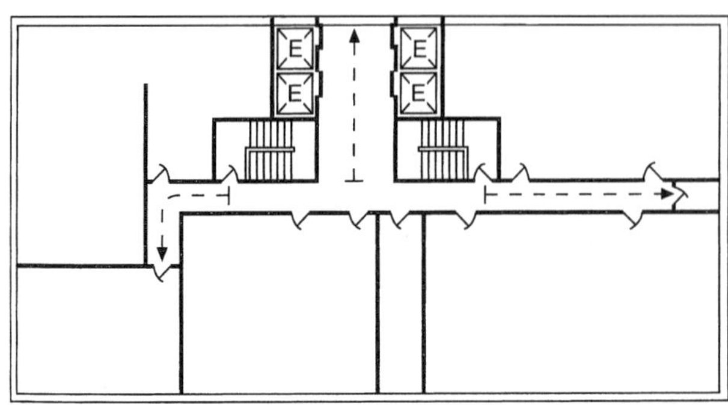

LSCH Figure 5-64(a)—Dead End Corridors. ©NFPA.

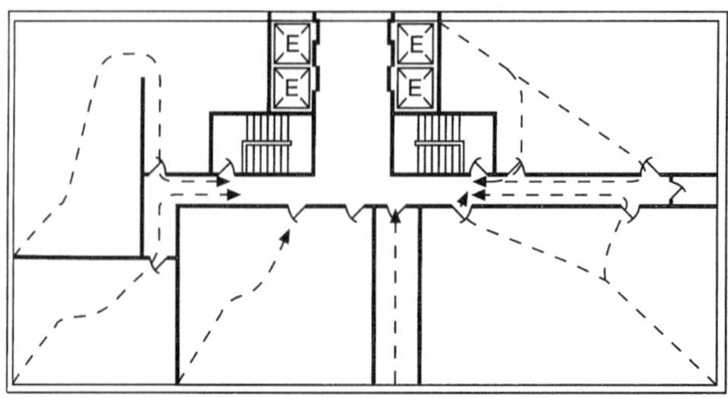

LSCH Figure 5-64(b)—Common Paths of Travel. ©NFPA.

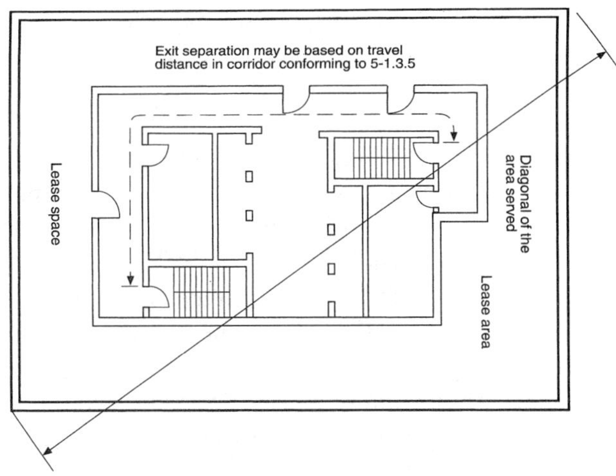

LSC94 A-5-5.1.4 (d)—Distance between exits is measured within the connecting exit access paths. ©NFPA.

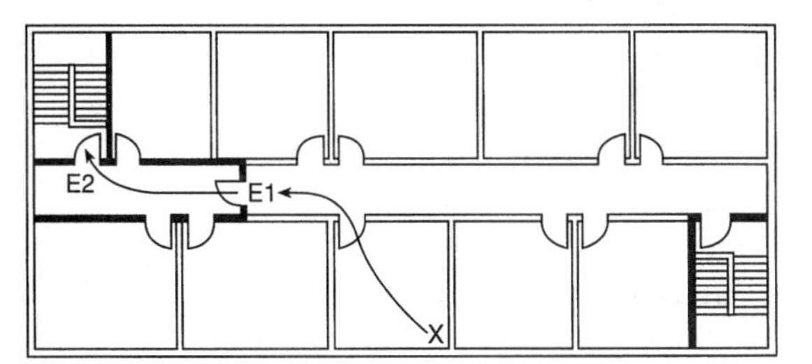

Figure 5-49(b). Exit passageway used to keep travel distance from becoming excessive. Travel distance measurement ends at entrance (d) to exit passageway. Distance from X to E2 exceeds allowed travel distance; distance from X to E1 within allowed travel distance.

LSCH Fig. 5-49(b)— Exit passageway used to keep travel distance within limits.©NFPA.

exits should be at least half the longest dimensions (usually measured on the diagonal) of the floor being served. If possible, more is better. Separation distance can be measured along the exit access corridor. This means that two exits built back-to-back could be kept a great distance apart by inserting a lengthy corridor between their doors [BOCA93 1006.4].

Optimizing Layout

Sprinklering Bonus

While distances between exits in an unsprinklered building must be at least half of the diagonal dimension of the space served, sprinklering brings a reduction to one-third from the Life Safety Code and one-quarter from BOCA.

Reach Of Exit

Exit distances can be reduced by extending exits, without changing the total distance from program space to vertical or horizontal exit [LSCH Figure 5-49 (b)].

Manipulating Common Paths

Within the allowed maximum travel distance to doors, fairly large interior distances may be acceptable when sufficient distance is provided between doors and alternate paths are made available.

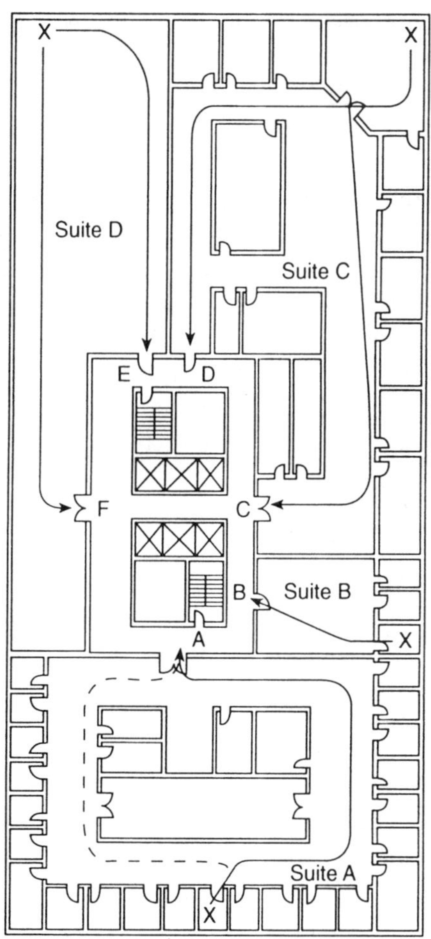

LSCH Figure 27-12 ©NFPA. Problematic egress paths. The path from Suite B is clearly a common path since there is no alternative. Although the path from Suite A provides a choice, both choices lead through the same point (the door). Even the paths from Suites C and D, while providing a choice and two doors, are nonetheless considered common paths of travel since their doors are not separated by at least one-half of their respective diagonal room dimensions.

Separation From Program Spaces

Whether or not they include aisle accessways or aisles, program spaces often must be separated from exit access routes by walls.

Unrated Partitions

Corridors that are useful or desirable but are not needed as exit access elements do not need to be fire rated, and therefore might be thought of simply as enclosed aisles. They can be designed using the requirements for aisles, with the added benefit that corridors are less likely to be blocked by furniture or personal items than are aisles.

Rated Walls and Doors

When separation is required, walls must meet the fire resistance required of fire partitions (See p. 243.) Doors in exit access elements are bound by the general design issues pertaining to doors, including rules for operations and hardware.

Outdoor corridors

One fairly effective way to separate exit access paths from program spaces, where weather permits, is by putting them outside. The advantage outdoor corridors provide is their extensive ventilation.

Outdoor corridors are often used in motels (especially in resort areas), low-income housing projects, and apartments, offices, and schools in warm climates. The Life Safety Code requires that they be more open than enclosed, and shaped to prevent smoke accumulation, so skirt panels hanging down from the ceiling are not generally a good idea unless somehow shaped to shed smoke. Design rules for outdoor corridors are similar to those for indoor exit accesses, including restrictions on dead ends and encroachments. They do have two unusual rules:

- Except at dead ends, the exterior walls of the building need not be rated.
- In regions that have four seasons, snow or ice buildup must be avoided. It is interesting to note that the Life Safety Code specifies a roof as the one acceptable way to do this , while the Life Safety Code Handbook is quick to urge readers not to interpret the rule too literally. On page 116, the handbook says, "Any method that accomplishes this goal and is acceptable to the local authority would be satisfactory. Examples of alternative methods are snow melting cables, heated pipes within the floor, or radiant heaters" [LSC94 5-5.3] .

Internal Conditions

The design parameters that influence the space within the exit access route are generally those that were noted under General Design Issues. Features unique to exit access elements are:

Maximum Length

Occupants can't stay in exit access elements for extended periods of time, simply because they are not required to be safe for extended periods of time. For example, a wheelchair-bound individual wouldn't want to wait in an exit access to be rescued, although an exit would be perfectly appropriate for just such an approach to egress.

Therefore, maximum lengths are mandated, varying substantially, depending on whether suppression systems are present.

Encroachments On Width

The primary problem with exit access elements is presented by door swings. People in the exit access moving toward exits could get hit by quickly opening doors from program spaces. It is for this situation that codes require designs that consider the doors' full range of motion, not simply position when open and when closed.

Slope (Ramps)

Rules for interior ramps used in means of egress are mostly the same as for outdoor ramps. (See p. 118.)

Optimizing Exit Access Path Rules

- **Length:** Use sprinklers.
- **Encroachments:** Recess all doors that lead into the exit access.
- **Exit signage:** Consider floor illumination strips for critical occupancies such as elementary schools, even though they're not yet required.

Areas Of Refuge

Background

Areas of refuge are conceptually somewhere between exit access elements and exits. They serve as temporary safe places during an emergency, allowing occupants to pause in the process of exiting. They are used in two distinctly different ways:

- as places where occupants can meet to assess their situation and decide what to do, and

- as places where the disabled and injured can rest or await rescue (also referred to in some codes as "accessible areas of refuge").

Although these two functions are quite different, the rules are the same, and the "area of refuge" terminology is common to both.

Not everyone in a building is capable of exiting on their own. While everyone in a building managed to enter it, conditions during normal entry differ in that speed is not a factor, and elevators are available. People who cannot use the regular exit path can wait in areas of refuge for rescue by the fire department.

Requirements

Location And Amenities

The two primary characteristics of areas of refuge are resistance to the effects of fire in general and smoke in particular, and direct access to exits. While access is a question of layout, fire resistance can be achieved in three ways [LSC94 A-5-1.2]. Areas of refuge can be provided by:

- Entire floors, even if undivided by rated barriers. This is sufficient for residential, mercantile, and business occupancies when the buildings are served throughout by sprinklers.
- Entire floors, when divided into two distinct areas by rated barriers. This is sufficient for assembly, educational, institutional, industrial, and storage occupancies when the buildings are served throughout by sprinklers.
- Individual rooms, when the rooms are separated by fire-resistant partitions or by location, in buildings that have partial or no sprinkler systems.

In addition, areas of refuge must have an intercom or other two-way communications system, so that occupants waiting there can make their presence known. The Life Safety Code applies this requirement to all projects [LSC94 5-2.12.2.5], BOCA applies it only to projects of over four stories [BOCA93 1007.5.3].

Spaces available for areas of refuge include the following, when properly designed:

- Elevator lobbies, which have the advantage of being easily accessible by fire fighters and paramedics by means of the elevators. Their disadvantage is that the elevator shafts and lobbies must be designed as smokeproof enclosures, except in sprinklered buildings and in cases where the area of refuge is over 1,000 SF and served by a horizontal exit [LSC94 5-2.12.2.4].
- Enlarged landings in exit stairs, which have the advantage of being even more protected than required, very visible due to the adjacency of equipment, such as standpipes and phone jacks, that fire fighters use, and already equipped with the required two-way communication system. Their disadvantage is that they require increased minimum stairs widths (55 in instead of 44 in) [LSC94 5-2.12.2.3].
- Other neighboring compartments accessible through horizontal exits, which have the advantage of being better protected than required for an area of refuge.

Keep in mind that none of this is necessary when buildings are fully sprinklered, since normal floor space can be used as an area of refuge [NFPA 5-2.12.3, BOCA93 1007.5].

Size

Areas of refuge may cover thousands of square feet. This is generally the case in fully sprinklered buildings. In unsprinklered buildings, where areas of refuge are frequently appended to exit enclosures, they are much smaller. Still, they must be large enough to hold 1 wheelchair (30 in x 48 in each) for every 200 occupants served by the area of refuge. This size must be in addition to any areas required for the exit path [BOCA93 1007.5.1].

For areas of refuge smaller than 1,000 SF, the project team must prove, by calculation or test, that conditions in the area will remain safe (codes call it "tenable") during a fire for at least 15 minutes.

Optimizing Areas of Refuge Rules

- As always, sprinklering yields huge benefits.
- Where areas of refuge are served by elevators, horizontal exits allow the elimination of elevator vestibules.
- The requirement for demonstrating "tenability" is waived when areas can be held above 1,000 SF.

EXITS

Layout

Background

Even on a ground floor, where exits can be as simple as doors in exterior walls, the fact that they must be linked to the building's circulation and at a fairly wide spacing means that their impact on design is substantial.

Quantity

Setting the number of exits is a very important decision, because its far-reaching implications for the construction budget and design efficiency (percentage of useable space). (See pp. 175–176.) A building with 4 required exits is very different from one with 3.

Rules don't stipulate types of exits. A requirement for 4 exits can be met with 4 stairs, or 2 direct exits, 1 stair, and 1 horizontal exit, at the discretion of the project team.

Spacing

As noted in relation to exit access layout, rules intend exits to be separated as far as reasonably practical to maximize their availability. Despite physical proximity, exits can still be made sufficiently distant by connecting them with circuitous exit access elements, as explored in the preceding section.

Relationship to General Circulation

Exits may be used as part of a building's general circulation, so long as their doors are rated and self-closing. Hold-opens are permitted for exits, and can help to make an exit seem more like a grand stair than a service element.

Optimizing Exit Layout Rules

QUANTITY: COMBINING STAIRS AND HORIZONTAL EXITS

When a large number of exits are required (more than 3) in a multistory building, one way to halve the num-

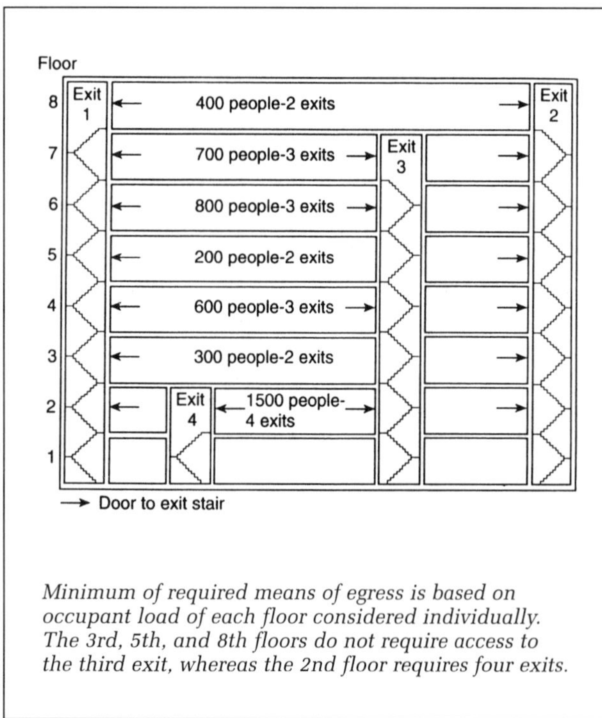

Minimum of required means of egress is based on occupant load of each floor considered individually. The 3rd, 5th, and 8th floors do not require access to the third exit, whereas the 2nd floor requires four exits.

LSCH Fig. 5-56—Exit riser diagram. ©NFPA.

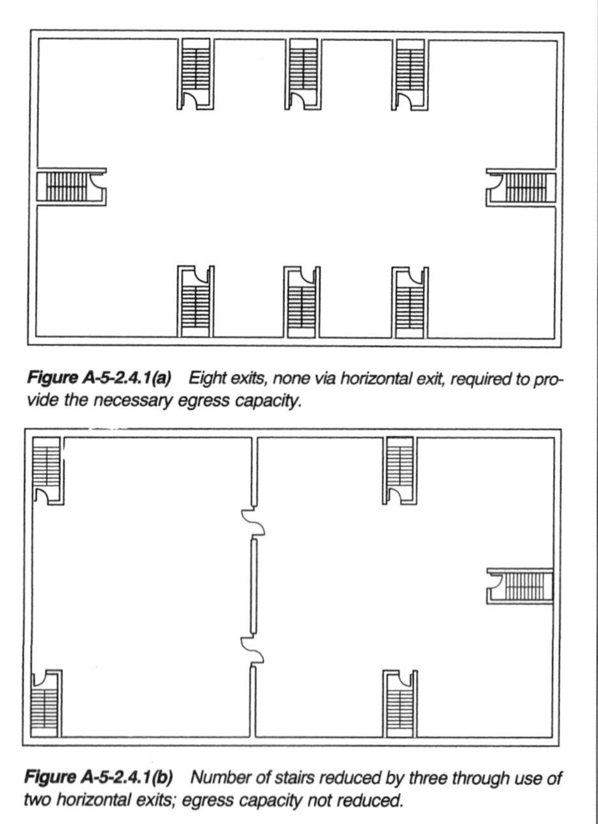

Figure A-5-2.4.1(a) Eight exits, none via horizontal exit, required to provide the necessary egress capacity.
Figure A-5-2.4.1(b) Number of stairs reduced by three through use of two horizontal exits; egress capacity not reduced.

LSCH Fig. A-5-2.4.1 (a, b)—Substitution of horizontal exit for two vertical exits. ©NFPA.

ber of actual exits while maintaining the same number of official exits is by dividing a building into compartments and substituting 1 horizontal exit for every 2 vertical exits after the first 2. In other words, 4 required exits can be provided by either:

- 2 vertical exits and 1 horizontal exit in a double compartment design.
- 4 vertical exits in a single compartment design.

Six can be provided by 2 vertical exits and 2 horizontal exits in a triple compartment design.

This is because exiting through a horizontal exit counts separately for each direction of travel. Moving to the east is an exit for those west of a separation wall, and moving to the west is an exit for those who are east. That counts as 2 exits. This strategy is limited by the rule that horizontal exits can't be used for more than half of the required exits from any given floor, except in medical facilities and prisons [LSC94 5-2.4].

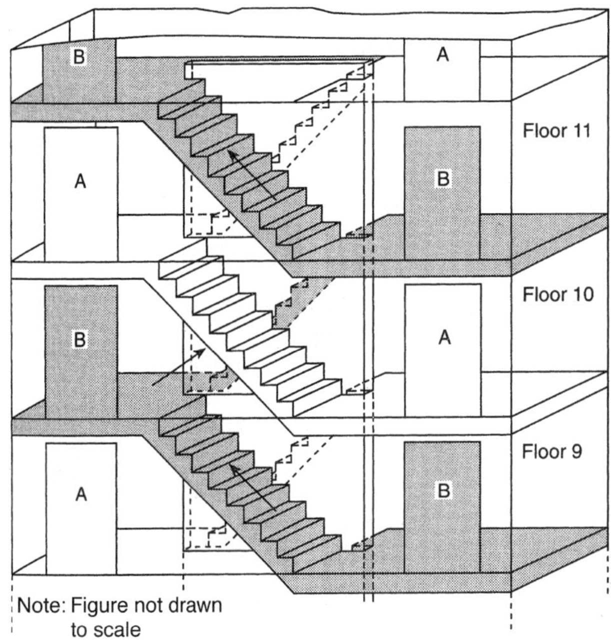

LSCH Fig. 5-60—Axonometric of scissor stairs. ©NFPA.

SPACING: SCISSOR STAIRS

A scissors stair is an arrangement whereby two completely independent stairways intertwine like a double helix and share a common set of enclosure walls. Each stair has a landing and door on each floor, but at opposite ends of the enclosure. They are the stair equivalent of double one-way ramps at a parking garage, the type stacked one above the other with cars on each ramp never seeing each other. The two stairs can serve in two possible roles:

- **As two separate exits:** Where two exits are required, one scissors stair may qualify for both. In this role, the two stairs sharing a common exit enclosure are nonetheless separated by a rated interior wall and rated stair assemblies so that smoke and fire cannot pass from one to the other.
- **As one exit of double width:** Where a very large occupant load must be handled, a scissors stair has twice the effective width of a conventional stair. In this case, both stairs can share the exit enclosure without internal separations, and are counted as a single exit.

Either way, the advantage of such arrangements is that they are more compact and that a single set of enclosing walls manages to protect 2 stairs.

What makes them work is the fact that each stair is designed with careful attention to headroom. In most scissor designs, each stair makes the trip from floor to floor in one run. This eliminates the intermediate landing which would otherwise create an obstacle to headroom. Calculating scissor stairs gets rather tricky. The following calculations explain the dimensional ranges possible:

- Because codes restrict runs to a maximum length of 12 ft vertically, scissor stair runs must be between the run maximum of 12 ft and a minimum comprised of headroom (6'-8") plus the thickness of the stair itself, which can be very minimal in the case of a steel plate stair.
- They therefore only work with landing-to-landing heights ranging from a minimum of about 8'-3" (the code required floor-to-floor height [7'-6"] plus structure), to 12'-0".
- Scissor stairs do not work for heights between 12'-0" and approximately 13'-4" (a double minimum run) and then they work again through 24'-0" (a double maximum run).

Vertical Exits

General

BACKGROUND

Stairs can be unnoticed, unused elements intended strictly for use during emergencies. They can also be

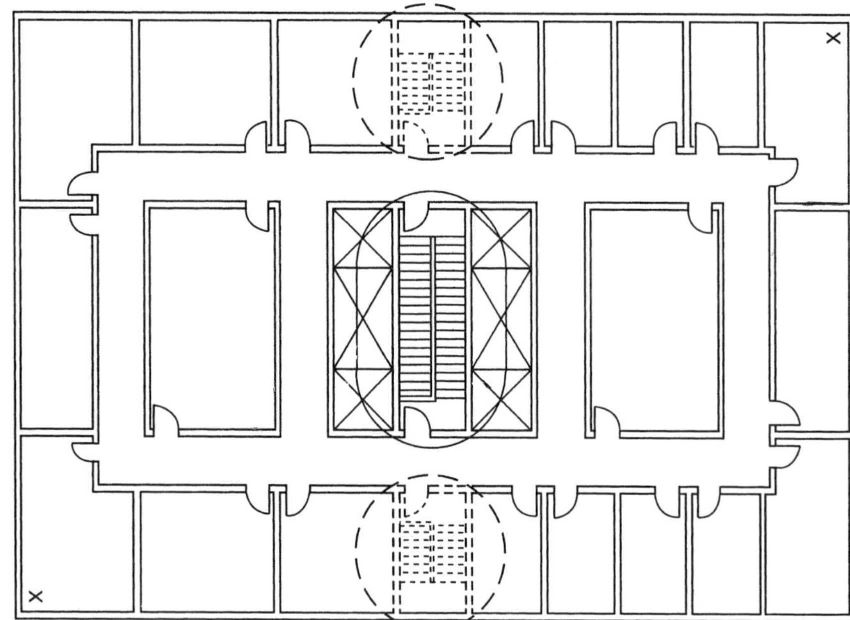

LSCH Fig. 5-61—Plan comparison of conventional stairs and scissors stairs. ©NFPA.

inspiring, well-worn, community-building, animating elements intended for use by a large proportion of a building's occupants on a daily basis. In either case, stairs can also serve as vertical exits in an exit path system. The descriptions in this section go a bit further than other sections to report code requirements. This is in recognition of their impact on the design options available.

Elevators, while generally taboo as exits, can serve as vertical exits in the context of areas of refuge [LSC94 5-2.12.2.4].

Shape

Many shapes of vertical exits can be designed, but not all are recognized by codes. The following list gives a brief evaluation of the requirements for each of the alternate exit designs [BOCA93 1014.6].

Forms That Are Generally Acceptable

- **Straight stairs**
  1. Step dimensions (4 in to 7 in riser heights, 11 in minimum tread depths). Common exceptions are made:
     - For houses and within individual apartments (4 in to 8¼ in risers, 9 in minimum treads)
     - To allow replacement of existing noncompliant stairs
     - For the steep stairs used in the balconies of theaters and stadiums to achieve appropriate lines of sight.
  2. All the steps in a flight must be the same size (within small tolerances), whatever that size turns out to be.
  3. Risers must be solid. Open risers are only permitted for dwelling units, and even then risers are limited to openings unable to pass a 4-in ball.
  4. Minimum grade change (12 in, requiring 2 risers). Smaller changes must be made with ramps [BOCA93 1005.6].
- **Circular stairs** are curved stairs whose total radiuses are at least three times their tread widths. There are no restrictions on where they may be used, and only minor ones on riser and tread design.
- **Ramps** are generally held to the egress requirements of corridors, with the addition of railings and occasional landings to reduce the danger of sustained downhill rolling in response to the sloping floor [BOCA93 1016].

Forms That Are Acceptable In Certain Situations

Those stair forms listed below are acceptable alternatives only in certain limited situations. Project teams that wish to consider any of these types as viable design solutions should check codes for further requirements.

- **Spiral stairs** are one type of curved stair. Whether they wrap around a center post or not, they have total radiuses of less than three times their tread widths [LSC94 5-2.2.2.7]. They are acceptable as exits only when they are at least 4'-4" diameter, with a maximum rise of 9½" and at least 6'-6" of headroom, and where they are:

1. In houses or individual apartments
2. Connecting a small mezzanine (less than 250 SF with no more than 5 occupants) to the floor below.
3. In small prison guard towers

Where a curved stair's total radius exceeds three times the width of its treads, the stair is not considered to be spiral, and may be used as part of an egress path.

- **Traditional winder stairs**, the ones with triangular treads frequently seen wrapped around newelposts in older houses, are now prohibited due to the hazard posed by the narrow end of the triangles. Modified winders, where the narrow end is at least 6 in deep, are permitted but only within a house or when serving a single dwelling unit [LSC94 5-2.2.2.8].
- **Alternating tread stairs**, which are hybrids halfway between stairs and ship's ladders, are generally allowed wherever ladders are allowed. They are also allowed as an alternative to spiral stairs at small mezzanine and prison towers, and for roof access. [LSC94 5-2.11]
- **Ladders** are prohibited from use as means of egress except as an alternative to straight or alternating tread stairs for unoccupiable low-pitch roofs of buildings more than three stories high. For occupiable roofs, the usual egress provisions relevant for the applicable use group must be followed [BOCA93 1026, LSC94 5-2.9].
- **Fire escapes** are prohibited from new buildings. They can be added to existing buildings if they do not use ladders, are accessed through doors rather than windows, and if the building is too close to its property lines to accommodate an exterior stair. Even then, at least half of the building's egress capacity must be handled by interior stairs [BOCA93 1025, LSC94 5-2.8].
- **Slidescapes** (straight or spiral people chutes) are allowed only for new hazardous uses and existing hazardous, educational, and institutional uses. They are an unusual form of exit device functioning like aircraft slides, but built as permanent fixtures, pitched between 24 and 42 degrees and enclosed by rated walls [BOCA93 1026, NFPA94 5-2.10].

An alternating tread stair. ©Lapeyre Stair Corp.

Separation From Exit Access Elements

WALLS

Under what conditions can exits remain unenclosed? Outdoor facilities that qualify as "smoke-protected assembly seating" (Use Group A-5), whether permanent stadiums or temporary parade review stands, do not require enclosure under some model codes [BOCA1014.11].

Owners of some historic buildings whose stairs were enclosed after original occupancy are now being given permission to demolish the enclosures and restore the buildings to their original open designs.

DOORS

Materials: See Chapter 10 (p. 249).

Locking: Clearly doors entering into exits must always be unlocked from the exit access side. The exit side, however, presents a conflict between the opposing

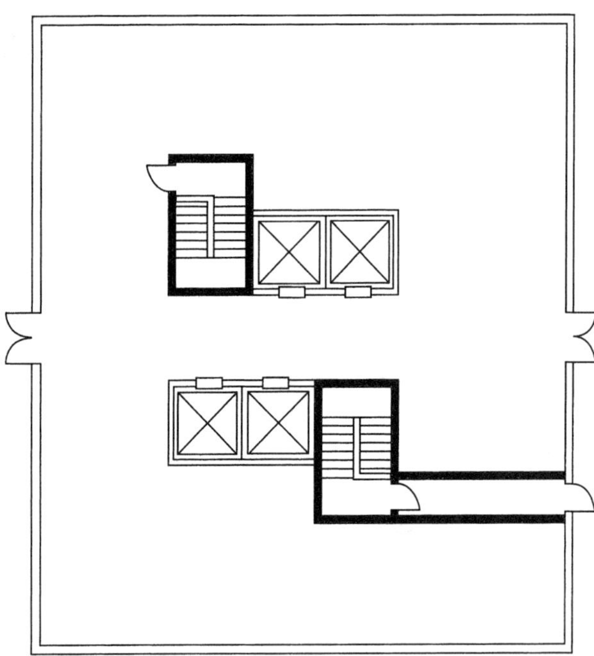

LSCH Fig. 5-49(a)—Exit Passageways make ground floor connections between exits and the outdoors. ©NFPA.

goals of safe egress and building security. If left unlocked, exits can allow unmonitored movement from floor to floor. If locked, people can get trapped in the unlikely event that the protections an exit affords are breached or otherwise compromised anywhere between the floor on which occupants entered and their discharge [BOCA93 1014.11.3, LSC94 A5.2.1.2.5]. (See also pp. 250–253.)

Offsetting Vertical Exits With Exit Passages

Exits may be shifted laterally through the use of exit passages. There is no material difference between an exit passage and an exit. The passage can be used as a connecting hallway so long as its enclosure is equal to those separating the rest of the exit from other parts of the building. In general, rules for the rest of the exit apply to exit passages.

Direction Of Travel

Exits may not reverse direction. A vertical exit leading downward from upper levels cannot go back upward, even for a few steps, before continuing down. Similarly, an exit from a basement level can not go down on its way up. Occupants get sufficiently confused while exiting without being confronted with reversals [NFPA App 5-7.1]. When an exit path needs to go up and down a few steps to get over, for example, a critically placed infrastructural element, move the element. This points to the importance of establishing the egress path very early in the design process.

Internal Conditions

Length

The total distance traveled within any one exit is unlimited. The only limitation listed pertains to the maximum length of vertical stair run (12 ft) or ramp run (30 in) between landings.

Awaiting Rescue

As explored in the previous section (Areas of Refuge), vertical exit enclosures can provide one of the safest places for disabled people to await rescue.

Fire Fighting Equipment

Requirements for standpipes are found in IBC 905. Phone jacks are required for connection to the Fire Command Station in buildings designed using high-rise strategies. (See pp. 187–188.)

Interior Finishes

Rules of thumb are of questionable help for this one. Finishes used in exits in sprinklered buildings vary by use group, ranging all the way from Class A to Class C to no restrictions (for use group U [Utility]). For uses that are given, and take the option of being built without sprinklers, finishes must generally be upgraded one Class.

Further, there are many highly specific exceptions allowed. For example, the IBC allows up to 1,000 SF of Class C paneling or wainscoting materials in ground floor lobbies of any use group. Wood ornamentation, trusses, and paneling is expressly permitted for churches, where finishes must otherwise be Class B. To maximize options, check the rules [IBC Table 803.4].

Horizontal Exits

Background

There is a substantial difference between horizontal exits in fire walls leading to other compartments (buildings or fire areas) and doors in smoke barriers leading to areas of refuge [LSC94 App 5-1.2]. (See pp. 180–185 for the significance of multiple compartments.)

Horizontal exits are openings that allow occupants to pass from one compartment to another without going outside. This is particularly useful in projects such as medical facilities and prisons, whose users can't or shouldn't simply evacuate the building.

Requirements

The simplest way to design with horizontal exits is to think of them as direct exits (see the next section).

People who come into a compartment through a horizontal exit must eventually get out of the compartment, since it, too, may not last forever if a fire is raging on the other side of the fire wall. Therefore, any horizontal exits used must be connected by means of exit access elements to other means of egress, such as stairs. For the same reason, as noted previously, only half the required exits from any given floor may be horizontal.

The opening itself and the doors which protect it must comply with the requirements of fire walls. (See pp. 241–243.)

Optimizing Horizontal Exit Rules

Walls that are being planned for functional or aesthetic goals can often be adapted for use as fire walls.

Direct Exits

A direct exit is a door in an exterior wall at grade. Under such conditions, stairs and areas or refuge are not necessary. They are particularly applicable to uses that have heavy ground level (or ground level and mezzanine) occupant loads such as convention centers, conference facilities, factories, and schools. By increasing the numbers of direct exits as needed, exit access corridors can be limited to the sizes programmed for normal functioning.

Exit Discharges and Safe Places

Exit Discharges

Exits must connect with safe places. At least half of them, with half of the egress capacity, must connect directly. Connecting via an exit passageway is considered to be a direct connection. The remaining half of the exits and egress capacity can connect indirectly, via spaces which qualify as exit discharges such as properly designed lobbies and rooftops [LSC94 5-7].

Arrangement

Interior Ground Floor ("Level of Exit") Discharges

Any interior space from which occupants can walk directly outside is potentially acceptable as an exit discharge if:

- Occupants can see the way to safety immediately upon coming into the ground floor space from the exit.
- It is designed as described in this section.

Exterior Discharges

It is not sufficient for exits to protect occupants only as far as the exterior walls. Codes generally take the position that occupants aren't fully discharged until they

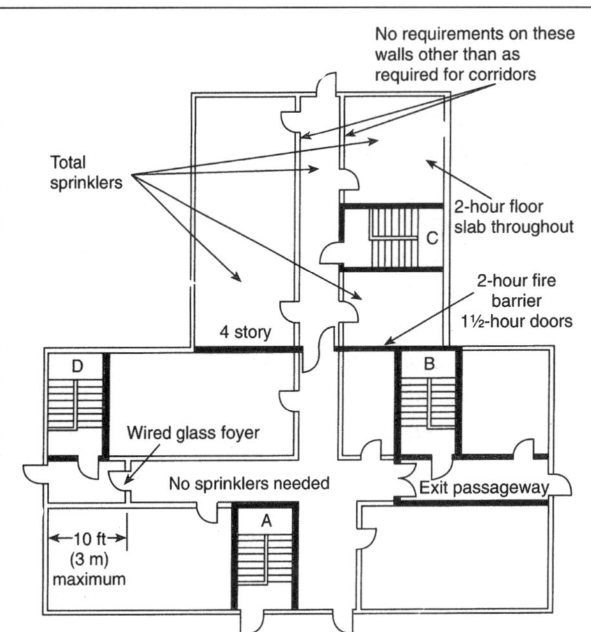

Figure 5-70. Exit discharge. The stairs provide four required exits from the upper floors. Exit stair A discharges directly outside. Exit stair B is also considered to discharge directly outside because its attached exit passageway affords protected passage to the door to the outside. The other two exit stairs, C and D, are permitted to discharge across the first floor because they do not constitute more than 50 percent of the number of exits from an upper floor or more than 50 percent of the egress capacity of any upper floor. Exit stair C discharges into an area on the discharge level that is sprinklered and separated from the remainder of the floor. The hourly fire resistance rating of the floor slab and the separating fire barrier are the same as required for the enclosure of exit stair C. Exit stair D discharges into a wired glass foyer in the nonsprinklered portion of the floor in accordance with the Exception to 5-7.2(b).

LSCH Fig. 5-70—Exit discharge. ©NFPA.

are in safe places (see below). Examples of exterior discharges include open ended courtyards, lawns, and surface parking lots. Exterior discharges must meet the same width requirements as interior discharges.

Rooftop Discharges

Although the model codes don't allow egress to roofs, the Life Safety Code does, under certain conditions. To qualify, roofs must be designed with at least two safe exit paths, so those who exit by climbing to the roof find an alternative to the stairs they just ascended. Helicopters are not considered to be sufficiently dependable to classify as an exit path [LSC 5-7.5].

Size And Separation

Ground Floor (Level of Exit) Discharges

In General: Clearly, discharges may be wider than the minimum required. Without wider discharges, columns could not occur in exit discharges because they would reduce actual width to below required width. Further, since it is not acceptable for the required portion of an exit path to narrow on its way out of a building, discharge width must be at least that of the exit leading to it.

Without Sprinklers: If the discharge is within 10 ft of the safe place, no more than 30 ft wide, and enclosed by wired glass, steel framed partitions or the equivalent, sprinklering is not required.

When Sprinklers are Provided: There are no size requirements other than the general width requirement. In terms of separation, however, the discharge will have to be separated from the rest of the ground floor and adjacent floors by assemblies rated as exit enclosures unless the adjacent areas are also sprinklered, or unless the adjacent ground floor and lower (but not higher) floors form a multistory atrium space with the discharge [LSC 5-7].

Rooftop Discharges

The roof underfoot must be built to the same fire resistance rating as the exit enclosure.

Optimizing Exit Discharge Rules

Use your imagination. So long as the design meets the 50 percent direct exit rule, there isn't much limit to what can be done.

Safe Places

These are the safe havens where exit paths end, beyond the limits of the affected building or fire area, and therefore beyond the scope of the design. The closest thing to a technical name for safe places is "public ways," but not all exit paths terminate in public ways. "Public ways" are spaces that are past a project's property lines and into adjacent streets or parks. The Life Safety Code accepts "other safe places" as well, an apparent reference to exterior areas of refuge allowed for jail and prison buildings (Endnote 17). Internal courtyards are not accepted as safe places since they, in turn, can be exited only by returning through the building.

It is the Code's intent that occupants be able to get to a safe place from which they can continue to move away from the burning building as necessary. At that safe point, Code requirements should cease to apply [LSCH p.35].

Of course, in some cases, the safe haven is another compartment of the same project, and therefore not beyond the scope of design for the overall project. Still, for the purposes of exit path design, it is probably best to think of, and design exiting for, each compartment individually.

Regardless of form or type, safe havens must be wide enough to accommodate the masses of evacuees arriving from all floors of a burning building, as well as fire department personnel, police, and bystanders. Codes offer no concrete way to account for these factors, but concerns for this issue could bring objections from a code official. Project teams should think about it long before applying for a building permit.

DOORS AND GATES

Not all doors are placed for the purpose of allowing people in and out. Some doors are more accurately described as shutters, or "openings protectives" in code terminology, and are intended to close off windows, edges of balconies, or other similar openings through which people don't pass.

For doors used by people it's simple: they've got to easily move out of the way when pushed. Easily, because some exiting people may be frail, due either to their general state of health or age, or due to fire-related causes such as exhaustion or injury. (See pp. 264–265.) Pushing is important because panicking people push whatever is in their way—they don't take the time to pull or slide when other people are behind them pushing. (For hardware, see p. 250.)

Background

Type Of Operation

Only a couple door types can easily be pushed.

- Hinged doors work, whether they use butt hinges, pivot hinges, "invisible" hinges or some other form of hinge [LSC94 5-2.1.4.1].
- Revolving doors work. This is not obvious, because a situation could occur where one person is pushing on one leaf (door panels), attempting to move the door counterclockwise, and another person is pushing on the opposite leaf, attempting to move it clockwise. In recognition of this, revolving doors are designed with "break-away" leaves that collapse when multiple leaves are pushed at once [LSC94 5.2.1.10].

Sliding doors (folding, pocket, patio, or industrial) and overhead doors (rolling or segmented), when built traditionally, cannot be pushed, and are prohibited by all building codes from use in a required exit path. They can be used when not part of an exit path, when used as openings protectives or when meeting the exceptions listed on p. 232.

Turnstiles and doors that must be locked for security purposes are also permitted under certain conditions.

Direction Of Operation

For doors that can be pushed, the push direction is critical. The flip side of pushing, pulling, requires more engagement (it doesn't happen by accident), more effort, and more time. Some of the most catastrophic fires in history, including the disastrous Triangle Shirtwaiste Company fire of 1911, attributed many deaths to the inability of panicking occupants to pull doors open, even when the doors were unlocked and operating properly.

As noted by NFPA [LSCH p.35], people who enter a building know at least one way out, simply by reversing their steps. Unfortunately, some design elements are not bidirectional; features that pose no obstacle to entry can become obstacles to egress. While direction of door

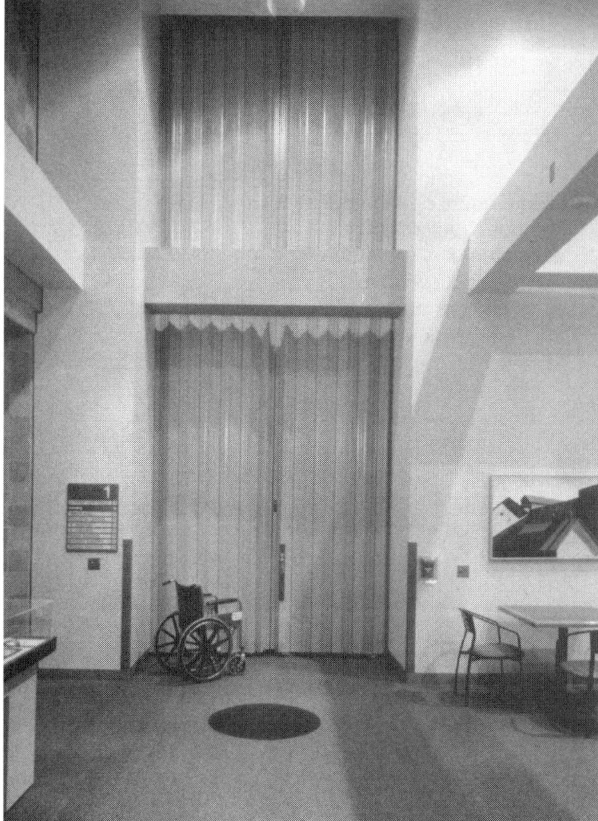

Push-slide door conforming with LSC94 5-2.1.14. ©Won-Door Corporation.

swing is very clearly one of these elements, it can easily be designed to avoid the problem of directionality by positioning it to pull on entering and push on exiting.

Visibility

It is important that people be able to find exit doors without having to look for them. Life safety rules prohibit designs in which egress doors blend into walls, are concealed behind drapes or other screens, or are intended to be undetectable.

Requirements

There are two requirements referring to door swing: the necessity of hinged doors over other types, and the direction in which they swing. Rules are written for doors that must be hinged, and that open by pushing in the direction of egress. Anything else requires an exception. The model codes offer different lists of exceptions, but include the following [IBC 1004.2.2, LSC94 5-2.1.4]:

1. Sliding, rolling, overhead, or power-operated doors are permitted when:

- In addition to being locked open during normal hours of operation, the condition is noted by a sign on the door, and they can be opened from the inside without particular knowledge or strength, and no more than half of the egress options use such doors [LSC exception #4].
- The doors serve an occupant load less than 10, and even then, only for private garages, and office, factory, and storage areas [IBC, LSC exception #6].
- The sliding doors are designed to operate when pushed in the direction of egress—that is, perpendicular to the direction in which they slide. They perform this neat trick with the help of electric motors. See the photos on p. 233 [IBC 1004.2.10, LSC94 exception #5].
- Used in prisons. This is allowed not because society doesn't care about the lives of prisoners, but because, as in health care facilities, there are trained staff members present at all times who can be expected to facilitate and control orderly exiting [IBC and LSC exception #1].
- The revolving door has collapsible-leaves, except in hazardous uses [IBC, LSC exception #7].
- The power-operated door can be operated manually in case of a power outage [IBC].

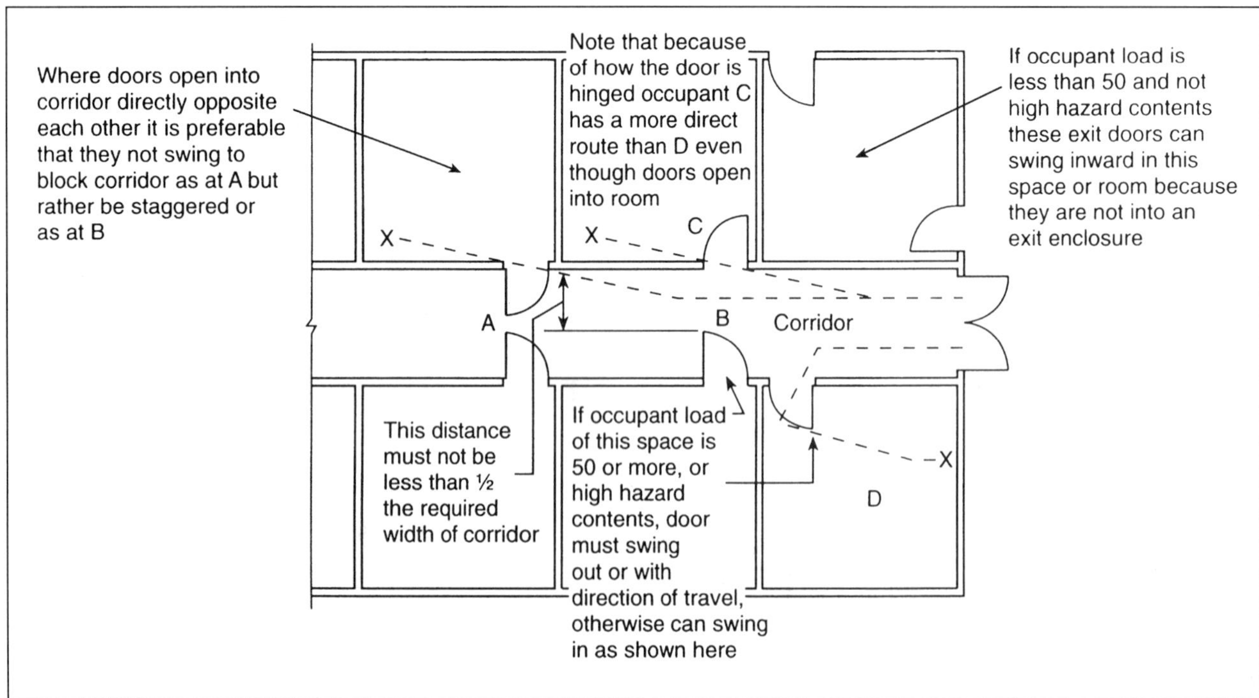

LSCH Fig. 5-12—Door Swing Considerations. ©NFPA.

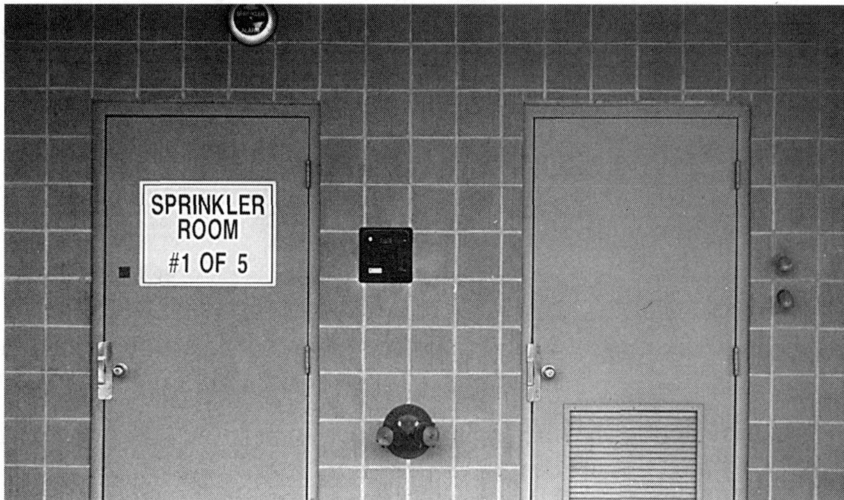

Required signage on doors. Photo by K. Wyllie.

2. Doors which pull in the direction of egress are permitted when:
   - Serving very low occupant loads. The IBC and LSC both require occupancies of less than 50. The LSC further precludes doors that lead directly into exits.
   - Serving as smoke barriers doors in existing health care facilities. This is permitted only because of the combination of hardship and the permanent presence of trained staff who can facilitate and control orderly exiting [LSC].
   - The reverse swing door is part of a pair whose other leaf pushes, and the condition is marked with a warning sign [LSC].
   - Serving as bidirectional horizontal exits in existing corridors less than 6 ft wide where double doors won't fit [LSC94 5-2.4.3.6].
   - In detention facilities [IBC].
   - When double doors are used in horizontal exits or smoke barriers, one leaf must swing in each direction.
   - Used within dwelling units [IBC and LSC].

For encroachments, see The System at the beginning of this Section [LSC94 5-2.1.4.3]. An exception is made for doors that intentionally block corridors, such as doors that are part of smoke barriers and horizontal exits.

Visibility: Although the intent is clear, there is some room for interpretation as to how much design blending constitutes concealment. The design moves that can make doors hard to find include use of Soss-style hinges, flush or recessed pulls, a matching of the finish between door and wall, or minimization of the door frame. Doors don't have to clash but they have to be obvious. The reverse is also true—something that isn't a door, such as a window, should not be designed to look like a door [LSC 5-2.1.1.2].

Optimizing Door and Gate Rules

There are two things worth avoiding in some design situations:

- Doors whose direction of swing inhibits normal use even though appropriate for egress. This is commonly the case for the disabled, who have trouble pulling, regardless of whether they are entering or exiting. It is also of frequent concern in hospitals, where doctors attending a patient on a gurney can't always stop to pull a door.
- The posting of required "Fire Command Station" or "Sprinkler Room" signs on otherwise unobtrusive or decorative doors visible in a public way.

One approach to avoiding either is to provide an exit alternative accessible from the push side of the door in question. While this may be an extreme measure, in some cases other design considerations may help justify the move.

One way to make a reverse swing work is to motorize it, so that users can open the door by pressing a switch or walking through a detector's sensor zone.

EMERGENCY: FIRE DEPARTMENTS

It isn't enough to have an efficient evacuation plan. There must also be some mechanism provided for alerting fire fighters to come to the scene and for fire department operations once they've arrived.

Automatic Detection And Notification By Building

Sprinkler systems detect fires early. They are actuated by heat sensitive capsules that shatter when their design temperature is reached. The capsules act as stoppers that keep water from flowing so long as they are there to hold it back. As soon as they shatter, water flows. It's simple. No human action or electrical power is unnecessary. Additionally, since codes require sprinkler heads to be located in concealed spaces as well as the more obvious ones, it is very difficult for a fire to escape detection.

Although sprinkler heads detect fires early and efficiently, they can't always do the whole job of controlling and extinguishing blazes. Thankfully, the same sprinkler system that knows enough to let loose with a shower of water at the right time can easily be made to notify authorities. Flow valves in sprinkler system water pipes can be fitted with electronic switches that send signals to annunciator panels within a building and to fire department personnel at a remote location.

While none of this has any noticeable impact on aesthetic design; the fact that such systems can eliminate the need for other, more visible, systems can prove to be of design benefit.

Manual Detection And Notification By Occupants

Pull stations are generally required to be located in the exit path. Obviously they have several drawbacks:

- Since they are so exposed, and are intended to be so, vandalism and false alarms take their toll.
- Exiting people may not take the time to activate the pull station, may be too shy to take such a bold step, or may not even notice that they are there.
- Pull stations can't be activated without people, and not all fires start when people are around.
- Critical time is lost while fires grow to the point that they are noticed by occupants and recognized for their severity. Fires starting in concealed spaces are even more elusive.

Fire Fighting And Rescue Operations

Measures required for high rise projects can just as efficiently help save lives in large low rise buildings (pp. 186–189).

Additionally, it is important to recognize that fire fighters use elevators to move about buildings during emergencies (see pp. 209–210).

10 DETAILING

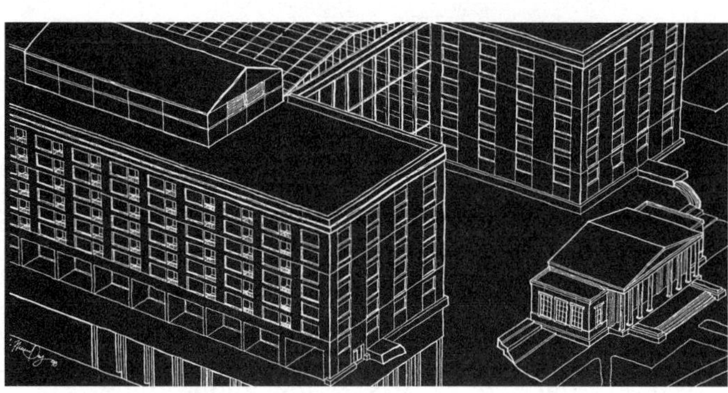

Selecting and Positioning Components

CONTEXT

Why This Issue Is Important

Eventually all conceptual designs must be thought through to the level of selecting individual materials and designing the details. A builder can't build without first ordering materials and once they arrive, understanding how they are to be prepared and assembled. The connection is, ultimately, what is recorded in contract documents at 1½ in scale. Details that are not developed by design professionals will be determined by workers in the field, who must make the building, part by part.

Why Regulatory Bodies Care

Materials and details are where forces exert themselves, where people physically touch the buildings they use, and where the authenticity of a historic design becomes manifest.

- Buildings that use inappropriate materials connected in inappropriate ways are firetraps. Materials must be chosen with regard to their combustibility and fire resistance, and assembled with regard to disaster containment.
- If forces are not properly handled, buildings tend to disassemble. Seams open, layers delaminate, fasteners unfasten. This can not only lead to premature failure of the building or parts of it but can also affect the users. Open seams can allow contaminants into the air; loose materials can fall on people or cause slippage.
- If people have difficulty handling the parts of a building physically, they slip and trip, have difficulty operating certain components, may not get the degree of physical support the building was intended to provide, and may not get certain messages the building was intended to communicate.
- When a building's details aren't historically appropriate, authenticity is difficult to determine, and a valuable cultural resource is compromised.

As far as zoning ordinances and historic preservation guidelines are concerned, the issue of authenticity is key, considering its effect on establishing the identity of a community and culture. For building codes and accessibility guidelines, the issues related to forces and people are critical, where health, safety, and welfare affect individuals.

How One Finds the Requirements

Model building codes dedicate several chapters to individual materials. Standards, such as those written by ASTM, are adopted by reference into many codes. Many manufacturers reference code requirements in their product literature, knowing that design professionals can only specify any materials known to meet code requirements for materials.

WALLS AND FLOORS

Context

Background

Ultimately it is walls and floors that define space. Rooms are not made by their contents but by their edges. In addition to providing physical, visual, and acoustic separation, walls and floors often also serve to keep people out of a fire's way and to reduce the quantity of available fuels through containment. (The discussion at pp. 180–185 explores the use of walls and floors to establish compartments when undivided areas are excessive or when the differing requirements of mixed uses must be met.)

This section, with its focus on detailing, explores wall and floor construction.

All building elements have spatial implications by the fact of their presence in or around built spaces. Rules generally classify them by other factors, which are:

- their importance as barriers to smoke and fire;
- their structural significance;
- their strictly spatial nature.

But owners and aesthetic designers (as opposed to technical designers) don't generally think in "code-speak." They tend to initially relate to building elements as shapers of space. This section uses that perspective for its exploration of rules.

Spatially, building elements are generally either one-dimensional (having length) or two-dimensional (having length and width). Of course, all elements have some thickness as well, but depth of assemblies is generally much smaller than length or width, and also more difficult for occupants to perceive. To simplify terminology, this section refers to one-dimensional elements as "poles" and two-dimensional ones as planes. "Poles" and planes combine to envelop and make three-dimensional space.

As design elements, building elements might be grouped as follows:

"Poles" (One-Dimensional)

Vertical

- Interior structural columns

Horizontal or Sloped

- Interior structural beams and trusses

Planes (Two-Dimensional)

Vertical Planes

Fire Enclosures

- Primary: exterior walls and building separations
- Secondary: fire separations
- Tertiary: fire partitions

Smoke Enclosures

- Smoke barriers

Strictly Spatial Enclosures

- Interior structural walls
- Interior partitions that are not required as barriers

Horizontal or Sloped Planes

- Floors
- Roofs

This chapter fits the codes' perspective to the owner and aesthetic designer's perspective.

Relatively speaking, although there are plenty of elements that need not be rated, there are also plenty that carry code-imposed restrictions. Project teams can't afford to ignore the regulatory demands made of these elements, or to postpone considering the effects of such demands while they develop designs that may prove to be noncompliant.

About Fire Resistance

Some of the worst fires of this century were made so by combustible materials used inside buildings. Codes have developed ways to assess the potential of various materials to contribute to conflagrations, thereby also determining their ability to resist them.

Characteristics are tested and regulated separately for wall and ceiling installations and for floor installations. Walls and ceilings tend to be much more of a problem, igniting and spreading fire with much greater rapidity than floors. Heat rises, and it is difficult to light a piece of paper with a match held above it.

Requirements

The matrix and descriptions that follow explore requirements for fire-resistive assemblies. (See also pp. 257–258.)

Optimizing Fire-Resistance Rules

The sprinkler exception works here as well. Floor finishes can be downgraded one category. Considering that there are only three to begin with, that can be a big change.

It is not clear that the following chart includes all possible categories of building element. For example, under which category would one include:

- Retaining walls beyond a building's exterior walls? Do codes simply ignore them as not being of concern due to their location? What is their position when the retaining walls are critical to the lateral load resistance of buildings' foundations?

- Roofs over underground vaults? Are such structures considered to be streets or sidewalks rather than roofs? Is there a difference in required fire resistance between an open garage structure and a parking lot located under a bridge?

These questions do not have fixed answers and may provide some degree of optimization room.

Where to Find the Requirements

Since the 1993 unified format, the model codes have put rules for achieving fire resistance in Chapter 7, "Fire Resistant Materials and Construction." The key to using the information contained in that chapter is found in matrices, contained in each model code, that correlate assemblies with hourly ratings. The proposed IBC has information for some assemblies in Table 601, but the user must hunt elsewhere in Chapters 3, 6, 7, and 10 for others. In BOCA, all assemblies are addressed by Table 602; in SBC they are split between Table 600 for structural elements and Table 700 for fire-resistive elements. A comparison of each of the model codes' assembly categorizations is found in Table 11-3.

Other codes are equally important. Many jurisdictions have adopted NFPA 101, the Life Safety Code, published by the National Fire Protection Association, to supplement their building codes. This code, whose format does not match the IBC standard, is concerned solely with fire protection and egress, with only a nod toward fire resistance. As such, it does not play a major role in regulating the design issues raised in this chapter.

"Poles"

One-dimensional elements cannot contain fire or smoke. The reason codes require fire ratings for them is to maintain structural integrity during fires until occupants can escape.

Vertical Structural Frame

This is a fancy way of describing columns, and includes both interior and exterior ones. Exterior bearing walls are classified as such, and described below as structural exterior walls.

Horizontal Structural Frame

The horizontal structural frame is defined as primary girders, trusses, and arches that are connected directly to the vertical structural frame (the columns). Structural members that can be supported by bearing walls are not considered to be part of a structural frame, as they carry distributed loads rather than concentrated ones. Where pilasters or embedded reinforcement are used to strengthen walls, they are defined as columns, and the horizontal members they carry as structural frames.

Structural Members Supporting Walls

This category is unique to BOCA. The elements it refers to are horizontal structural members used to provide midwall support to weak nonbearing rated walls. *Weak* means lacking sufficient strength to carry their own load to the floor below. This might be the case due either to excessively high slenderness ratios or insufficient compressive strength. In the former case, midwall supports resist buckling; in the latter they resist crushing.

Optimizing "Pole" Rules

Make sure they are structural: regardless of a component's form, if it isn't structurally significant, it need not be designed for fire resistance. Not all "poles" are columns or beams. Design elements appearing to be flying beams, pergolas, or even columns do not need to carry hourly ratings unless they are supporting other building components.

For the IBC, UBC, and SBC, the exemption from fire protection allowed for roof structures over volume spaces (see "Planes: Horizontal" later in this section) applies only to joists or other secondary structural members, not to the "structural frame" supporting them. It defines *structural frame* as beams, trusses, or girders ("poles") that connect directly to structural columns.

BOCA stands alone as specifically including the structural frame in this exemption.

Planes: Vertical

This section explores the rules for all walls, whether loadbearing or not, interior or exterior, and regardless of what they are enclosing. A primary source of information on required fire resistance is found in IBC Tables 601 (interior walls) and 602 (exterior walls).

As for maintaining the integrity of these walls at openings used for egress, see Chapter 9 (pp. 232–235). For other openings, such as windows, mechanical intakes and exhausts, and construction seams, see below (pp. 248–255).

General Concepts

Group	Subgroup	Class	Type	Description
POLES	VERT. & HORIZ.		Vertical Structural Frame	Columns
			Horizontal Structural Frame	Members connected directly to columns
			Mid-Wall Supports	Horizontal members supporting tall non-bearing rated walls
PLANES	EXT. VERT.	FIRE	Primary Fire Enclosures	Nonstructural Exterior Walls
				Structural Exterior Walls
	INTERIOR VERTICAL			Building Area Separations
			Secondary Fire Enclosures	Fire Area Separations
				Exit Separations
				Shaft & Hoistway Separations
				Other Separations
			Tertiary Fire Enclosures	Exit Access Separations
				"Tenant" Separations: Walls
		SMOKE	Smoke Enclosures	Smoke Barriers
		NONE	Strictly Spatial Enclosures	Structural Interior Walls
				Nonstructural Interior Walls
	INT./EXT. HORIZ.		Floor-Ceiling Assemblies	Basic Floor-Ceiling Assemblies (with associated secondary structure)
				"Tenant" Separation Floor-Ceiling Assemblies
			Basic Roofs (with associated secondary structure)	

General Comments

IBC Version (5/97) Table 601 & Misc.

Type	Part(s)
Structural Frame	Including columns, girders, trusses
See Table 602: Fire resistance for exterior walls based on fire separation distance	
Bearing walls	Exterior (see Table 602)
See Table 705.4: Building separation walls	
See Table 313.1.2, Article 706.3.4: Separation of uses, fire barrier walls	
See Article 1008.3: Vertical exit enclosures	
See Article 706.3.1: Enclosure of shafts	
See Table 1006.2.1: Corridor fire-resistance rating	
See Article 708.3: Fire partitions	
See Article 709.3: Smoke Barriers	
Bearing walls	Interior
Floor construction	Including supporting beams and joists
See Article 708.3: Fire partitions	
Roof construction	Including supporting beams and joists (adjusted for height to lowest member)

Only some requirements are in Table 601; for others (shown in gray), users must hunt.

BOCA Version Table 602

#	Type	Part(s)
8	Interior load-bearing columns, girders, trusses (other than roof trusses) and framing	Supporting more than one floor—Supporting one floor only or a roof only
9	Structural members supporting wall	
1	Exterior walls	Non-load-bearing
		Load-bearing
2	Fire walls and party walls	
		Mixed use and fire area separations
3	Fire separation assemblies	Fire enclosure of exits
		Shafts (other than exits) and elevator hoistways
		Other separation assemblies
4	Fire partitions	Exit access corridors
		Tenant spaces separations
6	Smoke barriers	
8	Interior loadbearing walls, and loadbearing partitions	Supporting more than one floor—Supporting one floor only or a roof only
7	Other nonloadbearing partitions	
10	Floor construction including beams	
5	Dwelling unit and guest room separations	
11	Roof construction including beams, trusses and framing, arches and roof deck	15' or less - 15' to 20' - 20' or more in height to lowest member

Very confusing regarding categorization of structural elements.

Comparison of assembly categories. Note that the CABO One and Two Family Dwelling Code has no such categories.

UBC Version		SBC Version Tables 600 & 700	
#	Type	Type	Part(s)
4	Structural Frame	COLUMNS [600]	
		BEAMS, GIRDERS, TRUSSES & ARCHES [600]	
3	Exterior Nonbearing Walls	EXTERIOR NONBEARING WALLS and gable ends of roof [600], EXTERIOR WALLS [700]	
1	Exterior Bearing Walls	EXTERIOR BEARING WALLS and gable ends of roof [600], EXTERIOR WALLS [700]	
5	Partitions—Permanent	PARTY AND FIRE WALLS [600], WALLS AND PARTITIONS [700]	Fire walls [700 only]
		OCCUPANCY SEPARATIONS [700]	
6	Shaft Enclosures	SHAFT ENCLOSURES (including stairways, exits & elevators) [700]	four or more stories—fewer than four stories —all refuse chutes [700]
5	Partitions—Permanent	WALLS AND PARTITIONS [700]	Exit access corridors [700]
		WALLS AND PARTITIONS [700]	Tenant spaces [700]
		WALLS AND PARTITIONS [700]	Smoke barriers [700]
2	Interior Bearing Walls	INTERIOR BEARING WALLS [600]	
5	Partitions—Permanent	WALLS AND PARTITIONS [700]	Within tenant spaces [700]
7	Floors—Ceilings/Floors	FLOORS AND FLOOR/CEILING CONSTRUCTIONS [600]	
8	Roofs-Ceilings/Roofs	ROOFS & ROOF/CEILING CONSTRUCTIONS [600]	

Small number of headings makes application uncertain.

Table suggests that hourly ratings do not vary depending on use group.

Primary Fire Enclosures

Nonstructural Exterior Walls

[IBC 704, BOCA93 705.2]

WHAT IS THEIR PURPOSE?

For buildings insufficiently distant from adjacent properties (see pp. 108–109.), exterior walls are critical to containing the spread of fire. Exterior walls prevent fires in adjacent buildings from attacking, a characteristic appreciated by investment teams. As an added bonus, such walls also keep fires within a building from expanding outward to neighboring buildings, a characteristic perhaps appreciated by other investment teams. Ratings for exterior walls are therefore related to distance, unlike any other building component.

HOW FIRE RESISTANT NEED THEY BE?

Nonstructural exterior walls must separate interior from exterior to a relative degree related to:

- The building's use: Higher separations are appropriate for uses that are either more likely to catch fire or that need to be better protected due to the nature of their occupancy. An example of the latter is an institutional building, whose occupants may be either incapacitated or forcibly restrained, and therefore at risk in case of fire.
- A building's distance from adjacent buildings, as required by IBC 602.
- A building's inherent fire resistance as a result of the materials and assemblies used. When there aren't many combustible components in buildings, fires in them are less likely to grow to sizes that threaten adjacent buildings.

For buildings facing streets, those public ways can provide sufficient fire separation distance. Side and often rear walls must compensate differently. Nonbearing exterior walls must provide separation equivalent to open space had it been available.

Structural Exterior Walls

In terms of separation required to compensate for proximity, there is no substantive difference between the construction required of structural exterior walls and nonstructural ones. Both require the minimum degrees of fire resistance appropriate to the distance separating adjacent buildings, described above.

The difference is this: When the minimum level of fire resistance appropriate to a wall's structural significance exceeds the level of fire resistance appropriate to

Fire wall after a fire. Note how little damage was incurred by the townhouses immediately adjacent to the one that burned completely. ©Gage Babcock & Associates.

Table 705.4.4
Allowable Protection Methods for Penetration Of Fire Resistant Assemblies[1]

Building Location			Conduit Pipes & Tubes		Ducts		Factory-Built Appliance Vents And Chimneys		Cables & Wires
			Penetrating Material Classification[4]						
			N	C	N	C	N	C	All
Penetration Through 3 Floors or More (Connecting 4 Stories or More)			A,E, C1[5], C2[5] or F	A or E	A	A	A	A	A or E
Penetration Through 2 Floors Maximum (Connecting 3 Stories Maximum)			A,B, C1[5], C2[5] E or F	A, B, E or F	A or B	A or B	A,B E, or F	A,B E, or F	A, B, E or F
Penetration Through 1 Floor Maximum (Connecting 2 Stories Maximum)	Monolithic Fire Rated Assembly		A,B, C2,E, or F	A,B, E, or F	A,B, or D2[3]	A,B, or D2[3]	A,B, or C2	A or B	A,B, C2, E, or F
	Fire Rated Assembly With Membrane Protection	Ceiling Membrane Penetrated	A,B, C1[2] E, or F	A,B, E, or F	A,B, D1[2] or F	A,B, D1[2] or F	A,B or C1[2]	A or B	A,B, C2 E or F
		Floor Membrane Penetrated	A,B, C2,E, or F	A,B E or F	A,B or D2[3]	A,B or D2[3]	A,B or C2	A or B	A,B, C2 E, or F
Penetration Throug Roof/Ceiling Assembly	Ceiling Membrane		A,B, C1[2] E, or F	A,B E or F	A,B D1[2]	A,B, D1[2]	A,B or C1[2]	A or B	A,B,C1[2] E or F
	Roof Membrane		No Requirement						
Penetration Through Roof Membrane Only									

C1 = Protection at the ceiling line.
C2 = Protection at the floor line.
D1 = Ceiling damper at the ceiling line.
D2 = Fire damper at the floor line.

SBC94 Table 705.4.4. ©SBCCI.

distance, that extra degree of fire resistance must be met. This is why structural exterior walls are listed separately from nonstructural ones in fire-resistance tables.

Building Separation Walls

WHAT IS THEIR PURPOSE?

As described in Chapter 7 (pp. 180–185), building separation walls separate building areas, for all purposes as effectively as do exterior walls and distance. They form one unbroken break from foundation to roof.

In some of the older codes still in use, the terms *fire wall* and *party wall* were used for building separation walls. *Fire wall* did not imply location, but *party wall* referred specifically to fire walls that were situated astride interior lot lines and serving two buildings. There were no rules for party walls beyond what was required for fire walls, and both terms have been subsumed in the IBC term *building separation wall.*

HOW FIRE RESISTANT NEED THEY BE?

Building separation walls must resist fire completely, in the sense that they must be structurally independent, with separate load-bearing conditions on either side of the enclosures. In many town-house fires, units burn to the ground without spreading the fire to immediately adjacent units simply because of the effectiveness of building separation walls.

RATED ASSEMBLIES

Building separation walls are the most fire-restrictive class of wall assemblies regulated by codes. The lowest acceptable rating is two hours, even for (or perhaps especially for) unprotected combustible classifications. Ratings of three and four hours are required for protected noncombustible construction. Like all ratings, these depend on the construction classification of the project [IBC 705, BOCA93 707].

MAINTAINING BARRIER INTEGRITY ACROSS OPENINGS

Doors are allowed in building separation walls and must carry fire protection ratings only slightly lower than the required ratings of the walls themselves [BOCA93 704.3, 708.2, 716].

Horizontal exits through building separation walls can be effectively hidden using automatic-closing doors, held open by electromagnetic releases and recessed into wall pockets. In noncombustible protected construction requiring doors rated for three hours, this option will work only with three-hour leaves. Although the option of two opposite-swinging one-and-a-half-hour leaves is generally allowed [BOCA93 716.3], it can't be used in this situation because some of the leaves would be swinging in the prohibited direction.

Secondary Fire Enclosures:

"Fire Barriers" [IBC 313, 706, and 707].

What Is Their Purpose?

Fire barriers can serve three purposes:

1. They are used to divide a project into "fire areas" when a design is compartmented.
2. They are also used to enclose areas that are notorious for their ability to spread fire when not separated, such as elevator hoistways, rubbish chutes, and duct shafts.
3. Last, they are used to protect those most needed of refuges—exits—maintaining their serviceability and life-saving integrity. This includes vertical exits and exit passages. Exit access corridors need only be enclosed by the less protective requirements of fire partitions.

How Fire Resistant Need They Be?

Fire barriers must contain fires for as long as people might still be in a building. They are the most fire-resistive elements of construction within a building's primary enclosure.

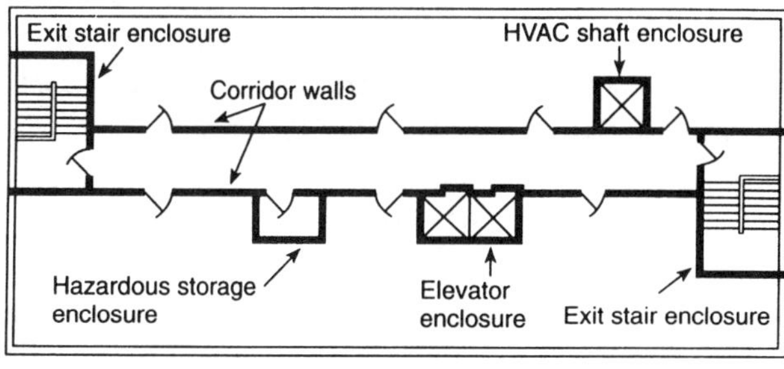

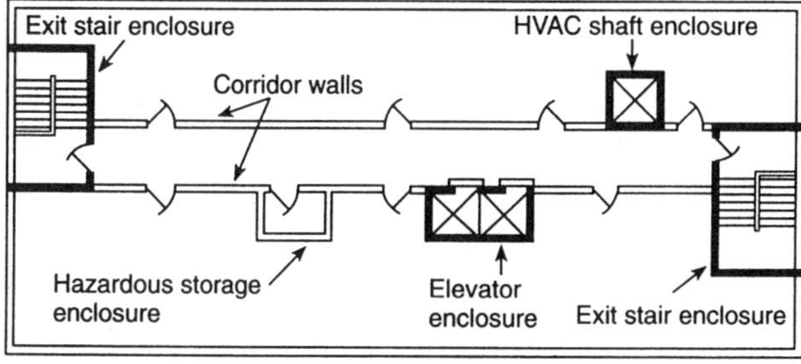

LSCH Figure 6-3: Fire barriers and rated structural supports. ©NFPA.

> **CASE STUDY**
>
> At Potomac Mills Mall in Dale City, Virginia, a single-story enclosed mall, Rippetau Architects intentionally stopped tenant and exit access partitions five feet short of the ceiling. This avoided compartmentation and created a huge smoke "corral" above head level to hold the smoke generated by any fire long enough for occupants to exit safely. By comparison, the traditional technique of enclosing spaces tends to trap smoke in the space that generates it, making that space much more dangerous even while sparing the occupants of other spaces.[48]

Tertiary Fire Enclosures:

"Fire Partitions" [IBC 708].

What Is Their Purpose?

Fire partitions are walls that enclose:

- **Exit access elements**
- **"Tenant" spaces**, areas controlled by single parties who may not be tenants in any conventional sense. They may be transient, short- or long-term, residential or commercial. "Tenant" spaces include conventional tenant spaces, dwelling units, and guest rooms. Rules control enclosure in both plan and section by regulating wall and floor assemblies. Further, they regulate walls between adjacent units as well as walls within units.

How Fire Resistant Need They Be?

Exit access enclosures are intended to hold heat and fire at bay long enough for occupants to get to exits, and provide a retrenchment space for firefighters.

Although walls between "tenant" spaces are no longer required to be built to the slab above, they still provide some degree of fire protection.

Smoke Enclosures:

"Smoke Barriers" [IBC 709].

What Is Their Purpose?

Smoke barriers are walls that enclose smoke to give people as much time to exit as possible before fire reaches them. They are found only in I-2 Incapacitated (hospital) and I-3 Restrained (prison) occupancies, where occupants must be brought to temporary refuge within a building. Since smoke isn't always sufficiently hot to activate fire detectors or fusible links, enclosing it can be a bit tricky.

How Fire Resistant Need They Be?

Smoke barriers are intended to keep smoke at bay. To meet this goal, their primary mission is to be airtight. Since the best gaskets won't do much good if a wall has burned down, codes require that smoke barriers have a one-hour fireresistance [IBC 709.3]. Where most fire dampers are activated by rising temperatures and are therefore not smoke resistant, dampers in smoke barriers are activated by smoke detectors.

Strictly Spatial Enclosures

Not Rated for Fire

Structural Interior Walls

Although they separate space, walls that are needed strictly as structural interior walls are not required to inhibit the passage of fire. They must, however, withstand it and are therefore addressed, if very minimally, in IBC Table 601 and Article 7.

Nonstructural Interior Walls

There are some elements found in buildings that are never required as barriers or as structure. Codes address them anyway. Although listed in some code tables, the only demands made on these elements relate to their use of fire-retardant treated wood [BOCA93 63.2, 2310].

Railings might be thought of as nonstructural interior walls. Their design, however, is highly regulated due to their importance in preventing falls. (See pp. 162–163.)

PLANES: HORIZONTAL

What Is Their Purpose?

Floors and roofs are critical to both fire containment and structural stability, slowing the spread of smoke and fire, holding together while occupants leave, and providing a secure surface for firefighters to walk on and under while they do their work. Roof assemblies have exterior exposure so must also resist fire spread to or from other buildings. If any function is compromised, a building's occupants, firefighters, and even occupants of nearby properties might be put at risk.

How Fire Resistant Need They Be?

Floor and roof assemblies are not as critical as vertical assemblies since they do not support other floors. When roof or floor assemblies fail structurally, only occupants and building components in the area of the failed assemblies are put at risk. Structurally failed floor assemblies jeopardize the spaces above them and below them, while structurally failed roof assemblies jeopardize only spaces below them. In comparison, when vertical components fail, everything above them that directly or indirectly rests on them is put at risk. This is not to underestimate the importance of horizontal assemblies, only to put them in context.

Rated Assemblies

There is some flexibility with respect to how much of a floor or roof assembly is considered in determining a rating. Between the floor of a room or a roof surface and the ceiling of the room below, there are, from bottom to top:

- (Optionally) A finish ceiling, mounted at some distance from the structural floor.
- (Optionally) The space above the ceiling, a plenum, which may contain such components as HVAC ducts, electrical wiring and light fixture housings, and sprinkler lines. It may or may not be rated, depending entirely on whether the selected assembly design included a plenum when it was originally tested. (See pp. 245–247.) If the design team added the plenum to the assembly design as published, it need not be rated.
- A structural slab and the secondary structural elements that connect it to the primary structure.
- The finish floor or roof membrane.

Some or all of these may be considered part of the rated assembly. Others, not integral to the rated portion, are unregulated, as they do not materially affect the performance of the rated portion [IBC 710].

Maintaining Barrier Integrity across Openings

FLOORS (AND ASSOCIATED SECONDARY STRUCTURE)

Floor openings are used for several reasons:

- To create multistory spaces, such as atriums. (See p. 202.)
- To allow unenclosed interfloor elements to pass, such as stairs, escalators, and inclined moving walks. For functional reasons, doors can rarely be used, so such openings must be protected while remaining open.
- To allow enclosed conveyances and infrastructural elements to pass, such as elevators, trash or laundry chutes, plumbing chases, and ductwork. (See p. 243.)

Roofs (and Associated Secondary Structure)

Openings in roofs are there for one of several reasons:

- Skylights.
- Smoke relief.
- Access to roof, generally from a stairwell.
- Penetrations for pipes and ducts connected to rooftop HVAC equipment.

(For both floors and roofs, see also pp. 248–255.)

Optimizing Horizontal Plane Rules

Raise the roof. When roof assemblies are sufficiently elevated from the floor immediately below, they are usually waived from fire-resistance requirements [IBC Table 601 note 3A]. The theory behind this "pardon" is that there is no fuel load sufficiently close to pose an immediate danger.

There are limits to this optimization, although good persuasive arguments based on the intent of this waiver have often resulted in the waivers' being granted. The limits include:

1 It doesn't apply to raised floors, only raised roofs.
2. Vertical or horizontal structural frames or bearing walls that support high roof elements would not be waived of their fire-resistance requirements.
3. When any part of a roof is less than the height required to waive the rules, the entire roof must fulfill required ratings. Examples of roofs that would not be permitted to omit the fire protection even for the portion above the "waiver line" would include the following:
   - A barrel vault 25 ft high at the peak but 19 ft high at the spring line.
   - The roof of a theater above the "waiver line" because the last two rows of a mezzanine in the back of the theater are too close to the roof to be waived.

Optimizing Wall and Floor Rules

Background

So far, none of the information discussed regarding fire resistance is of any direct use in building design. It isn't of much help to know that a particular wall in a particular location in a particular project must perform for at least three hours during a fire. How does that translate into a construction detail?

Assemblies need not be designed from scratch, and generally aren't. It is very difficult to try to predict the fire resistance characteristics of an assembly based on gut reactions to a paper design. Rules prohibit such experimental designs unless they have undergone thorough testing. Assemblies must be built as full-scale mock-ups, then ignited and observed. The entire process is described in ASTM E119 (see p. 63). Such testing can be done, but it is expensive and can delay a project.

Two alternatives to testing have been offered by some codes:

1. **Adopt a design:** The laboratories that test column, wall, and slab assemblies under controlled laboratory conditions publish their results. Lists and descriptions of successful assembly designs can be obtained:
   - Directly from publications of the testing agencies, which describe successful assemblies verbally and graphically.
   - From literature published by manufacturers of products used in successful assemblies. This literature does not usually describe the assemblies as thoroughly as the testing agencies' publications, but are often both verbal and graphic.
   - Through matrices found in some of the model codes that explain the required construction verbally but not graphically [IBC Tables 718.3A through C and UBC Tables 43A through C].

   Design teams are encouraged to use these pretested assembly designs, and can do so without fear of copyright infringement. Such designs must be used ver-

TABLE 3106.2A
TIME ASSIGNED TO WALLBOARD MEMBRANES[1,2]

Description of Finish	Time, Min.
⅜-in plywood bonded with exterior glue	5
15⁄32-in plywood bonded with exterior glue	10
19⁄32-in plywood bonded with exterior glue	15
⅜-in gypsum wallboard	10
½-in gypsum wallboard	15
⅝-in gypsum wallboard	30
½-in type X gypsum wallboard	25
⅝-in type X gypsum wallboard	40
Double ⅜-in gypsum wallboard	25
½ + ⅜-in gypsum wallboard	35
Double ½-in gypsum wallboard	40

1. These values apply only when membranes are installed on framing members which are spaced 16 inches o.c.
2. Gypsum wallboard installed over framing or furring shall be installed so that all edges are supported except ⅝-inch Type X gypsum wallboard may be installed horizontally with the horizontal joints staggered 24 inches each side and unsupported but finished.
3. On wood framed floor/ceiling assemblies, gypsum wallboard shall be installed with the log dimension perpendicular to framing members and shall have all joints finished.

TABLE 3106.2B
TIME ASSIGNED FOR CONTRIBUTION OF WOOD FRAME

Description of Frame	Time Assigned to Frame, Min.
Wood studs 16 inches o.c.	20
Wood floor and roof joists 16 inches o.c.	10

1. This table does not apply to studs or joists spaced more than 16″ o.c.
2. All studs shall be nominal 2x4 and all joists shall have a nominal thickness of at least 2″.
3. Allowable spans for joists shall be determined in accordance with 1706.3.1 and 1708.1.1.

TABLE 3106.2E
TIME ASSIGNED FOR ADDITIONAL PROTECTION

Description of Additional Protection	Fire Resistance, Min.
Add to the fire resistance rating of wood stud walls if the spaces between the studs are completely filled with glass fiber mineral wool batts weighing not less than 2 lb/cu ft (0.6 lb.sq ft of wall surface) or rockwool or slag mineral wool batts weighing not less than 3.3 lb/cu ft (1 lb/sq ft of wall surface).	15

Excerpts from SBC91—Table 3106.2A, B, & E. ©SBCCI.

batim as published, but may be concealed behind other construction if in conflict with aesthetic intent.

2. **Calculate a design:** As recently as 1991, the SBC contained a lengthy guide to the fire-resistance ratings of individual components of assemblies [SBC94 709, SBC91 701.2 and Chapter 31]. With this technique
   - design teams selected individual components. They then
   - where required, turned to matrices and graphs to find numeric values related to the type of material selected and the intended thickness. Next, they
   - used these values in formulas provided to find out how many minutes the individual component could have been expected to resist fire. Last, they
   - totaled the fire resistances of each component to get the expected fire resistance of the entire assembly.

It was a lot of work but was the only process of those allowed that permitted a combination of materials never before tested to be approved for use without fire testing a full-scale mock-up. Although it is not included in more recent versions of the SBC or the IBC, project teams using this approach might be able to demonstrate the performance of alternative assembly designs.

Testing Laboratories

The most widely used of testing lab publications is Underwriters Laboratories *Fire Resistance Directory.* This publication, currently filling two volumes, is a compilation of hundreds of tested assemblies, grouped by location within a building (walls, floor/ceiling assemblies, etc.), by materials used, and by hourly fire-resistance rating. Each assembly design is numbered, allowing architects to specify assembly designs by their UL numbers. Some jurisdictions prefer to see entire UL design drawings and descriptions inserted as is within a set of contract documents.

The designs included are very specific. Each design includes a specification for materials that have been tested in that assembly, and are therefore acceptable. Each also lists products by brand name and manufacturer. If the product desired isn't on the list, it can't be used.

Designs are listed once they pass their fire tests. The tests are usually commissioned by private concerns. Commonly, the concerns are manufacturers who want their products used more often. Less frequently, requests come from project teams who want to do something innovative. Due to cost, tests cannot reasonably be commissioned for smaller or even moderately sized projects.

Other laboratories that conduct full-scale fire tests include the University of California and Architectural Testing Labs. While assembly designs published by UL seem to be universally accepted, the recommendations of other labs may not be. Design teams are urged to confirm the acceptability of other labs' findings with local code officials before adopting particular assembly designs.

Manufacturers' Guidelines

Unlike the UL directory, manufacturers' lists usually include tests by other independent testing laboratories. Some of these guides report, in addition, the results of tests other than those for fire resistance, such as those for acoustic performance.

The IBC and UBC Matrices

With the IBC and UBC, everything needed is right in the code. Tables A through C describe a series of numbered assemblies. Design teams using these assemblies can specify them by their numbers. They differ from testing lab and manufacturer guides in that they are not illustrated and do not specify manufacturer. Table A describes structural fire protection, B describes wall and partition assemblies, and C describes floor and roof assemblies [IBC 718.3, UBC91 Chapter 43].

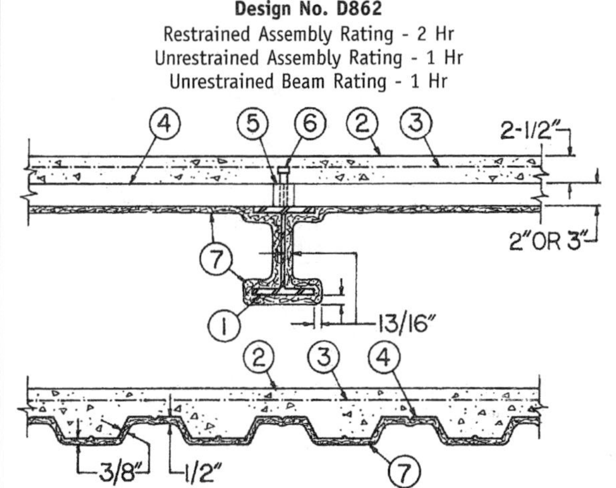

1. **Beam**—W8x21 min size.
2. **Lightweight Aggregate Concrete**—Lightweight concrete, expanded shale, clay or slate aggregate by rotary kiln method; 102 + or - 3 lb. per cu ft. unit weight, 4000 psi compressive strength, vibrated.
3. **Welded Wire Fabric**—Min. 6 x 6, W1.4 x W1.4.
4. **Steel Floor and Form Units***—Composite 2 or 3 in. deep, 24 or 36 in. wide, galvanized steel, all fluted units. Min gauge is 22 MSG. Units welded to supports not over 12 in. OC. Side joints of adjacent units button punched or welded together 36 in. OC.
   Gens Metals Inc.—24 or 36 in. wide Types LF2, LF3. Types LF2, LF3 units may be phos/ptd.
   United Steel Deck, Inc.—24 or 36 in. wide, Types LF2, LF3. Types LF2, LF3 may be phos/ptd.
   Vulcraft, Div. of Nucor Corp.—24 or 36 in. wide, Types 2VLI, 3VLI. Types 2VLI, 3VLI units may be phos/ptd.
   Wheeling Corrugating Co., Div. of Wheeling-Pittsburgh Steel Corp.—24 in. wide Types SB-200, -300; 24 or 36 in. wide Types P20LF, SB-P21LF, -P31LF. Types P20LF, SB-200, -300 may be phos/ptd.
5. **Joint Cover**—(Optional) 2 in. wide, pressure-sensitive cloth tape.
6. **Shear Connector**—Studs, 3/4 in. diam by 3-1/2 in. long, headed type or equivalent per AISC Specifications. Welded to the top flange of beam through the deck.
7. **Spray-Applied Fire Resistive Materials***—Applied by spraying with water in one coat to a final untamped thickness as shown above, to steel surfaces which are free of dirt, oil, or scale. Min average untamped density of 13 lb per cu ft. on beam and floor units with min individual of 11 lb per cu ft for Types II or DC/F. Min avg and min ind densities of 22 and 19 pcf, respectively, for Type HP. For method of density determination refer to Design Information Section.
   Isolatek International—Type D-C/F, HP, or II, Type EBS or Type X adhesive/sealer optional.

*Bearing the UL Classification Marking

An assembly as shown in the UL Fire Resistance Directory. ©UL.

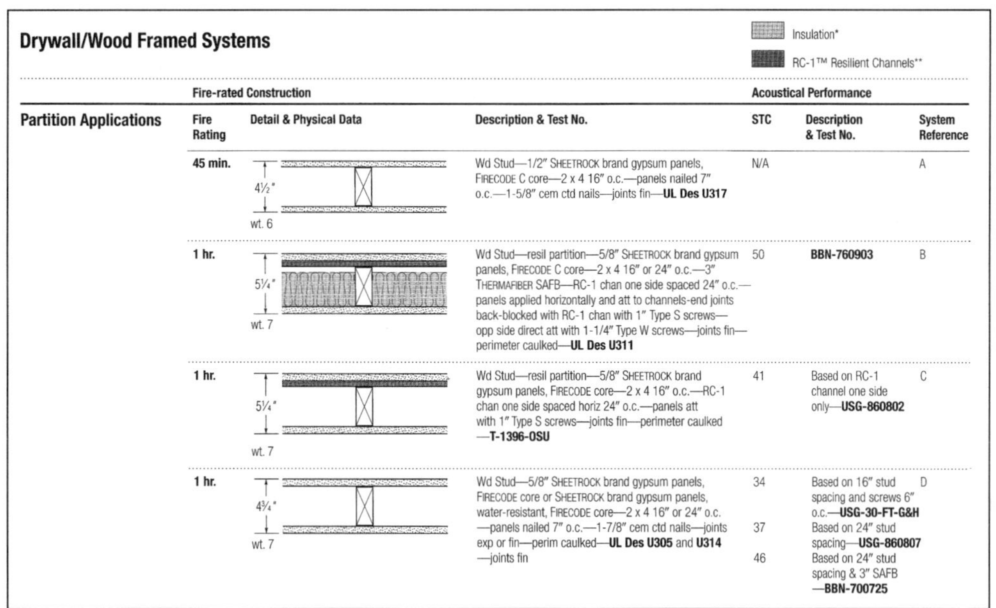

Drywall/Wood Framed Systems

Insulation*
RC-1™ Resilient Channels**

Partition Applications	Fire-rated Construction			Acoustical Performance		
	Fire Rating	**Detail & Physical Data**	**Description & Test No.**	**STC**	**Description & Test No.**	**System Reference**
	45 min.	4½" wt. 6	Wd Stud—1/2" SHEETROCK brand gypsum panels, FIRECODE C core—2 x 4 16" o.c.—panels nailed 7" o.c.—1-5/8" cem ctd nails—joints fin—**UL Des U317**	N/A		A
	1 hr.	5¼" wt. 7	Wd Stud—resil partition—5/8" SHEETROCK brand gypsum panels, FIRECODE C core—2 x 4 16" or 24" o.c.—3" THERMAFIBER SAFB—RC-1 chan one side spaced 24" o.c.—panels applied horizontally and att to channels-end joints back-blocked with RC-1 chan with 1" Type S screws—opp side direct att with 1-1/4" Type W screws—joints fin—perimeter caulked—**UL Des U311**	50	**BBN-760903**	B
	1 hr.	5¼" wt. 7	Wd Stud—resil partition—5/8" SHEETROCK brand gypsum panels, FIRECODE core—2 x 4 16" o.c.—RC-1 chan one side spaced horiz 24" o.c.—panels att with 1" Type S screws—joints fin—perimeter caulked —**T-1396-OSU**	41	Based on RC-1 channel one side only—**USG-860802**	C
	1 hr.	4¾" wt. 7	Wd Stud—5/8" SHEETROCK brand gypsum panels, FIRECODE core or SHEETROCK brand gypsum panels, water-resistant, FIRECODE core—2 x 4 16" or 24" o.c. —panels nailed 7" o.c.—1-7/8" cem ctd nails—joints exp or fin—perim caulked—**UL Des U305** and **U314** —joints fin	34 37 46	Based on 16" stud spacing and screws 6" o.c.—**USG-30-FT-G&H** Based on 24" stud spacing—**USG-860807** Based on 24" stud spacing & 3" SAFB —**BBN-700725**	D

Assemblies as described in the Selector, published by United States Gypsum. ©USG.

TABLE 7 - RATED FIRE-RESISTIVE PERIODS FOR VARIOUS WALLS AND PARTITIONS [a, 1]						
			MINIMUM FINISHED THICKNESS FACE-TO-FACE [2] (inches) 25.4 for mm			
MATERIAL	ITEM NUMBER	CONSTRUCTION	4 Hr.	3 Hr.	2 Hr.	1 Hr.
Concrete masonry units	6-1.1[11, 12]	Expanded slag or pumice	4.7	4.0	3.2	2.1
	6-1.2[11, 12]	Expanded clay, shale or slate	5.1	4.4	3.6	2.6
Noncombustible studs— interior partition with plaster each side	14-1.1	3¼″ (82 mm) by 0.044 inch (1.12 mm) (No. 18 carbon sheet steel gage) steel studs spaced 24″ (610 mm) on center. ⅝″ (15.9 mm) gypsum plaster on metal lath each side mixed 1:2 by weight, gypsum to sand aggregate.				4¾[6]
	14-1.2	3⅝″ (92 mm) by 0.055 inch (1.4 mm) (No. 16 carbon sheet steel gage) approved nailable[16] studs spaced 24″ (610 mm) on center ⅝″ (15.9 mm) neat gypsum wood fibered plaster each side over ⅜″ (9.5 mm) rib metal lath nailed to studs with 6d common nails, 8″ (203 mm) on center. Nails driven 1¼″ (32 mm) and bent over.			5⅝	

An example of the assemblies described in the IBC and UBC, ©ICC. The referenced footnotes (a, 1, 2, 6, 11, 12, and 16) allow for the acceptance of other recognized published designs; allow substitution of equivalent staples for nails; limit some assemblies to nonbearing situations; provide a methodology for calculating masonry thickness to compensate for plastering, core percentages, and other factors; and explain how steel studs can be made nailable.

OPENINGS IN WALLS AND FLOORS

GENERAL ISSUES

Context

Fire must be prevented from traveling through openings in or penetrations through assemblies, and it must be prevented from compromising the performance of critical building components.

Openings are of several types and can be grouped according to their purpose and what they are filled with:

- Openings in walls for occupant circulation or to admit light, air, or view, and filled with doors and windows
- Openings in walls or floors for duct or chase penetrations, filled with shutters and dampers (openings protectives)
- Openings in floors, intended to create multistory spaces or to allow an escalator through, and left open
- Openings in floors or walls for pipe penetrations, enclosed by partitions
- Construction gaps in walls or floors, including cracks and joints

Shaft openings, those in horizontal separations, are of much more concern than openings in vertical separations. The latter allow smoke and fire to spread across a single floor. Except in the case of building area separations, fires that are confined to individual floors can be brought under control relatively efficiently. In contrast, openings in horizontal separations are of more strategic importance. Fires generally don't move quite as fast horizontally as vertically. This is because of "chimney effects," whereby rising columns of hot air create negative air pressures that suck smoke and flame upward. Openings in horizontal separations can cause chimney effects and must therefore be carefully protected.

Leaving Openings Open

Openings can be left open where doing so is expressly allowed by rules. A complete list of such situations is found in IBC 707.2, and includes:

- Covered mall buildings (see pp. 189–190)
- Locations where escalators pass through floors (see p. 211)
- Mezzanines (see pp. 203–204)
- Unenclosed openings four or fewer stories high in individual residential units
- Parking garage ramps
- Unenclosed openings connecting only two stories that are outside required egress paths and not concealed within construction
- Properly packed piping or duct penetrations

General Requirements and Implications

Maximum Size, Number of Openings, or Percentage of the Required Separation

For separations requiring substantial fire-resistance ratings, rules limit the size of individual openings. Limited openings are allowed even in building separation walls and fire barriers, so long as they are protected [IBC 705.8 and 706.6]. The total width of openings in building separation walls and fire barriers must be held below a maximum percentage (25 percent) of the wall's length, apparently without exception. The size of each individual opening is unrestricted if the whole project is sprinklered, but strictly limited (120 SF) if it isn't [IBC 705.8]. BOCA allows an intermediate level of restriction for first-floor spaces when at least those floors are sprinklered [BOCA93 708]. Openings in fire partitions and smoke barriers are not limited other than being required to use openings protectives.

Some penetrations are also allowed for building components that have to get from one side of a building separation wall or fire barrier to the other, such as ducts, pipes, and wires, so long as both the penetrating component and the space between it and the fire wall are designed to minimize discontinuity [BOCA93 707.7, 707.8].

DOORS

Doors are an obvious choice for openings protection, and when used in that capacity must be specially designed to maintain the fire resistance of the separation assembly.

Material Selection

Primary Material

Doors in rated partitions are required to be fire resistant. Manufacturers use a system of labels developed by Underwriters Laboratories to certify the ratings of their doors. (See p. 252.) Fire performance depends mostly on the materials used for the cores of the doors, so does not have much impact on design other than through budgetary concerns. Wood doors carry only the lower hourly ratings, but manufacturers produce some reasonably effective wood veneered doors that carry the higher ratings.

Glazing Material

Fires love openings, and windows in doors can unfortunately provide them. In contrast, windows can also be useful in alerting people on either side of the door to traffic and danger coming their way. Codes have worked with this issue by allowing windows based on two considerations:

- The size of the opening, with permitted openings getting smaller as ratings get longer [IBC 714.2.4].
- The glazing material. For years, wire glass was the only material accepted. Now intumescent glass that works without wires has joined the list of options. Acrylics (such as Plexiglass®) and polycarbonates (such as Lexan®) are prohibited in any rated door [IBC 714.2.4 and 714.4].

Finishes

Codes require doors used in means of egress to look like doors. Designing doors to blend in with the walls around them may create a sleek look, but it can also lead to lost lives. People fleeing fires must be able to find their way out. The *Life Safety Code* specifically cites the dangers of mirrored doors, doors that are overhung by banners or fabrics, and doors that are paneled or trimmed to visually "disappear" even when in plain view, whether part of the initial design or added later as a function of a building's operations and maintenance [NFPA 5-5.2.2 and 31-1.2.3].

Optimizing Materials Selection Rules

If a partition doesn't need to be rated, neither do doors within it. Where rated doors would conflict with aesthetic or functional intent, look for ways to eliminate the rated partition requirement, possibly by reducing the overall construction classification (although by the time design teams are selecting door materials, it's generally too late to rethink classification) or by using the techniques noted at the beginning of this section for leaving openings open.

Hardware

Background

The term *hardware* refers to any piece of equipment attached to a door or the wall opening it occupies, either fixed in place or operable. Proper use of hardware can help resist the passage of smoke.

At the Door

Locksets

Lock mechanisms: A locked door in an egress path that can't effortlessly be unlocked is a death trap. This fact doesn't generally alleviate the security concerns of occupants. Rules allow doors in an egress path to be lockable when one of the following measures is taken:

- Specify locksets that lock only in the direction of ingress, not egress. For small occupant loads, this can be accomplished by specifying the simple type of lockset commonly used in classrooms, bathrooms, and closets.
- For larger occupant loads, panic hardware is required. Panic bars release locks without requiring the use of hands, or even consciousness.

  Where security from the inside is also a concern, locks and panic hardware can be wired to alarms or indicator lights to inform security personnel of their use. This doesn't keep them locked, but does allow for monitoring [LSCH p. 48].
- Use "captive key" hardware that retains the key in the lock unless it is in the unlocked position [LSC94 5-2.1.5.1 #3)].

Knobs and levers: Protruding knobs are invitations to personal injury, particularly in the case of recessed doors that are normally left open and designed to be unusually inconspicuous. For doors that are frequently closed, push plates are usually sufficient for ease of maintenance. (See p. 253.)

Making Doors Close

Fire-rated doors are useless unless closed. All doors in rated partitions are required to be fitted with closers. Closers are mechanical devices that exert pressure on open doors to cause them to swing shut. The force needed to open them is limited by rules. In addition, special "delayed action" closers that wait a few moments before swinging closed are available for use in accessible egress paths, giving occupants more time to get through doors after getting them open [IBC 714.2.5, LSC94 5-2.1.8].

The most common type of closer consists of a fairly heavy spring mechanism packaged in a metal box attached to the top of the door and connected to the door's frame with a metal arm. The spring pulls the door and frame together. Unfortunately, it is fairly large, and because it is surface mounted, usually looks like an afterthought. Where the design intent is to minimize the visibility of such mechanisms, alternative closer models are available in three forms:

- Recessed closers may be installed in the head frame, with an arm below that attaches to the top edge of the door. These are quite undetectable but can handle doors of medium or light weight only.

Function/Application	(ANSI) List Nos. Mortise Locksets	(ANSI) List Nos.Cylindrical Locksets	(ANSI) List Nos. Mono-Locks
Passage Latch • For doors that do not require locking. • Either knob operates the latchbolt at all times.	(F01) Knob Lever 8701 8701FL 8601 8601FL 4601 4601FL	(F75) 5401 5501 5401L 5301	(F36) 6201
Privacy Lock • Latchbolt by knob either side. • Anti-Panic operation. Turning inside knob retracts latchbolt or latchbolt and deadbolt simultaneously. • Emergency release from outside.	(F19) Knob Lever 8702 8702FL 8602 8602FL 4602 4602FL • Deadbolt by Thumbturn inside. • When the deadbolt is projected outside knob is automatically made rigid.	(F76) 5402 5502 5402L* 5302 • Button automatically releases when inside knob is turned or door is closed.	(F37) 6202
Patio Lock • For exit doors with limited entry. • Deadlocking latchbolt. • Either knob operates latchbolt unless knob is locked by pushbutton inside. • Button automatically releases when inside knob is turned or door is closed. • Inside knob always active.		(F77) 5403 5503 5403L* 5303	
Storeroom or Closet Lock • For use on storeroom, utility, exit doors. • Deadlocking latchbolt. • Outside knob rigid at all times. • Latchbolt by key outside, knob inside.	(F07) Knob Lever 8705 8705FL 8605 8605FL 4605 4605FL	(F86) 5405 5505 5405L* 5305	(F44) 6205

A partial list of possible lockset combinations. ©Yale Security, Inc.

A door fitted with a panic device connected to surface-mounted flushbolts, a surface closer, and an electromagnet hold-open. Photo by the author.

- Recessed closers may be installed within the thickness of the door itself, with an arm that attaches to the frame above. This type of closer conceals its spring mechanism but not its arm.
- The most inconspicuous solution is a floor closer, recessed in a concrete floor and concealed by a cover plate that is flush with the floor or threshold. This type of closer works with doors mounted on pivots but not those on hinges. The closer works by rotating the pivot's spindle. It is also the most durable type of closer, usually made of cast iron, and quite expensive. Finally, while floor closers for medium and lightweight doors can be as thin as 1½ or 2 in, floor closers for heavy doors (including exterior ones) may be up to 4 in thick and can be installed only in fairly thick floor structures. Still, for heavy doors or where concealment or ease of maintenance is a priority, they meet the intent of the rules well.

Holding Doors Open

Most doors can normally be left in the closed position. Such doors do not need to be released, as there is nothing holding them open. Codes refer to such doors as simply "self-closing."

In some cases, it is convenient to prop open a door. In other cases, doors that are programmatically unnecessary or even inconvenient may be installed specifically for the purpose of containing fire. When occupants want to keep a door open, they usually find a way. Too often, people use doorstops to hold self-closing doors open. Then during a fire, when everyone forgets about the doorstops, the doors remain open and fail in their critical role as barriers.

Such doors can be released by automatic actuators. "Automatic-closing" doors are held open by means of electromagnets, relatively small devices that act as magnets as long as they are being powered by electricity. Fusible links, although permitted by some past codes, are no longer permitted for this function. They don't release until they melt, and they often don't melt in time to prevent the passage of smoke. Conversely, electromagnets allow doors to self-close when power is turned off by signals from fire-detection systems, swinging shut in response to the force exerted by closers.

At the Door Frame

Frame Stops

Single doors seal against the passage of smoke and fire through the use of frame stops (not to be confused with doorstops), with supplemental gasketing when necessary. Double doors can use frame stops except for the joint between the paired doors, where an astragal and a coordinator are needed.

Astragals

Astragals are interlocks on the meeting edges of double doors that prevent smoke from blowing through when the doors are closed, so they take the place of a single door's strike-side stop. They may be built into the edges through rebating, or they may be surface-applied.

Coordinators

Where closers with astragals are used, coordinators are needed. They see to it that the door to which the astragal is mounted stays open long enough to let the other panel swing shut first. This way, the astragal doesn't prevent the doors from closing.

At the Floor

Doors must be designed to seal against floors. The simplest approach would be to leave only a hairline space between the door bottom and the floor, but there are two problems with this.

- Construction tolerances are a bit too loose for this to work. If floors aren't perfectly level and clean, doors bind and won't close at all.
- Gaps may be required for air transfer, to let stale air out of closed rooms and into return registers located in corridors.

Therefore in most situations a gap must be allowed, but must be capable of being sealed when the door is closed. This can be done in several ways, some more visible and tactile than others.

Sweeps

These follow a very simple idea. There is only one moving part, but it's constantly moving. Sweeps are like brooms attached to the bottoms of doors that drag along the floor as the doors are opened and closed. They tend to wear out quickly, make a faint sound when operating, and are usually visible since they are attached to the door face at the bottom edge. In addition, they don't seal as positively as the seals found on thresholds and automatic door bottoms, whose seals are pressed between parallel meeting edges.

Thresholds

Thresholds take the opposite approach to sweeps. They don't move, but simply sit on the floor under the

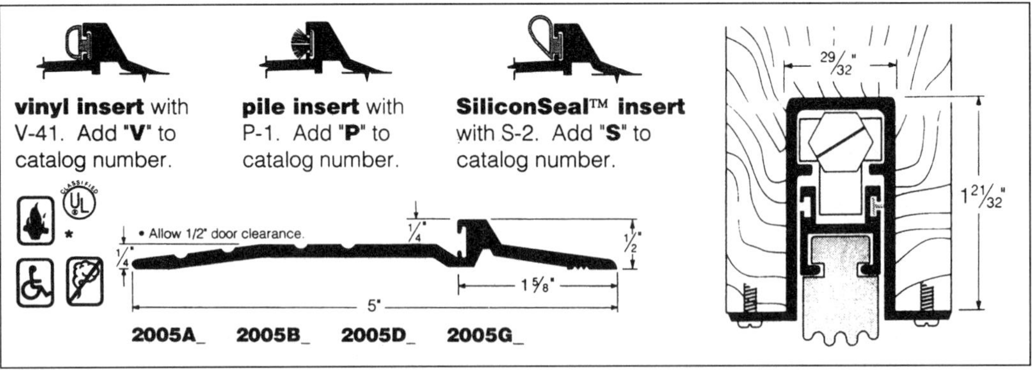

Sweeps, thresholds, and automatic door bottoms. ©Pemko Corp.

closed position for the door. When the door is opened, the threshold is out of the way, leaving the door to swing easily. When the door is closed, the threshold fills the gap. Unfortunately, thresholds can't be concealed, and their shape and position pose tripping and accessibility problems. Accessibility guidelines generally allow thresholds up to ½" high, but no threshold is even better.

Automatic Door Bottoms

These are the most complex and expensive, but also the most discreet solution. Automatic door bottoms are gasketed bars that fit into slots in the bottom edges of doors. When a door is moving, the bar is retracted up into the slot. When the door closes and makes contact with the jamb, the bar is forced down against the floor by a cam. These devices need no power to operate and require no corresponding floor treatment.

Requirements

Door Ratings

The fire-resistance ratings required of doors, like partitions, are listed by the time they are expected to last [IBC Table 714.2]. They are either the same or slightly lower than those of the walls in which they are located. The highest required rating, 3 hours, is achieved either by installing a single 3-hour leaf, despite its weight and expense, or by installing two 1½-hour doors in the same frame, hinged on the same jamb, but opening in opposite directions, despite its complexity. Doors between rooms and exit access corridors require only a 20 minute rating, even though exit access enclosures may require as much as a 1-hour rating. Any material that can meet these requirements is permitted.

It wasn't always this way. For many years, the rules referred to rated doors by letters indicating their intended function, and consequently were not terribly precise. The system is found in Appendix E of NFPA 80: "Fire Doors and Fire Windows." As this system is still widely used in reference, it is worth including. Its designations are as follows (parenthetically indicating the current terms that most closely correspond to the identified components):

- **Class A:** For use in fire walls (building separation walls) and in walls that divide a single building into fire areas (fire separation walls). This class has generally been applied to 1½ hour doors intended for use in two-hour partitions. As with the other classes, Class A doors are more commonly referred to as "A Label" doors.
- **Class B:** For use in enclosures of vertical communications through buildings (fire separation walls at shaft enclosures) and 2-hour partitions providing horizontal fire separations (again, fire separation walls). The Class B designation has generally been applied to ¾-hour doors intended for use in 1-hour partitions.
- **Class C:** Openings in walls or partitions between rooms and corridors having a fire-resistance rating of 1-hour or less (fire partitions and smoke barriers). The Class C designation has generally been applied to 20-minute doors intended for use in nominally rated partitions.
- **Class D:** Openings in exterior walls subject to severe fire exposure from outside of the building (no direct comparison).
- **Class E:** Openings in exterior walls subject to moderate or light fire exposure from outside of the building (no direct comparison).

Permeability

The degree to which a door and its frame seal against smoke [IBC 714.2.1] is generally addressed through performance requirements for volume of air leaked. This is expressed as so many CFM (cubic feet per minute) of air leaked per square foot of door opening. The requirements are mandated for 20-minute doors in corridors, so affect many doors. While these doors require gasketing, doors in other enclosures generally don't. Normal construction tolerances for door frames and doors are considered to be sufficiently tight for most partitions.

Locks and Latches

Regulations require that doors in the egress path be opened easily and generally without use of a key when moving in the direction of egress, with exceptions made for individual dwelling units, prisons, and for low-occupant-load assembly uses where doors remain unlocked during normal hours of use. They go on to regulate the use of manually operated flushbolts, locks tied to alarm systems, and mounting height [LSC94 5-2.1.5, IBC 1004.2.6]. Panic hardware is addressed separately [IBC 1004.2.7].

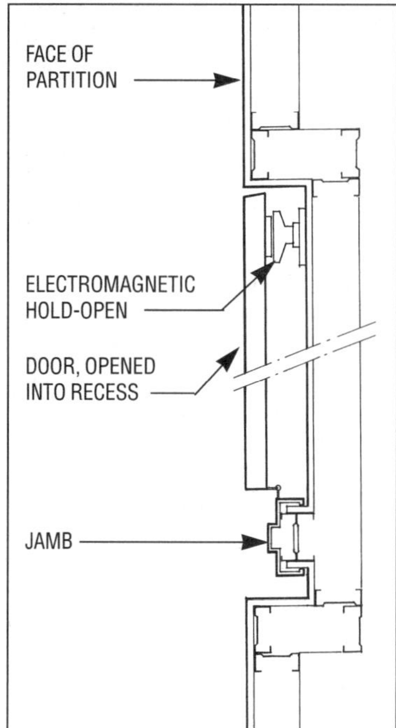

Plan detail of recessed door.

Optimizing Door Rules

Recessed Doors

Fire separations tend to divide space functionally and aesthetically. This is a common problem where corridors pass through fire separation assemblies. There, the intersection of the two must be designed to permit closing off the corridor to smoke and fire without closing it off to people. The trick is to do it without having the closure interrupt the visual or functional continuity of the corridor.

Concept

Some sort of fire curtain that drops down out of the ceiling would seem possible. Unfortunately, it is hard to reverse—once closed, it could ensnare panicking people, even if only temporarily while they try to push it aside. The best closures are those that move out of the way when pushed but that then return to their closed position when the pressure is relaxed. Hinged doors fit the description well. Unfortunately, even glass doors create more of a visual barrier than no doors at all.

The key to making doors invisible is to recess them when open and automatically force them closed when a fire is detected.

Technique

To recess a door, the frame must be recessed into the corridor wall. Since the wall most likely must be fire rated, reducing its thickness is not an option. One can achieve the same effect by building up the wall everywhere except where the door is when open. Through the use of swing-clear hinges, normally used to refit existing narrow doors for wheelchair clearance, the door face and jamb face can be brought into alignment, and both can be set to align with the surrounding wall face. Frame stops, unfortunately, cannot be eliminated, as they are needed to prevent the passage of smoke between the door and the frame.

Mechanism

Electromagnetic hold-opens, as described above, are used in conjunction with closers to automatically force the doors closed. Anyone who wants to use the doors has only to push them open [IBC 714.2.5.3].

Achieving Required Width

Swing-clear hinges can decrease the difference between the door width and the clear width of the doorway. (See pp. 214–215.)

Extra-Wide Openings

What about openings too wide to span with a set of double doors? An 8-ft opening is the widest allowable for two doors under the current 48-in maximum rule for individual doors. When three doors are needed, and a design calls for doors that are all glass, intermediate jambs spoil the sight lines and make the fire separation assembly obvious. In terms of construction

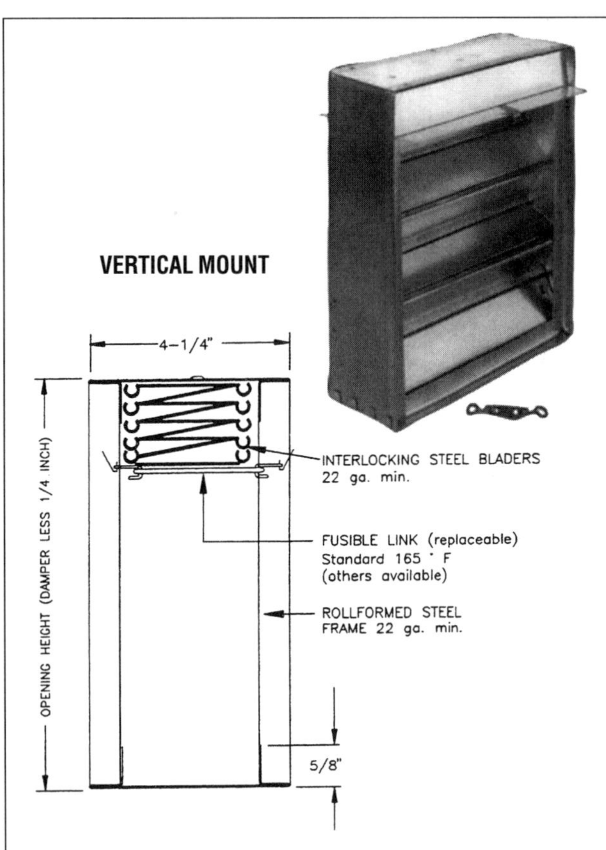

A typical damper. ©Lloyd Industries, Inc.

technology, the only way to avoid such jambs is by using pivot-hinged doors instead of the typical butt-hinged variety. These are common with all-glass doors anyway. In terms of code optimization, officials might consider allowing such a solution if hinge-side astragals were used, assuming that such astragals could be achieved effectively and in a visually discreet way.

Another solution is the use of fire-rated accordion doors. At least one manufacturer offers ratings of up to 3 hours with such doors, although it sets height limits of about 11 ft for labeled openings and 23 ft for certified openings, and sets width limits of about 12 ft for labeled openings. Certified openings are offered in unlimited widths.

SHUTTERS AND DAMPERS

Not all openings protectives are doors. Shutters are fire-resistive panels that cover openings, and dampers are gates that block ducts where they pass through rated assemblies.

Options beyond those discussed for doors are possible because people do not exit through these openings. Interior windows and the space above atrium railings are examples of such openings. For them, protectives can be stored in pocket walls or coiled in dropped ceilings, waiting only to be released by fusible links or magnetic hold-opens before sliding to their closed positions.

Codes allow two types of approaches to protection:

- Either exposed or concealed panels that extend automatically. The mechanisms that cause such panels to extend must be able to function without human participation, and without reliance on electricity. As discussed for doors, electromagnetic or electrically operated mechanical hold-opens work fine.

  In addition, fusible links can be used where smoke is not an issue, for instance where dampers are required inside a duct. Since smoke isn't sufficiently hot to make fusible links melt, dampers held by them stay open in the presence of smoke, waiting for temperatures to rise.
- Panels that are permanently fixed across openings, but that are not activated until exposed to fire. Intumescent glass is commonly used in such installations. Such glass is normally clear, but foams, thickens, and becomes visually opaque and fire resistant upon exposure to excessive temperature.

Code requirements can be found in IBC 714.

PLUGGED GAPS

Since the detailing of concealed spaces, joints, and penetrations has very little, if any, impact on conceptual design and the planning of building projects, an in-depth exploration of this issue or discussion of optimization strategies is beyond the scope of this book. It is important, however, to understand the roles played by these components in overall fire separation. These are the weak links in the chain of fire resistance.

Concealed Spaces

Definition

One of the hallmark advances seen in construction over the last century and a half is the use of layered materials. Walls are no longer generally made of single materials. Solid stone castles and uninsulated log cabins have been replaced by projects that use sophisticated veneers and backups. The layering results in better buildings, since each layer can be selected to achieve just one design goal, but achieve it well. Facings need not carry structural load; room can be found for infrastructure.

The downside is that the air spaces between the layers, even if unintended and residual, can provide passage for sparks or smoke.

Another hallmark is the increasing amount of infrastructural circulation used in our buildings. Building floors and walls are now not only pierced by electrical wires and sanitary plumbing, they may also contain computer raceways, recycling chutes, graywater distribution networks, pneumatic delivery systems, medical gases, and passive solar antistratification, thermal storage, or convection passages. With each of these penetrations comes the potential to spread fire.

Barriers inserted into concealed cavities within required enclosures are called draftstops and fireblocks [IBC 715.2 to 715.4]. The materials that fill them (until recently called safings) are generally expansive and sufficiently resilient to move without damaging themselves or the adjacent materials.

It is worth noting that, in addition to stopping fire, it is necessary to fill gaps in exterior walls to preclude rodents. See "Vermin Control," below.

Requirements
Fireblocks and draftstops must be located at the point where the curtain wall passes closest to the floor slab. For further exploration, see IBC 715.

Joints

Definition
Joints occur where partitions meet exterior walls or where expansion joints allow for expansion and contraction of a building. Smoke and flames could pass through joints between materials if they weren't also made fire resistant.

Requirements
"Fire-resistive joint systems" (joint fillers) are required to be as fire resistive as is required for the assemblies into which they are installed [IBC 712]. They are not required in floors within individual dwelling units or spaces that are not separated, such as malls and atriums.

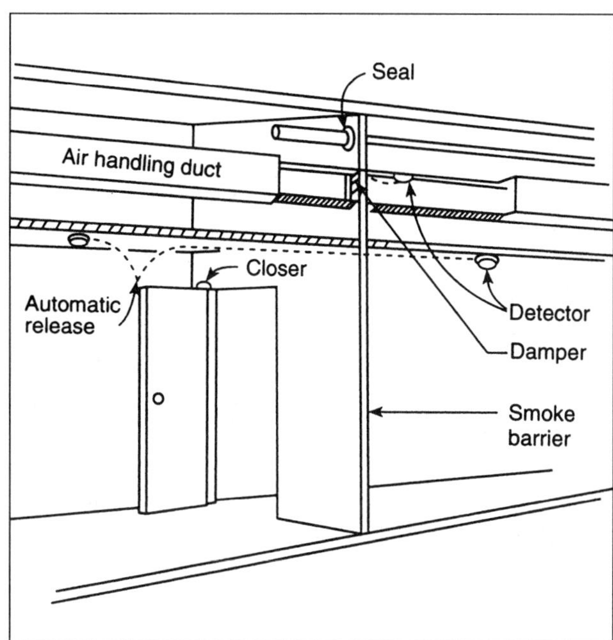

LSCH Figure 6-24—smoke barrier penetrations. ©NFPA.

Penetrations

Definition
There are two kinds of penetrations:

- Through penetrations that puncture an assembly from one side to the other
- Membrane penetrations that just break the surface

Either way, annular spaces occur where penetrating components such as fixtures, ducts, or pipes are recessed into or pass through partitions or floors. Since holes in partitions or floors must be at least a little larger than the components passing through them, spaces remain between the outside surface of the components and the edges of the corresponding holes made in the assemblies. Regardless of how fire resistant one makes a penetrating element, smoke and flames could pass through these annular spaces if they weren't also made fire resistant.

Requirements
Rules require that through penetrations be plugged with "penetration firestop systems" consisting of fire-resistive materials [IBC 711.3.1 at walls and 711.4.1 at floors]. The rules for membrane penetrations are similar [IBC 711.3.2 at walls and 711.4.2 at floors].

GENERAL DESIGN CONCERNS

MATERIALS

Context

"Base Building" Materials

(For exterior materials, see pp. 146 and 150.)

Structural materials are commonly concealed behind finishes, but their impact on a design is substantial. For example, the range of options possible in a post-tensioned concrete building is different from those possible with a heavy timber building. While structural configuration affects seismic performance (see pp. 150–153), structural materials affect span, column locations, and many other very visible factors.

Interior Materials

Since the most intense contact people have with buildings is by interacting with interior materials, such materials are a focus of regulatory concern. Unlike other forms of regulation, zoning ordinances are mute regarding choice of interior materials. This is because the choice doesn't affect community identity or property values, unlike exterior materials selection.

Building and accessibility codes do regulate interior materials, however, since these materials can affect the health, safety, and welfare concerns expressed by building codes and the opportunity concerns embodied in accessibility codes. They do this by influencing:

- Fire resistance, by regulating performance when exposed to fire. (See pp. 238–239.)
- Durability, by regulating material strength and workmanship.
- Acoustics, due to its impact on sound levels.
- Air quality, by regulating use of insulation in ducts and filtration systems.
- Electrical conductivity, due to its effect on electronic equipment and flammable gases, especially as related to medical services.

Durability

Background

Durability is a composite characteristic, embodying such material properties as hardness, compressive strength, and density, and such assembly characteristics as expansion control, attachment, and construction tolerances.

It can be argued that nothing says quality so much as durability. This is why banks were traditionally made of stone and why homeowners like brick houses. Durability is something we'd all like to have but that more of us would rather pay for later. That attitude works until it causes us to incur unexpected losses. If it weren't for the inherent danger, rules could leave us to suffer the economic repercussions of our short-term planning. For better or for worse, durability, or lack of it, carries implications for safety. There are, however, no mandated warranties.

To some degree, durability is simply a cost item for investment teams, and therefore should not be of concern to building codes. However, codes recognize that high maintenance costs resulting from low durability often translate into poorly followed maintenance schedules that impact that key concern of codes—building safety.

For historic properties, existing materials may pose particular durability problems. Although they've been durable enough to still be there, they may no longer be in acceptable condition. Existing materials can't be chosen since they're already there, so ways must be found to work with them, augmenting their durability through careful integration of newer technologies.

Requirements

DURABILITY ROOTED IN DESIGN

Optimizing durability starts with choosing the right assemblies for the right locations with the right materials. These are decisions typically made by the design team. Building codes typically describe:

- Situations in which various assemblies would be appropriate, including special measures to be taken in difficult situations. For example, codes might list requirements for ice shields where roof shingles will likely be subject to ice dams.
- Quantities of materials to be used, such as minimum thicknesses for load-bearing masonry walls and minimum thicknesses of plaster finishes.
- Type, size, and spacing of mechanical fasteners, adhesives, or other attachments.

DURABILITY ROOTED IN MANUFACTURING AND FABRICATION

Many rules establish quality standards for materials. Considering that virtually no materials are used as they are found in nature, these rules ultimately regulate manufacturing and fabrication processes. The "ball drop test" for safety glass is but one example of a material standard. Most materials that can be used structurally, such as masonry, steel, and wood, are required to meet minimum requirements for compressive strength.

Each of the chapters in the model building codes related to materials (Chapters 19, "Concrete," through 31, "Special Construction") sets certain rules for durability. Frequently they are tied to standards, such as those published by ASTM. For example, among the IBC's requirements for roof assemblies are several under the heading "Performance Requirements." They specify minimum levels of durability, defined as "physical integrity over the working life of the roof based upon... accelerated weathering tests conducted in accordance with ASTM G23, G26, or G53."

DURABILITY ROOTED IN WORKMANSHIP

Workmanship is regulated by installation codes and standards, of which there are many. Although recognition of these rules is the responsibility of the design team, fulfillment is typically the responsibility of the construction team. Building codes generally describe:

- Required installation techniques and preparation of surfaces and substrates.
- Conditions under which old materials can be reused in a new assembly.
- Postinstallation testing.

Maintenance procedures are needed when even durable materials start to show their age. In the model building codes, maintenance is addressed in Chapter 34, "Existing Structures." BOCA93 [3401.2] puts responsibility for maintenance and repair, regardless of the age or condition of the building, on the owner. The model codes reference property maintenance codes such as the BOCA National Property Maintenance Code, and fire protection rules, such as NFPA 101, The Life Safety Code, and NFPA 72, *Installation, Maintenance, and Use of Protective Signaling Systems.*

Optimizing Durability Rules

Is this a portion of the rules that you really want to skirt? Regulatory standards are minimal at best, and reducing quality of materials below code-mandated minimums may be risky. Keep in mind that a claim of quality based on "meeting every single requirement of the codes" is about the same as saying, "If we had skimped on even one issue, this project would have been illegal."

Fire Resistance

Walls and Ceilings

Fire resistance for walls and ceilings is generally defined as performance under the Steiner Tunnel Test. This process goes by two names, ASTM E84 and NFPA 255, each of which uses separate labeling as noted below. Until the upcoming IBC resolves this bifurcated system, the ASTM version will continue to be referenced by BOCA and the UBC, while the NFPA version will continue to be referenced by the SBC and the Life Safety Code.

The test measures two critical factors, for both of which high numbers translate to greater danger:

FLAME SPREAD

This describes the speed with which fire travels along a material. The findings are grouped into three categories and a noncategory:

1. 0 to 25: Class I (ASTM) and Class A (NFPA)
2. 26 to 75: Class II (ASTM) and Class B (NFPA)
3. 76 to 200: Class III (ASTM) and Class C (NFPA)
4. Over 200: Unclassified, but allowed under some codes with the permission of the local officials

In use, codes correlate class with use group and with location along the exit path. As a rough summation:

- Houses and storage uses may include materials with the lower ratings.
- While program spaces are allowed to use materials of Class II/B or III/C, exit access elements are restricted to Class I/A or II/B, with exits limited to Class I/A only.

For specifics, please check the applicable codes and amendments [BOCA93 Table 803.4].

SMOKE GENERATED

This part of the test measures the amount of smoke emitted by burning materials, strictly due to the smoke's potential to reduce visibility. Regardless of class, a test result of 450 has been determined to be the threshold of acceptability; products with higher ratings are prohibited.

Ensuring that selected materials comply with these ratings is as simple as looking for class listings in manufacturers' literature before specifying or ordering interiors products.

OTHER FACTORS

Two historical asides are noteworthy. First, the Steiner Tunnel Test originally had a third part, which tested for "fuel contributed," the amount of combustible product in a material. As the measurement methodology of this test proved not to be valid, it was dropped in 1978. Second, editions of the Life Safety Code prior to 1988 had language allowing rejection of materials due to "the character of the products of decomposition," a reference to irritants and toxins that are potential byproducts of incomplete combustion. While the issue seems to be valid, the provision was difficult to enforce because its human factor made assessment inconsistent.

Floors

Floor finishes generally ignite as a result of exposure to heat rather than exposure to flame. As a result, the Radiant Flux test, another one published twice as ASTM E648 and NFPA 253, tests the amount of heat it takes to ignite a floor. Its results are measured in watts, as is the energy blasting out of a hair dryer or room heater. Unlike the Steiner test, here higher numbers are achieved by materials with greater fire resistance.

Another test is used for carpets. The "pill test" does not directly test fire resistance, but rather the tendency of a carpet to let go of its fibers. Since fibers that are tightly bound into a carpet are less likely to ignite than individual yarns, this test is an indirect predictor of fire resistance. It is published as DOC FF-1 by the Department of Commerce's Consumer Product Safety Commission because carpeting is much more likely to be bought by consumers than are other materials regulated by building codes. When carpets are used on walls or ceilings, they must follow the rules that pertain to walls and ceilings.

Findings are grouped into three categories:

1. Radiant flux over 0.45 W/cm^2: Class I
2. Radiant flux of 0.45 to 0.22 W/cm^2: Class II
3. Carpets complying with the "pill test"

As with walls and ceilings, codes correlate class with use group and with location along the exit path. As a rough summation:

- In the exit path, most use groups other than hospitals must use at least Class II floorings. Hospitals must use Class I. Factories, hazard, utility, and storage uses can use any of the three categories.
- Outside the exit path, any of the three categories is acceptable for any use group.

For specifics, please check the applicable codes and amendments [BOCA93 Table 805.3].

As with wall and ceiling materials, ensuring that selected flooring materials comply with these ratings is as simple as looking for class listings in manufacturers' literature before specifying or ordering interiors products.

Acoustics

Background

REGULATORY CONCERNS

For an exploration of the need for sound controls, see Chapter 8 (pp. 198–199). Sound travels through buildings through air and through construction materials. It also bounces around within individual spaces. Airborne transmission of sound is measured as STC rating, material-conducted sound is measured as IIC, and sound reflection or absorption within a space is measured as NRC.

STC

STC, or Sound Transmission Class [ASTM E90], describes the ability of walls and floor/ceiling assemblies to inhibit the passage of airborne sound. When partitions carry low STC ratings, people complain that "you can hear every word." Various codes set different standards for this factor.

IIC

IIC, or Impact Insulation Class [ASTM E492], describes the ability of walls and floor/ceiling assemblies to absorb blows without translating their force into construction materials. When partitions carry low IIC ratings, people "wonder what the people upstairs are doing." As with STC, various codes set different standards for this factor. IIC ratings are often achieved through lease requirements for tenant-supplied carpets.

NRC

NRC, or Noise Reduction Class, describes the ability of wall, ceiling, and floor coverings to inhibit the reflection of sound within a space. When NRC ratings are high, sound generated within a space is absorbed, making the environment more pleasant for occupants who don't want to be subjected to the sound. Over the last decade or two, restaurants have moved toward lower and lower NRC ratings on the theory that a noisy dining room is an exciting one, where customers will eat more and are more likely to come back.

Requirements

Building codes require minimum acoustic suppression only in residential occupancies, and then only for walls or floor-ceiling assemblies that enclose individual dwelling units.

STC: The IBC and BOCA come in at 45 [IBC 1206], and UBC holds out for 50. To put this in perspective, an uninsulated single-layer drywall partition carries an STC rating of about 38. With 1½ in acoustic insulation batts and perimeter caulking, the rating increases to the 45 required by the IBC. By adding a second layer of gypsum board to one side, or adding to both sides without the insulation batts, STC increases to meet the UBC criterion.

IIC: The IBC and BOCA come in at 45 [1214.3], and UBC holds out for [50].

NRC: As noted in Chapter 8, noise within a space is not addressed by building codes, but is by OSHA, in 29 CFR 1910.95. Still, this regulation has no direct bearing on NRC and therefore no direct influence over materials selection.

Air Quality

Background

Many of the factors related to air quality (explored in pp. 197–198), can be traced to materials. Careful selection can eliminate many of the direct and indirect sources of pollutants and irritants. That being said, it is also worth noting that some air quality issues are best handled by making changes in behavior—getting occupants, visitors, and maintenance/repair teams to use the building in a different way.

MICROBIAL SUPPORT

Some construction materials, particularly those made of organic materials such as cellulose or held together with organic binders or adhesives, can provide sustenance to some microbes. State hygiene codes seek to control such materials in settings where microbial support is particularly problematic:

- Restaurants
- Hospitals
- Inside ductwork and other HVAC equipment

ABILITY TO BE CLEANED

In addition to the normal deposition of airborne particulates and skin oils (if within reach), vandalism provides a good incentive to use materials that can be easily cleaned. Key to this is hardness and impermeability, since smooth materials don't give such coatings as spray paints much opportunity to grab on [SBC 1204].

AIR CHANGES

Air freshness is partially regulated by specifying minimum numbers of air changes per hour. More changes mean less chance of inhaling poor-quality air. In some cases, the facilities whose occupants need it most are those whose products would be most damaged by it. Pharmaceuticals laboratories need fairly calm air, but their employees benefit from increased numbers of changes. Some of this conflict can be resolved with appropriate use of localized environments of the type established by incubators, cabinets, or possibly fume hoods. Some codes, particularly older ones, may specify minimum levels of ventilation by formulas using CFM per SF foot of floor area.

Requirements

Rules for air changes are found in ASHRAE 90 for common projects, and in standards published by the National Institutes of Health (NIH) for laboratories and hospitals. They commonly recommend or require, depending on the authority of the enforcing agency, anywhere from three to thirty air changes per hour, or even more in very rare situations. For a better sense of what changes mean, it might help to translate from number of air changes to number of minutes in a cycle. For example, ten air changes per hour means that all the air in a building is replaced every six minutes.

Optimizing Air Quality Rules

POLLUTING MATERIALS

When existing materials or operations are found to be sources of indoor air pollution, three strategies are available:

- **Remove them:** When a loading dock is located adjacent to an intake grille, delivery trucks can be required to shut off their engines while loading or unloading. Smokers can be limited to certain areas or banned altogether, although the way things are going, state and federal laws will eliminate the need for individual properties to act. Ask cleaning services to do their work in off hours.
- **Substitute other products:** Use HCFCs instead of CFCs. Use linoleum in place of carpeting.
- **Encapsulate them:** Seal friable asbestos with a layer of approved sealers. Put polyethylene sheeting down in an earthen crawl space to contain molds and mildews.

Coating and Encapsulation

When the surface characteristics of needed or desired materials are inadequate to fulfill regulatory requirements, coating and encapsulating them with more acceptable materials may be a reasonable design response. When materials are coated, a thin layer of paint, sizing, or other such material is applied that binds the base material, holding it together. When they are encapsulated, they are completely enveloped in materials that are acceptable in exposed situations. Friable asbestos can be encapsulated with special coatings. Before this solution is attempted, encapsulation coatings must be checked for incompatibility problems with the materials to be encapsulated.

Electrical Conductivity

Background

Figurative and sometimes literal sparks can fly when:

- Amassed static electricity is spontaneously discharged.
- Ferrous metals strike one another. This can be a deadly combination in hospital operating rooms if steel scalpels are accidentally dropped onto steel carts in the presence of flammable medical gases such as oxygen.
- "Live" conductors come in sufficient proximity to one another for electricity to jump between them.
- Lightning hits a building or near a building.

Regardless of source, the discharge of electricity can wreak havoc on electrical equipment, not to mention its effect on occupants when they ground themselves to doorknobs. Sparks can cause substantial problems in medical, computer, electronics, and some manufacturing settings.

Static electricity accumulates in response to indoor humidity levels, the electrical characteristics of finish materials, and activities occurring in a space. It is discharged in response to grounding, which happens when charged materials come in momentary contact with electrically conductive materials, including people.

The problems associated with static electricity can be managed with:

- Materials that don't generate sparks. Electrically speaking, glass is highly insulative (nonconductive). Wood and masonry materials are also good insulators. Some plastics and any ferrous metals tend to generate sparks. Review this issue, where relevant, with product manufacturers.
- Materials that don't support static buildup. Manufacturers' literature lists those standards with which their products comply.
- Use of grounding straps to safely conduct static electricity to a safe ground before it has a chance to accumulate. Some products, such as static-control pure vinyl flooring, are available with grounding straps supplied as part of the material system. After installation, electricians connect the straps.
- Treatments that can be applied to problem materials that reduce their ability to accumulate a static charge. Such treatments are available for carpets and most fabrics, and provide a base level of protection for electronic equipment.

The problems associated with lightning are still managed with a rather old technology: the lightning rod, or arrester. These are metal conductors mounted in atmospherically prominent positions and attached to grounding straps that, when properly used, provide paths of least resistance to the ground, where electrical energy can be dissipated.

Requirements

Rules pertaining to electrical conductivity may be found in state health codes as related to hospitals, and in the National Electric Code.

Assemblies

Context

Assemblies are applications of materials to specific locations and functions within buildings. Assemblies may be made from single materials or groups of materials—what makes them assemblies is the task they are expected to serve. For this reason, the types of qualities we look for in assemblies are different from those we want in materials. Key characteristics that must be evaluated in choosing assemblies include:

- Safety, a result of carefully controlling the spacing of components and of positioning assemblies to avoid burns and scalds
- Fit, a question of providing sufficient width and minimal obstruction, proper support, proximity for reach, shaping or texturing for grip, and managing the force required to operate movable assemblies
- Aesthetics, involving choices of color and luster, and fidelity to historic detailing
- Communication, designing to help users orient themselves, to warn users of potential hazards, and directing them, when necessary, to safe egress
- Vermin control, choosing materials and tolerances to preclude entry by animals and insects

Safety

Spacing

Definition

When openings in grids, gratings, and the like are inappropriately sized, safety can be compromised. People and objects can fall through openings that are too big. Openings that are too small don't allow undesired materials to filter away.

Guardrail Panels

Rules for guardrails have tightened significantly in the last few decades. If the only issue were restraining people at waist height, almost any design could work. But that wouldn't handle all situations.

It wouldn't restrain balls rolling or flower pots being accidentally pushed over the edge, either of which could pose a hazard to passersby below. For this reason, some rules require toe- or kick-boards, solid barriers that extend from the floor plane upward several inches.

Guardrails must also contain middle-size objects that wouldn't necessarily be restrained by toe-boards or rails positioned to stop adults of average height. Children and pets could easily slip below top bars, in between the vertical supports needed to support the top bar. To this end, all codes require that the area between the top bar and the toe-board be spanned by some kind of infill panel. The panels need not be solid (not yet, at any rate—wait for a few more cases of personal injury to go through the courts), but any openings in them are limited.

Requirements

Codes have developed a ball-passing test to judge the suitability of guardrail panels [IBC 1003.13.2, LSC94 5-2.2.4]. If a ball of whatever size the code specifies is able to be pushed through the panel, in any place or way, its openings are rejected as being too big. Current building codes are requiring a 4 in ball, reduced from 6 in as recently as 1991. In the not-too-distant past, some codes required only one or two intermediate rails.

The Life Safety Code Handbook has an interesting commentary on this[49]:

The change from a 6-in . . . to a 4-in sphere was . . . based on a proposal that received the backing of the American Academy of Pediatrics. Approximately 950 out of 1,000 children under age 10 can pass through a 6-in wide opening.

Burn and Scald Hazard

Many, but not all, of the building components that are sufficiently hot to burn or scald on contact with skin are inaccessible to occupants. For example, hot-water pipes are located within plumbing chases or floor slabs for most of their length. Where they can potentially be exposed to occupants, rules address the safety hazard.

Three building elements have the potential to burn and scald:

- Hot pipes
- Water coming out of hot pipes
- Elements intended to heat things, such as radiators and stoves

For the first two, protection or control can help prevent mishaps. Occupants can be protected from dangerous elements either by locating such elements beyond reach or by screening them with protective materials. Many of the rules contained in ADAAG define what is reachable and regulate the way hot components in such areas are screened. They can also be controlled through use of balancing valves that restrict the flow of water when temperature surges are detected, as happens sometimes in showers when toilets are flushed.

For the third, occupants tend to be cautious, since they expect them to be hot. Even so, accidents can happen. Injury prevention is addressed by manufacturers in the design of such equipment and by project teams complying with occupational rules established by OSHA.

Fit (Accessibility)

Fit is the notion addressed by the science of ergonomics. It is all about designing our environments to fit us, rather than forcing us to adapt to use environments designed with insufficient consideration of human physiology. Fit is a question of sizing, support, operability, reach, and operating force.

Passage

DEFINITION

Elements in the built environment must be sized for access. Spaces must be big enough to accommodate our passage, and smooth enough not to obstruct it. Making spaces big enough is a question of circulation, as explored in Chapter 9. Making spaces smooth enough is a question of detailing.

Examples of design features that can pose passage problems include:

- Curbs or thresholds that are too high.
- Floor expansion joint covers that catch cane tips or the kinds of thin tires found on strollers, suitcases, and wheelchairs.
- Counters, sinks, and drinking fountains that are too low to pass the knees of wheelchair users.

To a large degree, compliance is a question of specifying properly designed components. This is fairly easy because, in general, manufacturers comply with passage rules, knowing that noncompliant products won't sell.

Still, not everything that goes into the built environment is premanufactured. Architects must choose moldings and trim pieces carefully and watch the thickness of finish materials relative to the height of substrates.

REQUIREMENTS

For the problem of low counters, sinks, and drinking fountains, ADAAG figures 8(c1) and 8(d) illustrate required clearances.

For curbs, thresholds, and joint covers, the most ubiquitous requirement is to keep all transitions to ½ in total height or less, in ¼-in increments. That dimension appears, with a bit more subtlety, throughout accessibility regulations like ADAAG and ANSI. The apparent simplicity of that dimension is misleading. Required surface elevations can be met only by adjusting the thickness of materials or by modifying substrate elevations. Either can have far-reaching consequences for design. Many materials come in standard thicknesses that can't easily be adjusted, and many flooring systems do not easily lend themselves to localized adjustment. Consider, for example, the problem of maintaining a ½-in vertical transition in a floor that combines brick pavers with vinyl composition tile.

The scale of the transition becomes an important factor as well. When areas finished in different materials are very wide, whole sections of floor can be installed using different parameters. However, when areas are small, as when an accent strip two bricks wide is specified, every joist may need to be notched, or an entire slab may need to be poured with greater thickness to maintain its minimum thickness where the thickest materials are used.

OPTIMIZING PASSAGE RULES

Modify the Materials

Finish materials now come in an amazing range of installation modes. Brick pavers need not be 2¼ in thick. Wood strip flooring is now made as resilient flooring, with real veneers laminated between layers of acrylic film. Terrazzo can be thinset. Even marble and granite are now available in sheets ¼ in thick, using fiberglass backing to provide the needed flexural strength.

Modify the Substrates

For wood floor structures, joist depths could be trimmed, joists could be installed at different elevations, or additional layers of sheathing could be installed over joists that are kept at the lowest needed elevation.

For steel and concrete structures, formwork could be made with the necessary complexity or topping slabs could be considered. Cellular floor decks become almost impossible because some of the knockouts wouldn't be able to reach the surface.

Awareness

DEFINITION

People can get hurt when they aren't aware of the presence of potentially dangerous objects. As an example, people who aren't being careful or whose vision is blocked by smoke might run headfirst into wall-mounted exit signs hung at inappropriate heights. The blind frequently navigate with the aid of a cane swung from side to side in front of them. It is too low a sensory technique to detect objects more than waist high. Ensuring that components are effectively placed can be critical.

REQUIREMENTS

ADAAG prohibits any objects from protruding more than four inches from walls when such protrusions start more than 27 in from the ground. This is in response to the difficulty of detecting such objects with a cane. In a seemingly unavoidable and totally absolute design conflict, ADAAG also requires that countertops and drinking fountains have clearances to the floor of at least 27 in, allowing wheelchair users to get sufficiently close by maneuvering their knees into such undercounter spaces. These two rules are mutually exclusive by the thickness of a hair, but neither can be modified.

OPTIMIZING AWARENESS

RULES

One way out of the awareness problem is by recessing walls from which objects are protruding, letting the blind go by with no awareness of, or danger from, the object. The blind need not be aware of protruding objects as long as they don't protrude more than the four inches mentioned.

Another technique is to locate an obstacle at the ground to warn a cane user of something hanging higher up, in the danger zone. The space under escalators is frequently ringed by railings or occupied by planters for this reason, extending back from the bottom of the escalator to the point where the sloping escalator doesn't pose a headroom problem.

Support

DEFINITION

Proper support of a worker's body is critical to health. Back trouble and repetitive-motion injuries (or more technically, CTDs: Cumulative Trauma Disorders) cause substantial productivity losses. CTDs alone generate $20 billion in workers' compensation claims annually.[50] Solutions often come from the science of ergonomics.

In general, support is provided by furniture rather than building elements. Building components involved in providing support include handrails and seating surfaces such as might be incorporated into planters and retaining walls.

HANDRAILS

General

Handrails provide critical support for anyone using a stair. Even young, healthy, coordinated people occasionally lose their balance. For the very young and the frail, they are physically and psychologically critical. Handrails must be properly positioned and able to sustain the impact or tug of a heavy person moving suddenly and with some speed. Handrails are quite different functionally from railings (guardrails). As the name suggests, handrails are for gripping when support or guidance is needed. Railings serve as barriers at edges of walking surfaces (floors, ramps, or stairs) to prevent falls. (For railings, see pp. 162–163.)

Position

Handrails must be properly positioned vertically and horizontally for optimal function. Rules are very restrictive on both, requiring exactly 1½-in clearance between handrail and wall, and a height of 32 in to 34 in above the floor or a stair nosing.

The clearance rule is unusual in allowing no margin. In this case, both larger and smaller dimensions have been found to be problematic. People, sighted or not, often find handrails by reaching for walls, then sliding their hands down until their palms find the handrail. When clearance is too wide, they are more likely to find the handrail with their wrists, which makes the handrail harder to use.

The height is likewise set for availability. Although many people do not walk with hand on handrails, they may grab for one when they trip or lose their balance. Since people vary in height, the distance from their hands to the ground varies. Still, the 2-in range required has, apparently, been found to work for most people.

For handrail size, see "Operability: Getting a Grip," below.

Structural Capacity

Handrails are not guardrails. Even if one came loose, there is no possibility that anyone would fall over an edge. Still, people can lose their balance and fall down a flight of stairs. Therefore, handrails must be properly anchored to floors or walls and have almost identical requirements as guardrails—200-lb point loads in any direction and 50-lb uniform loads in any direction [ASCE7 Section 4.4, BOCA93 1615.8.1]. They differ only in uniform load resistance, since railings must handle 50 lb horizontally and 100 lb vertically. When support walls are framed, blocking must be provided or handrail brackets must be attached directly to studs. When support walls are masonry or concrete, anchors with sufficient pull-out strength must be used.

SEATING SURFACES

Seating surfaces built into planters and retaining walls, or used as window seats, are generally not occupied by the same person for very long. Proper support is consequently not critical. Proper support is important for furniture, but since that moves beyond the scope of building design, it is beyond the scope of this book. Suffice it to say that workplace furniture is regulated by OSHA's "general duty clause," which requires businesses to "maintain a safe and healthy workplace."

Reach

DEFINITION

Rules for reach address the range of motion limits exhibited by different users. Simply put, users who can't reach operable components can't use them. Items typically controlled for reach include:

- Elevator controls
- Toilet accessories, such as paper towel dispensers
- Light switches and outlets
- Fire pull-boxes

Too low can be as much of a problem as too high. Rules generally assume that it is easier for a tall person to stoop than for a young, short, or wheelchair-bound person to find a step stool. While maximum height rules have been enforced since the first accessibility guidelines were written, minimum height standards are more recent. Such rules respond to the difficulty elderly or wheelchair-bound users have bending down to reach outlets.

The rules themselves are so clear, complete with dimensioned graphics, that it would be redundant to explain any of it here. Besides, the rules are very specific and quite extensive, beyond anything that would be reasonable to summarize here. Readers are urged to look through ANSI, ADAAG, or whichever accessibility standard applies.

OPTIMIZING REACH RULES

One useful drawing to include in a set of contract documents is a mounting height schedule. This is a type of interior elevation that shows the heights for hanging each type of wall-mounted fixture. In such a drawing, dimensioning must show distance from the floor or other convenient reference to the rough opening or centerline for the piece of equipment. This tells the worker what to do, since much of the construction may need to be completed before the fixtures are delivered to the site. Determining the appropriate distances for these dimensions is usually a function of the operable element's position relative to that installation position. For example, to install a feminine napkin dispenser so that its coin slot and actuating levers are within 44 in of the floor, one must know where the slot and levers are relative to the edge of the fixture.

Operability: Getting a Grip

DEFINITION

Some building components, such as doorknobs, appliance controls, and thermostats, must be moved, turned, slid, or otherwise operated and manipulated. This can be very difficult if the user has arthritis, small hands, or hands that are temporarily wet. A component's ability to be gripped is determined by its size, shape, and texture.

Two other factors are related. To be gripped, the component must be within reach. Once gripped, it must be manipulated with sufficient force. These characteristics are addressed below.

REQUIREMENTS

Handrails are required to be round and 1¼ in to 1½ in diameter, as this size and shape has been found to be most easily grasped by most populations. Designers should note that 1½-in nominal pipes have an outside diameter exceeding these limits.

"Growing Older." ©Roger K. Lewis

Knobs, which will not turn without sufficient friction between knob and hand, are now prohibited in commercial installations, replaced by a requirement for levers that can be operated with the push of an elbow.

In some situations, rules require metal operating surfaces to be knurled so as to provide some degree of grip. A good, though not perfect, way to evaluate the appropriateness of a design is to check that mechanisms can be operated without fingers, by pushing or pulling with fists or elbows.

Operability: Required Force

Definition

To ensure that doors can be opened by any occupant, codes dictate minimum amounts of pressure required to operate them. For exterior doors, the minimums set must be low enough to allow occupants of limited physical capacity to exit, yet high enough to prevent winds from blowing the doors open without human participation. This latter situation would pose a hazard to occupants and create problems for heating and cooling systems.

When exits are pressurized to keep smoke from entering, the force needed to open doors leading to them can exceed acceptable levels. Pressure regulators may need to be designed into the system [LSCH p. 47].

Finally, operating force is an issue for doors in the egress path that incorporate collapsible hardware. Such doors include revolving doors and some sliding doors [IBC 1004.2.8.2].

Operating force may reach unacceptable levels when a device is:

- Too short to allow leverage. This is a particular issue with smaller devices.
- Sufficiently heavy that lifting or just overcoming inertia becomes a problem.
- Sufficiently dampened that human efforts at manipulation are thwarted. Dampening is a technique used to prevent extremes of movement, such as when the wind slams a door open. It can be provided by mechanical governors or by the viscosity generated by oil or air bladders.
- Subject to wind loads that pose an additional force to be overcome.

Requirements

Operating force for doors is regulated with some specificity. Different values are provided for the task of opening latches, which are relatively low, and of pushing doors open, which are relatively high. Different values are also provided for doors without closers, for doors with closers or nonhinged doors, and for collapsing the leaves of revolving doors. Except for the revolving doors, at 130 lb (in egress path) or 180 lb (nonegress), the highest of these values is 30 lb [IBC 1004.2.2, LSC94 5-2.1.4.4].

Optimizing Operating Force Rules

This often requires hunting for a different device. Most manufacturers are aware of the rules for force, and design their equipment to comply. For projects with demographically extreme populations, such as elementary schools or assisted living housing, a higher standard may be reasonable.

Aesthetics

Style

BACKGROUND

What is style? This may be the most subjective question in the building industry. Can style be legislated? Should anyone other than a property owner have the right to influence construction simply based on style?

Exterior details have a powerful effect on establishing community identity. Santa Fe's citizens support its strict enforcement of adobe as the material/color/style of choice. Many suburbs require all new projects to be built with brick front facades, even when side and rear walls are faced in vinyl siding. Some people feel that by adhering to a certain community architectural identity, property values will be maintained.

Abundant examples exist to argue the effectiveness as well as the intrusiveness of this solution. This is one of the issues over which property rights advocates grapple with the idea of governmental intrusion. It is also the issue over which, for better or for worse, neighbors have successfully fought the imposition of "ugly" projects.

> **Shutters**
>
> Objective/Goal: To select and install shutters that are harmonious with, and will enhance, the architecture of the house.
>
> The DRB will review your application for shutters based on the following general guidelines:
>
> - The architecture of the house and the style of the shutter must be compatible (typically, pre-manufactured shutters are found only in designs suitable for "traditional-style" homes.
> - To provide visual continuity, there should be other approved shutters in the cluster [group of houses], and the approved shutters should be compatible in design with them.
> - The proposed color of the shutters should be harmonious with the exterior colors of the house and the cluster.
> - Shutters should be applied to all windows on an elevation and should be architecturally logical, that is, installed on both sides of a window, matching the size and configuration of the window, and not cut down to fit a space that is too small for a full shutter.

An example of the influence over style that covenants can have. Example excerpted from the Covenants of Reston, Virginia. ©Reston Association.

REQUIREMENTS

CC&Rs have a lot to say with respect to materials selection, but sometimes also with respect to historic style, arrangement, and proportion.

Historic preservation rules might be interpreted as very liberal with respect to style, requiring that new work in historic buildings deliberately avoid styles already present. This helps building users understand what is authentic and what was added. The rules have been criticized for preferring contemporary aesthetics as the style of choice for new work. In response, historic preservationists have maintained that the addition of another period style, such as Georgian or Gothic, to an existing property would confuse rather than enlighten.

The Secretary of the Interior's Standards for Rehabilitation [1996] directly address the historic fidelity of details through the following mandates (numbers listed are the citation references):

- 4. Most properties change over time; those changes that have acquired historic significance in their own right shall be retained and preserved.
- 5. Distinctive features, finishes, and construction techniques or examples of craftsmanship that characterize a property shall be preserved.
- 6. Deteriorated historic features shall be repaired rather than replaced. Where the severity of deterioration requires replacement of a distinctive feature, the new feature shall match the old in design, color, texture, and other visual qualities and, where possible, materials. Replacement of missing features shall be substantiated by documentary, physical, or pictorial evidence.
- 7. Chemical or physical treatments, such as sandblasting, that cause damage to historic materials shall not be used. The surface cleaning of structures, if appropriate, shall be undertaken using the gentlest means possible.
- 9. New additions, exterior alterations, or related new construction shall not destroy historic materials that characterize the property. The new work shall be differentiated from the old and shall be compatible with the massing, size, scale, and architectural fea-

tures to protect the historic integrity of the property and its environment.

- 10. New additions and adjacent or related new construction shall be undertaken in such a manner that if removed in the future, the essential form and integrity of the historic property and its environment would be unimpaired.

Note that although *The Secretary of the Interior's Standards* do not address interior work, some local amendments do. Further, in jurisdictions that assign preservationists to represent government interests, the preservationists help determine the rules that apply.

OPTIMIZING RULES RELATED TO STYLE

CASE STUDIES

All of Santa Fe, Mew Mexico, works under zoning guidelines that require all construction to either be adobe or at least look like it. Although this ordinance has resulted in a few notable stretches, such as the gas station with the "adobe" canopy cantilevered twenty feet, it has also managed to preserve a remarkable sense of place. Note that these requirements are part of the local ordinances, and not the historic preservation laws. They are aimed at visual consistency, not historic authenticity.

A house faced in oiled plywood designed by Santa Monica architect Frank Ghery was rejected by community boards, but not based on style. Apparently, the board could not be convinced that the project wasn't industrial, a prohibited use under the district's zoning.

Maywood is a small community in Arlington, Virginia, designated as a historic neighborhood. Historic preservation rules there prohibited use of window shutters made of anything but wood. Upset at not being allowed to use any of the lower maintenance alternatives, one homeowner painted black rectangles next to each window, directly on the siding boards of his house. Although this solution clearly violated the intent of the preservation guidelines, it was not expressly prohibited by them.

Shape

Shape is at the heart of detailing. The differences between one shape alternative and another affect the perceived scale of a building as well as its perceived quality, since complex shapes make for a finer, more delicate design and are generally more expensive to produce. This kind of thinking affects historic preservation guidelines, which require that "new construction shall be compatible with the massing, size, scale, and architectural features" of the existing property, while not being sufficiently similar to be mistaken as original work. (See also pp. 145–146.)

It is worth pointing out that there isn't much else to say about the impact of rules on the project team's decision regarding the shape of exterior wall details. This reinforces the notion that rules tend to ignore issues that are beyond their areas of concern.

Color and Luster

Color is an issue with no real impact on accessibility, but quite a bit on community identity, safety, and efficiency of energy and lighting. Perhaps surprisingly, color has little impact on federal historic preservation rules, which are neutral on the issue, and leave color selection to project teams. Where those soft, faded colors so often associated with colonial architecture are mandated, it is by local rules, and generally seen only in projects that market themselves as "living museums," such as Mount Vernon and Colonial Williamsburg, Virginia.

Color can have a significant impact on energy costs, since light colors reflect radiation away while dark colors absorb it. The color difference between light gray terne-coated steel standing seam roofs and almost-black slate shingles has a substantive effect on fuel bills. Color is one of three factors considered in calculating heat gain, along with R-value and shading factor. Numerical coefficients for various colors, which are used in energy calculations, are listed in the ASHRAE handbook. However, significant as color is, energy codes currently do not address it directly.

Color can be used for warning, as typified by OSHA regulations. Color is used to indicate low headroom and location of buried pipes, and to delineate crosswalks. These regulations mandate certain colors to indicate various levels of danger. Predictably, the color scheme tends toward yellow and red [29 CFR 1910.144]. Red is required for fire protection equipment, including any alarm bells mounted to building facades. It is also required for danger signs, emergency

"Design Review in Action"
©Roger K. Lewis.

stop buttons, and for containers of low-flash-point flammable liquids. Red lights must be provided at construction barricades. Yellow indicates caution and is required when physical hazards must be marked.

For direction, color is mandated for egress signage by BOCA and NFPA. Mechanical codes mandate a specific color palette for marking building piping. While pipes and their color tags are not usually exposed in finished spaces, there's no rule prohibiting it.

Color and luster are both valuable for controlling lighting levels and glare. Interior designers and lighting designers must work together to coordinate selection of finishes with use of natural lighting and selection of artificial lighting.

Communication

Tactile signs, audible signals, and visual emphasis.

Orientation

DEFINITION

A building's details can play a substantial role in helping the sensory-impaired find their way. It can also help reinforce wayfinding for those who are easily confused. Considering the aging of our population, this may be a significant design issue in coming decades. With some complicated designs, it may be necessary reinforcement for the general population. How many people never experience confusion when relocating a car left in a parking garage?

This issue is also critical for the general population during emergency conditions, such as fires. When smoke may obscure visibility and people are in a hurry, aids to orientation become critical. (See p. 188.)

Even though the deaf can see, and most of the cues provided by our buildings are visual, special design attention is appropriate for them as well. Some of our most critical cues are audible, in the form of alarms, and therefore imperceptible to those with severe impairments.

Tactile cues: These include the shape of a railing, the texture of a pavement. Tactile range is severely limited compared with sight; canes can only increase it by a few feet at best, and depend on direct contact. A cue that goes untouched remains unread. Braille is mandated for some locations.

Audio cues: Echolocation depends on acoustic reflectivity and therefore affects selection of materials and their method of attachment. Also critical are audio signals, such as chimes indicating the direction of an arriving elevator (one for up, two for down).

Visual cues: Flashing lights can be used to announce incoming telephone calls, and for fire alarm systems. Oversize lettering is mandated for some locations. For use of color to aid in recognition and identification, see the previous topic.

Olfactory cues: Human beings have the ability to locate the source of distinctive smells. This can be of help in orientation and warning.

Aiding memory: Sound and smell can help tremendously with orientation, particularly if distinctive, even at low "dosages." Parking garages have used some very clever cues to help their customers remember where they parked, including "theme songs" played over sound systems, theme cities or cultures (a "French area" complete with France's national colors, a mosaic of the Eiffel Tower, etc.), and the obvious letter and color designations (Row F, the "Blue Level," etc.).

Requirements

OSHA stipulates minimum sizes, distances, and illumination for signage. Building and fire codes stipulate sizes and illumination of lettering for exits and elevators. See NFPA 72, The National Fire Alarm Code, for the most widely used regulations.

Warning

Definition

Many times, the need for warning is related to a piece of equipment rather than to a building design in general. Manufacturing equipment must warn users of impending dangers.

There are also situations where an environmental danger is integral to public architecture. Examples include motorized gates, areas of low headroom, reflecting pools, steep roofs over walkways in snow zones, and platforms at subway and rail stations.

Requirements

OSHA has many requirements in this area. Most are related to the presence of certain types of potentially dangerous equipment or machinery. Projects teams planning such projects are well advised to refer to OSHA to determine specifics.

Accessibility guidelines require the use of textured paving at curb cuts and at rail platforms to warn visually impaired pedestrians of nearby hazards.

Direction

Alarm systems are intended to direct occupants during an emergency. The onset of panic is delayed when occupants feel that the situation is being controlled, and directive communications are an effective way to foster such feelings.

They are traditionally audio, but now must be visual as well. Audio alarm systems simply make noise, alerting people to the need for special behaviors, such as exiting the building. Voice alarm systems "talk" to occupants, either prerecorded or live, telling them what to do. Visual alarms use strobes to attract attention.

Requirements

For alarm systems, see Chapter 9 (pp. 218–219) and refer to NFPA 72, The National Fire Alarm Code.

Use

Definition

Some components designed into a building must be used, and in some cases, use requires communication. Examples include elevator controls, ATM machines, and some types of dispensers. In such cases, if the component can't get a message to the user, the user won't be able to operate the component. Blind people aren't able to tell elevators to hold their doors open if they can't figure out which control is which. Deaf people can't know to open the door if they can't tell when the doorbell is ringing.

Accessibility guidelines stipulate sizes for braille and for raised letters and pictographs (including international symbols).

Requirements

In an interesting apparent contradiction, ADAAG 4.34.4 and 4.1.3(20) do not exempt drive-up ATM machines from the general requirement to be outfitted with braille controls. One may wonder why a drive-up machine should be fitted for the blind when there isn't much chance that blind people would be driving. There are potentially two answers to this:

- Blind people can sit in the rear driver's side position when driven by others, affording them a degree of confidentiality and dignity that would not be possible were they dependent on others to make their transactions for them.
- Some legally licensed drivers, especially in states with older populations and loose license renewal laws, might be driving yet have sufficiently poor vision to need oversize lettering.

Vermin Control

Background

General

This is an issue that is closely tied to region. Termites can quickly destroy an unprotected building in the southern United States, while they are not known at all in the northern-most states. Many codes contain maps delineating the geographic areas of concern. Rules for rodent control generally apply to all occupiable rooms as well as places where food or animal feed is stored.

ANIMALS
Rats, squirrels, raccoons, birds, and bats can cause problems ranging from mild inconvenience to structural damage and rabies. Many animals have an amazing ability to squeeze through tight spaces. Adult rats can flatten their rib cages to pass through openings as small as ½-in wide. Woodpeckers attack EIFS (Exterior Insulated Finishing Systems), vinyl siding, and other products that sound hollow.

INSECTS
Termites are the main infestation concern, although carpenter ants and wood borer bees can also cause structural damage in a matter of months. Many types of bees can inflict occasionally fatal stings, and cockroaches, flies, and mosquitoes can spread disease.

Requirements

For rodents, building codes commonly require [BOCA93 1215, IBC 1202.3, 1210]:

- Metal screens or grilles over foundation vents, with openings no larger than ½ in. For birds, rules require screens or grilles over attic vents, with openings no larger than ¼ in.
- Sealing of gaps at foundation penetrations.
- Size limits on the gap between a door and its frame at the jambs and sill.
- Metal screens over any windows near the ground. Where higher windows are near wires, they'll accept either screens or wire-mounted guards.
- That floors at grade be either concrete or, if wood, have metal shields or masonry skirts.

The need for or prohibition of garbage disposals is often a question of relative importance: Is the danger posed by undisposed solid waste more or less of a threat than the danger posed by inadequate capacity of available sewage treatment facilities? In some jurisdictions, health codes mandate disposals to minimize the amount of garbage kept in dwelling units. In others, they prohibit disposals because their infrastructure can't handle such appliances.

Rules identify certain tree species whose heartwood is considered to be "naturally durable" and therefore appropriate for use in contact with soil. For insects that cause structural damage, redwood and eastern red cedar are listed as termite resistant.

To control insects that spread disease, some local building codes or building code addenda require screens in all operable windows during the summer months. Health department regulations frequently mandate screens in specific types of facilities, such as commercial kitchens.

Optimizing Vermin Control Rules

When screens or barriers are needed, it is best to incorporate them into the design rather than look for ways to avoid their use. There is nothing aesthetically pleasing about the damage vermin can cause.

Structural Capacity

Background

A lot of ink in codes is expended trying to establish adequate structural capacity. Tables and formulas are provided for sizing simple structures. Codes seem to provide more guidance on smaller scale projects than on the big ones. The CABO One and Two Family Dwelling code provides extensive structural guidance. Specific structural issues addressed by codes include:

- Designing for normal and severe planetary forces such as wind (including hurricane), rain, snow, flood, soil, and earthquake, as already explored [IBC 1608 to 1613]
- Designing for use, with special focus on methods for calculating live load [IBC 1606, 1607]
- Designing for temporary measures, particularly as needed during construction
- Structural testing and inspections [IBC 1700]
- Designing for deflection so that components dependent on the structure for support get it

Certain building assemblies arising from either unique design features or unusual situations call for special structural properties. Rules recognize the following for special attention:

- Pedestrian tunnels and elevated walkways [IBC 3104]
- Temporary structures [IBC 3103]
- Tents, canopies, and awnings [IBC 3102, 3105]

- Moving existing structures [IBC 3406]
- Demolition and excavations [IBC 3303, 1803]

Requirements

Many of the issues noted above either have been covered elsewhere in this book or are too specialized for its scope. One that is neither is allowable deflection [IBC 1604.3.6, BOCA96 1604.5]. Key considerations include the following:

- Rules limit deflection relative to the length of the component being deflected—the longer the piece, the greater an absolute deflection one can expect.
- Formulas are given for live load conditions, snow loads, and wind loads.
- Some rules specifically warn design teams to consider the effects of ponding should allowable deflections be sufficient to cause any.
- Some formulas are dependent on the finish to be applied, with brittle finishes such as plaster requiring structural designs that deflect less.

Allowable deflections range from a high of l/60 for cold-formed steel roofing and siding panels to a low of l/360 for structures supporting plaster.

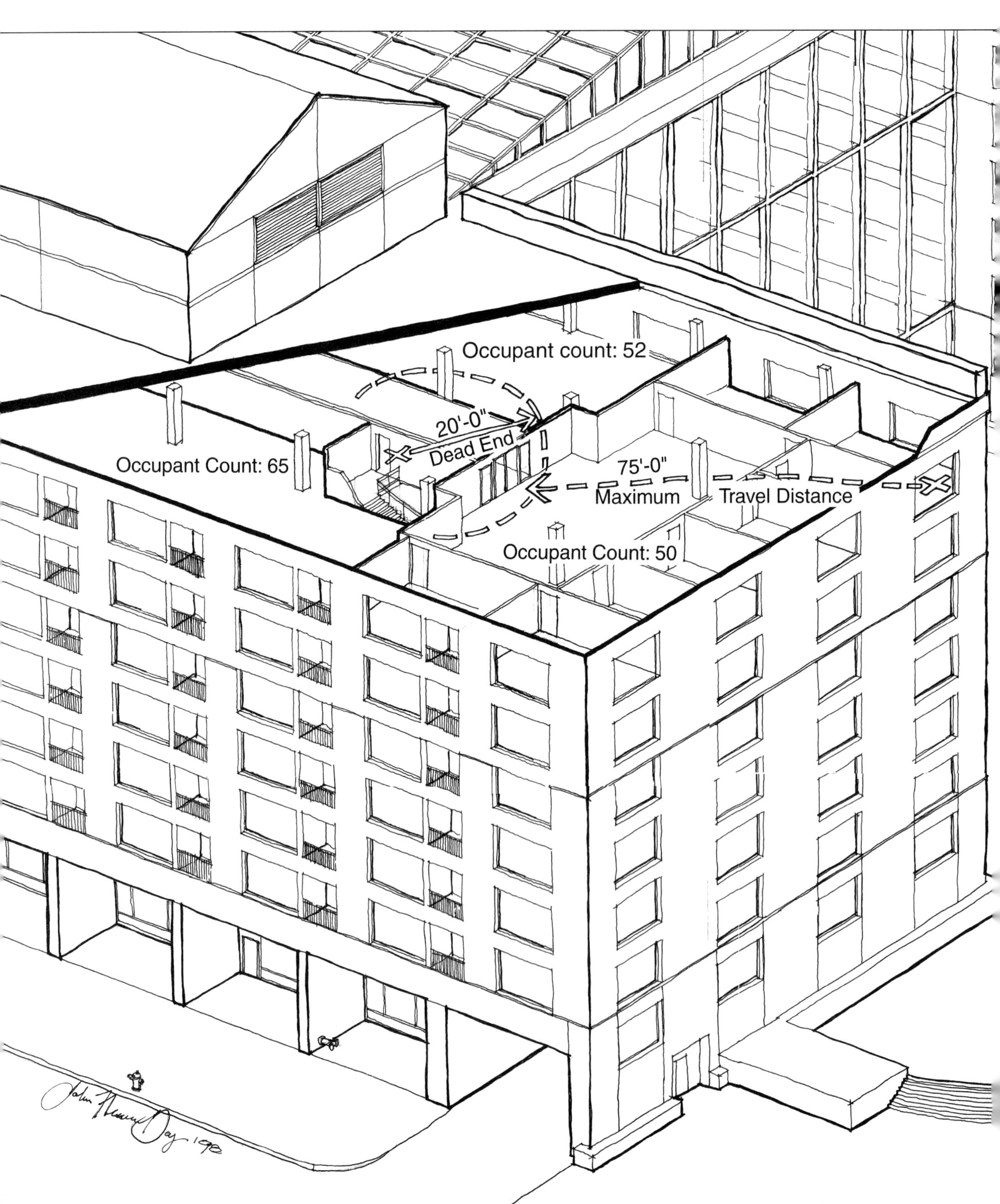
Occupant count: 52
20'-0"
Dead End
Occupant Count: 65
75'-0"
Maximum
Travel Distance
Occupant Count: 50

PART 

PROCESS AND APPENDIX

We have now discussed the origins of rules for building and have explored the major regulatory aspects of project design. You may still be wondering how to use this knowledge in practice situations. This section explores the process of regulatory analysis and documentation, and provides two additional practice tools. It includes:

- Suggestions for conducting, diagramming, and recording a project regulatory analysis
- A matrix of the subchapters of this book listing where related rules can be found in many of the widely used codes, regulations, and guidelines. This will help those who want to check the wording of the actual rules.

11 METHOD

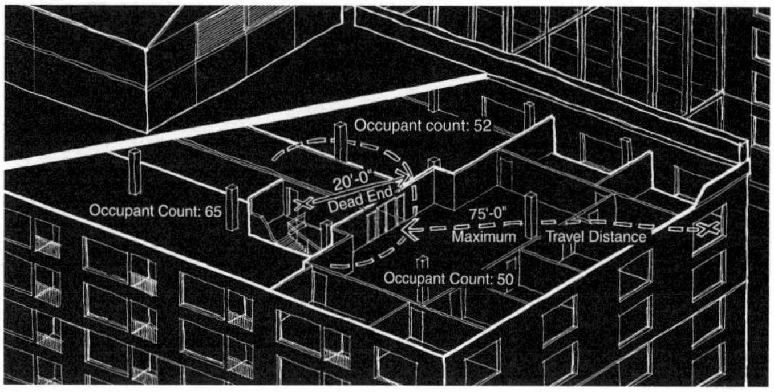

Analysis and Documentation

Determining and diagramming zoning compliance and life safety

CONTEXT

Why This Issue Is Important

As introduced in Chapter 2 (pp. 56–61), it is sometimes hard to know where to start. It doesn't make much sense to do much design work before getting some sense of the accumulated knowledge stored in codes. One does this by conducting code analyses, the scope of which should be related to the size of a project and its stage of development at the time that any particular set of code-influenced design questions are being researched.

The teams that develop, design, and build large projects require better information and better channels of communication for sharing that information. For small projects, where the aesthetic designer may also be the technical designer, formal procedures for recording and disseminating analyses may not be so critical, but the analyses themselves are no less important. There are, however, fewer variables in simpler work such as houses than in more complex projects such as community centers.

After going to all the work of analyzing rules, and them complying with them, there is nothing so frustrating as having a project get stalled during the permitting process because the plans reviewers can't (due to incomplete documentation) or don't have the time to (just as likely) figure out the extent to which your design complies with regulatory requirements.

There are three components to the work a project team must do to achieve a set of design documents and a built project that complies with rules. Each component has its own methodology:

- **Analyzing rules:** Ensuring that rules are read and interpreted properly.
- **Recording the design:** Documenting the configuration of the final design. Rules set the formats for some aspects of construction documents.
- **Demonstrating compliance:** Showing how the design fulfills rules. These are diagrams or other graphic analyses of the proposed design showing how it fulfills code mandates.

The Content

Code analyses are studies of rules, usually documented in reports that contain:

- Lists of alternative approaches (not definitive solutions) for those project issues that are informed by codes, such as site placement, massing, layout, and facade development. There should be several alternatives for each issue so that there is a range from which to choose.
- Lists of the facts that support each of the alternative approaches, with section citations from the relevant codes.

When a code analysis is done well, it answers several questions:

- What do the codes say? The architect gets an acquaintance of codes by reading them. Nobody is born with this knowledge.
- How can codes guide my design work? The architect finds out how code guidelines can help design better buildings and reduce guesswork.

- How can my design work take creative advantage of the codes? The architect reads for alternatives, exceptions and premiums that will maximize his or her design choices.

ANALYZING RULES

PERSPECTIVE

There are two perspectives one might have in using codes. One might be a member of an investment or design team who starts by analyzing rules in order not to waste time developing noncompliant design approaches. Alternatively, one might be a code official reviewing a more-or-less completed design to determine whether to issue a building permit. The official must start by analyzing the proposed design to see if it fulfills regulatory requirements. Either way, the same issues must be examined, although their order will be somewhat inverted.

How one does either of these is a personal matter. For the project team, it is important to understand that code officials look only at the design attached to the permit application to see if it complies; they don't care about the thoroughness, accuracy, or even existence of the analysis work on which it is based. Without any requirements from the officials, you are free to follow any methodology that works for you. For code officials, the government for which you work may require you to use specialized checklists for such analyses. In either case, a few methodologies are suggested below.

SUGGESTED PROCEDURES

Briefly, the BOCA and NFPA procedures have the following common elements:

- Determine the use group that applies to the project's program.
- Determine whether codes consider the building to be existing or new.
- Determine site position with respect to separation distances and their implications for exterior wall design.
- Determine height and area options.
- Determine a range of possible construction classifications. Decide on a single combination or a narrowed group of combinations of construction type, allowed fire area size, and exceptions used (such as those allowed with sprinklering) that are feasible for the project.
- Determine occupant load based on the expected sizes of program spaces. Expected sizes are usually determined by market studies conducted by the investment team.
- Determine hazard level of contents.
- Assess demands on fire performance, interior environment (including materials, arrangement, and accessibility issues), exterior envelope, structural performance, and infrastructure. Read through rules that are related to the use group of your project, noting those that apply to your project. The extent to which you note them is up to you. While your investigation can be more cursory during the early planning stages of a project, the appropriate depth is likely to vary from issue to issue. For instance, it would be prudent to take careful notes regarding permitted use and massing from the beginning while preparing predesign feasibility studies. Detailed analysis of exiting requirements could probably wait. It is easier to know what to analyze once some work is under way, since work tends to focus the analysis on only those design issues that apply. Just don't wait until you find yourself having to throw out efforts due to noncompliance.
- Determine whether any parts of the project count as special uses, and note their requirements.
- Determine requirements for the submitted contract documents—scale, type, and quantity of information to be shown (including any analysis diagrams required), and number of copies to be submitted with the permit application.

SEQUENCE

Although a proper code analysis must include the things listed above, the things listed below are just suggestions. This is by no means the only way of doing a code analysis.

Step 1: Discover

Discover what the codes want and what they offer as incentives relative to your project.

Follow These Steps:

1. Read through the parts of the applicable zoning codes, building code (including the sections on handicap accessibility, and plumbing code that apply to your project.
2. Write down rules that apply, and how they apply to your project, citing paragraph numbers to help you if you need to check them later. You will need to have a rough idea of how big and high the building you design might be to do this part.
3. List the abstract facts (e.g., "Maximum FAR is 2.0"), and the applied facts (e.g., "Therefore, the 8,000 SF site will permit a 16,000 SF building").
   - Look for exceptions and alternatives. Every code lists incentives that encourage developers and architects to explore options that the public feels are worthwhile. Codes generally do not limit a designer to one option. If you think you're required to design in just one particular way "because the code said so," you haven't read carefully enough.
   - Explain how the rules you found apply to your project (e.g., "I could use a construction classification of 3A, since that will allow for a 30,000 SF floor, and my program will probably need that much").
4. Recheck for options: Go over the facts you just wrote down to make sure you've found and understand the flexibility inherent in the guidelines. You won't find options for every rule, but you'll find many, especially for the more basic decisions. Do this in two ways. First, look for listed exceptions. Second, think about what the codes don't say. It's easy to get stuck on what they do say and forget about all the options that aren't restricted.

Look for These Things:

ZONING CODE

Find Out

Check if any additional requirements apply, so you don't get caught short when it's too late to handle it in an optimal way.

How many permits will be needed, and with what filing requirements and fees? Separate permits are commonly required for storm water management, sediment control, transferable development rights, and forest conservation.

Ask

Is your anticipated use permitted in this location? If so, is it permitted as a standard use or a permitted exception?

What are the required bulk (setback, lot coverage, height, and FAR) restrictions? What are the corresponding premiums or exceptions?

How much off-street parking must you provide? What are the corresponding premiums or exceptions? Are you permitted to put part or all of the parking required for your project on another site? Under what conditions?

How much off-street loading must you provide? What are the corresponding premiums or exceptions?

BUILDING CODE

What is (are) the occupancy(ies) and use group(s) of your project?

What are the construction classifications from which you can choose for this project, assuming that you don't use sprinklering (fire suppression) or compartmentation? What additional options are possible if you add sprinklering? How about if you compartmentalize?

What is the number of occupants in your project for purposes of calculating egress, room by room, and total? If there are several ways that you could calculate this, show each way.

What egress requirements must you meet? Make sure you check for permitted exceptions. The answer to this question is quite long, involving hallway widths, number of stairs, distance between stairs, and many other factors that all affect schematic design.

For each space listed in the program, what minimum levels of light and ventilation are required? Are mechanical ventilation and artificial lighting acceptable, or must natural sources be available?

What special requirements must you meet to provide for handicap accessibility?

PLUMBING CODE

What is the number of occupants in your project for purposes of calculating plumbing fixtures, room by room, and total?

How many fixtures (toilets, drinking fountains, etc.) will you need for this population? What options are available here?

Step 2: Analyze

Analyze what it all means to you ("What If" Analysis). This step is the most critical part but will be meaningless if you haven't done a thorough job of the preceding parts.

Zoning Code Options and Recommendations

If your use is not listed as a permitted use, what zoning change would be most likely to allow your project to proceed?

Prepare a series of site diagrams examining bulk (massing) options. There are often quite a few choices based on different ways of combining code premiums.

Prepare a series of site diagrams examining parking options. There are often quite a few choices based on different ways of working with the site and using code exceptions.

Building Code Options and Recommendations

What are the implications of the choices you've identified regarding construction classification? What other decisions will this decision affect?

What are the implications of each of the occupancy counts you calculated? Comment on which method you will use and why. What are the implications of providing a complete fire-suppression system (sprinklers)? Would it be worth the cost and complexity?

RECORDING THE DESIGN

How should contract documents be drafted and written to assure code-compliant construction?

General Scope

The general tone is set in Chapter 1, "Administration," of each of the model codes.

Mandated Information

[SBC 104.2]

... Drawings and specifications shall contain information, in the form of notes or otherwise, as to the quality of materials ...

With the following language, they preempt design teams from simply noting that construction should follow codes:

Such information shall be specific, and the technical codes shall not be cited in whole or in part, nor shall the term "legal" or its equivalent be used as a substitute for specific information.

Further, they force design teams to take responsibility for their work:

All information, drawings, specifications, and accompanying data shall bear the name and signature of the person responsible for the design.

It is noteworthy that the word *design* used here refers to the general discipline of preconstruction planning, rather than the narrow sense of aesthetic design.

For Particular Features

Many chapters of the model codes contain sections that describe minimum information to be conveyed in construction documents. The following examples are typical and make requirements quite clear.

Asking for Architectural and Structural Information

The IBC asks for a very specific, and not very brief, set of structural information to be included in the contract documents [IBC 1603], but most is information intended to demonstrate compliance rather than simply document the design. Documenting the design requires member sizes, sectional areas, and locations, fully dimensioned, for all structural members.

Rules mandate specific information for certain architectural features as well [BOCA93 2103]. The second paragraph is identical in SBC 2101.2:

The construction documents shall describe in sufficient detail the location, size, and construction of all masonry fireplaces. The thickness and characteristics of all material and the clearances from walls, partitions, and ceilings shall be clearly indicated.

The drawings and details shall show all the items required to be shown on the drawings by this code including the following:

1. *Specified size, grade, type, and location of reinforcement, anchors, and wall ties;*
2. *Reinforcing bars to be welded and welding procedure;*
3. *Size and location of all structural elements; and*
4. *Provision for dimensional changes resulting from elastic deformation, creep, shrinkage, temperature, and moisture.*

Asking for Use Patterns

[BOCA93 1203]

Drawings . . . shall designate the number of occupants to be accommodated in the various rooms and spaces . . .

Notifying Design Team of Required Correspondence

[BOCA93 1003.2]

. . . *The fire prevention code official shall be informed* in writing *of the calculated occupant load.*

Impact

This and other similar language outlines a fairly clear, if very basic, directive as to the scope of documentation and communication expected. Although compliance with these requirements does not influence the eventual design, it certainly affects the scope of work required of, and the liability incurred by, architects and engineers (some of whom may be employed by the builders). It also affects the roles of the builders' site crews, which are clearly set as following the designers' documents. Last, although not necessarily by intent, it helps set a minimum level of documentation below which results are difficult to predict. Owners would do well to note the extensive nature of required documentation when considering the fees requested by their design teams.

DEMONSTRATING COMPLIANCE

BACKGROUND

Where and When Would a Project Be Documented?

Zoning Ordinances and CC&Rs

Unless the project is very small and you can "walk it through" the permit process, your project's drawings will have to speak for themselves. Also, if a variance is sought, hearings require good graphics and thorough research.

Building Codes

As with zoning issues, your project's drawings will have to speak for themselves. How will code officials know what the occupant load of each room is? Either they will have to calculate the area of each room and do their own analysis, something that the taxpayers aren't paying them to do, or they can find it on a drawing you prepare. This is not information that you would generally put on your contract documents as they are of little concern to the builder, and the drawings would be crowded enough with notes, dimensions, and targets that there would be precious little room for code diagramming.

Accessibility Rules

You may only need documentation of compliance in situations where a building is being retrofitted for compliance. In such cases, the documentation may be needed for a court appearance.

Historic Preservation Rules

If your site is registered, some documentation may be required, but others may be available. As a starting point, look for the Historic Structures Report (HSR) prepared for a property. It documents the historic and stylistic significance that justified its original designation.

How Does One Find the Documentation Requirements?

As usual, look in model codes and in local amendments to the model codes. Call your local building office for up-to-date requirements. If these types of documentation are required, you'd better find out what information they should contain. After all, these diagrams serve only one purpose—to expedite the permitting process. Make sure you give them what they want.

Governments also differ in the number of sets they require, and in their requirements for professional seals. This last topic involves a whole catalog of rules by itself, with some wanting ink seals, some accepting only embossed seals, some wanting each sheet signed, others expecting only the cover sheet to be signed. Some want calculations and diagrams as well as contract documents to be sealed; others want only contract documents sealed. Ask.

What are The Documentation Requirements?

Most jurisdictions do not require formal code analyses, but having them can reduce the time the project spends in a permit office. The next few pages explore some ideas for documentation related to each type of rule.

- It is generally helpful to list quantities required by rules as well as quantities provided in the project as designed. This is as true of FAR calculations as it is of egress path width, exhaust capacity, or travel distance.

ZONING

Zoning envelopes can show a variety of compliance issues. If drawn in isometric or axonometric form, they can convey not only such two-dimensional concerns as setbacks and lot coverage, but also such three-dimensional concerns as height, number of stories, upper-floor setbacks, and FAR compliance.

Graphic Presentations

Zoning diagrams can be prepared for two distinctly different purposes:

- Before a schematic design has been developed, to give the designers an idea of the parameters to which they must design.
- After a schematic design is prepared, to demonstrate compliance to the zoning reviewers.

One methodology useful for establishing zoning graphics before a schematic design has been developed examines the interaction among the various formulas influencing the zoning envelope. To begin, it must be recognized that the building area allowed by zoning is defined by the smallest of either . . .

- **FAR formula:** Floor Area Ratio x Area of property
- **Lot Coverage formula:** Maximum number of floors x Lot coverage x Area of property
- **Setback formula:** Maximum number of floors x Area within setback lines

. . . where the maximum number of floors equals the maximum height divided by the programmed floor-to-floor height. To review any of these concepts, see Chapters 4 and 5.

The FAR formula says nothing about the building's location on the property or its massing. The Lot Coverage formula is almost as bare, but does establish, if indirectly, the maximum area allowed per floor. The setback formula sets some hard constraints with respect to position and massing, so is worth further consideration. Even when it is not the most restrictive in terms of allowable building volume, it is usually the most restrictive in terms of building shape and position within the property. It is often explored through use of zoning envelope diagrams. These diagrams establish the size and shape of the maximum permitted mass. They are generally drawn in axonometric, but may also be drawn as isometrics or perspectives.

1. Maximum height can be expressed with a plane drawn at the proper distance above a drawing of the property, as shown below.
2. Next, planes representing setbacks can be added.
   - Fixed setback formulas can be expressed with vertical planes.
   - Step-variable formulas can be expressed with corrugated planes.
   - Continuously variable formulas can be expressed with angled planes.

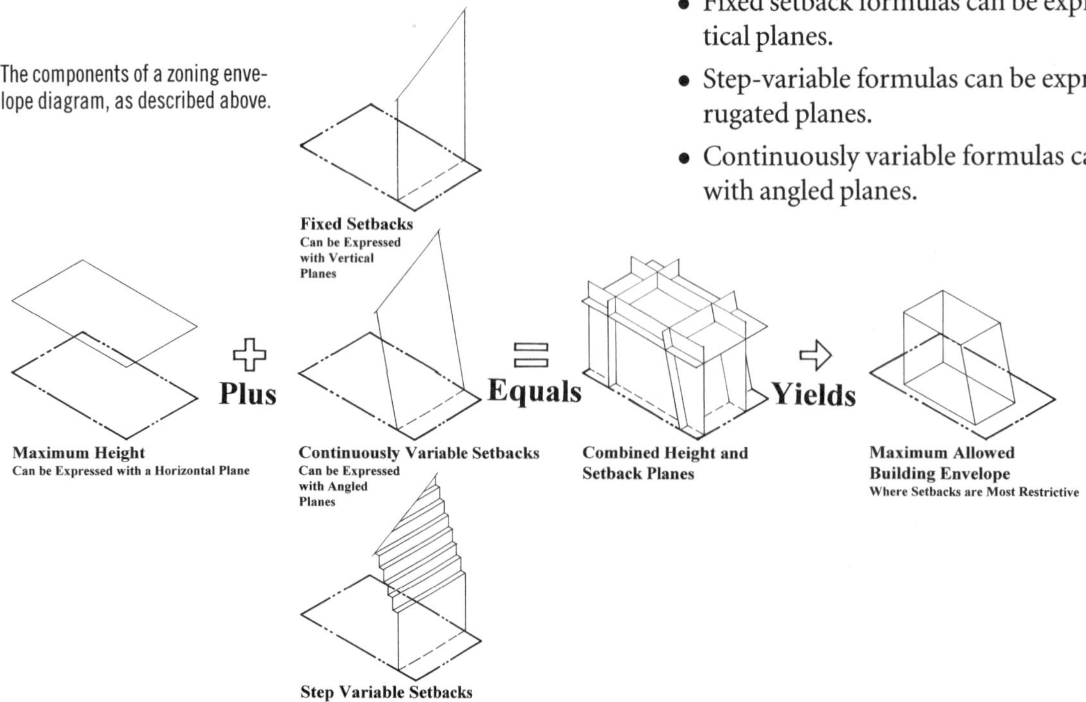

The components of a zoning envelope diagram, as described above.

3. Together, the setback and height planes define the maximum building envelope allowed by the setback formula.

While buildings are never required to fill such building envelopes, envelope diagrams are useful in helping visualize the limits. Such diagrams might be thought of, conceptually, as representing molds of the maximum allowable mass. Proposed designs need not fill these molds. Even if FAR and Lot Coverage formulas are more restrictive, building envelope diagrams indicate permitted locations for the building volume. Designs will meet regulatory requirements so long as they stay within the maximum zoning envelope (except for permitted projections and encroachments as discussed later) and comply with FAR or Lot Coverage formulas when either is more restrictive.

Of course, zoning analyses can also be used to clarify just how effectively a particular design optimizes zoning provisions. In this case, since the conceptual design is already determined, the analysis need not be graphic.

District: B7-6	
Maximum Dwelling Units Allowed 1 multiroom unit allowed for each 115 SF of lot area, 1 efficiency unit for each 75 SF of lot area	246
Dwelling Units as Designed	240
Off-street Parking Required	155 stalls
Off-street Parking as Designed	186 stalls
Base Allowable FAR	12.00
FAR Premium 0.15 per floor for setbacks at 25 residential levels	3.77
Total Allowed FAR	15.77
Building Area as Designed	499,620 SF
Total FAR as Designed	15.36
Total Unused FAR	0.41, equaling 10,420 SF

After schematic design: proposal for Chestnut Place. © Yatt Architecture, Ltd.

CASE STUDY: MULLET HOUSES

A project proposal presented by KCF Architects of Washington, D.C., several years ago demonstrates these issues well. The investors wanted to maximize the amount of space they could add to a group of existing buildings. The architect presented four different massing schemes before starting schematic design. The design of only one was fully within the envelope allowed by the zoning ordinance. The other three required some degree of optimization, with no guarantee that such optimization would be achievable. Nonetheless, the architect presented all four schemes to its client, clearly indicating what was and was not within the rules, and the extent to which optimization need be achieved.

A photograph of the existing property. Photo by the author.

CASE STUDY: MULLET HOUSES (continued)

Memo

From: KCF Architects

The following summarizes four approaches to the development of the Mullet Houses site. For the purposes of this study, the Mullet Houses site is looked at independently of the rest of the development at 2501 Pennsylvania Ave. To determine compliance with the zoning rules, it is assumed that the lower two floors are of nonresidential use and the upper five floors are residential. In all cases it is assumed that the front of the Mullet Houses will remain intact. The four approaches are as follows:

SCHEME A: Development within the limits of the existing building envelope. This scheme would require no zoning relief.

SCHEME B: Development of the full site area for four floors, with the top floor, behind the mansard roof, remaining in the current configuration. This scheme would require zoning relief for both lot occupancy and rear yard.

SCHEME C: Development of the full site for five floors, including a full floor behind the mansard at the fifth floor. This scheme would require zoning relief for both lot occupancy and rear yard.

SCHEME D: Development similar to Scheme C, except that both lot occupancy and rear yard rules would be met. This scheme would require no zoning relief.

A memo from KCF to the investors

SITE DATA

Zone: C-2c (High-Density Community Business District)
Site Area: 6,508 Square Feet

LOT OCCUPANCY

Residential Portion (Nonresidential allows 100%)

Allowed	A	B	C	D
80%	66%	**99%**	**99%**	80%

REAR YARD SETBACK

Less than 20 ft above grade (measured from center of alley)

Allowed	A	B	C	D
15 ft	37 ft	**5 ft**	**5 ft**	15 ft

More than 20 ft above grade (measured from property line)

Allowed	A	B	C	D
15 ft	32 ft	**None**	**None**	18.89 ft

Comparative tabulations, with violations in **bold**

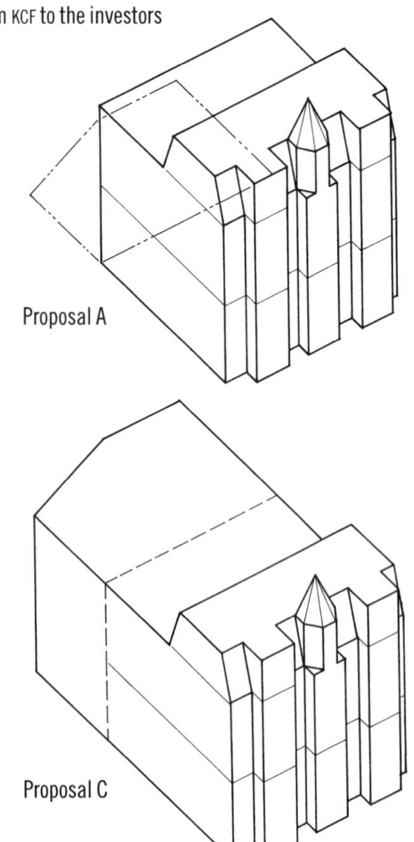

Proposal A

Proposal C

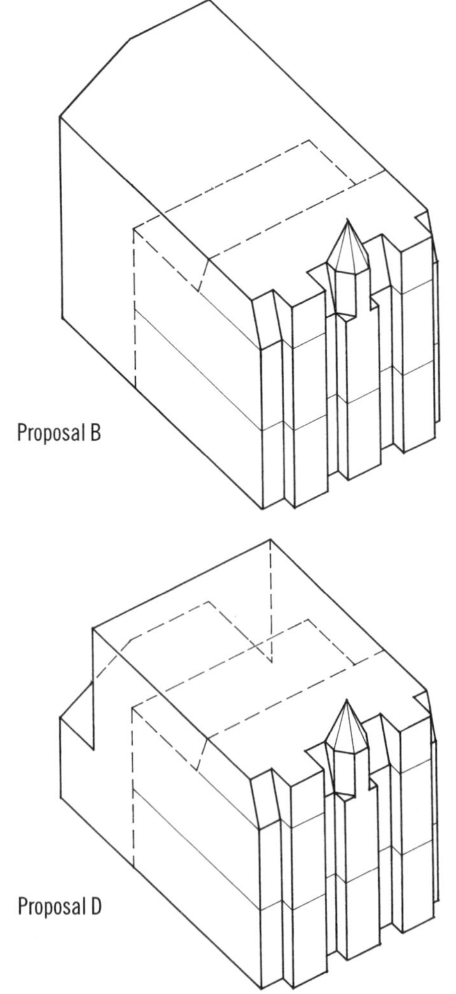

Proposal B

Proposal D

Verbal Presentations

The purpose of diagrams can also be achieved with carefully considered verbal analyses. As an example of a well-crafted opening statement in defense of a proposed variance, consider the following:

Testimony of Arthur H. Fawcett, Jr., in Board of Zoning Adjustments Hearing for variances related to the construction of a roof deck at 1763 R Street NW.

Good Morning.

I am Arthur H. Fawcett, Jr., resident and owner, with my wife, of the house at 1763 R Street NW.

We are seeking variances from the zoning rules (1) to permit enlargement of a nonconforming structure and (2) to vary the floor area ratio requirements for a small structure for access to the roof of the house.

This property is a single-family house with a small apartment to the rear. The house was built in 1909, many years before the enactment of the zoning rules. It is now in an R-5-B district; about 25 ft to the east is an SP-1 district along New Hampshire Avenue. The property across P Street is also zoned SP-1.

1763 R Street contains about 6,850 SF of floor space, where the current FAR rules permit 4,271. The house covers 1,712 SF of the lot, where the rules permit 1,425. Most, if not all, of the buildings on the R Street frontage from the SP district along New Hampshire Avenue to the C-3-B district along Connecticut Avenue, which were built around the turn of the century, are nonconforming structures in the R-5-B district.

We are seeking this variance to permit construction of a roof deck on the building. All of the outdoor space on the lot is needed for the two off-street parking spaces that exist there. These spaces reduce parking congestion on the streets nearby. In order to create livable open space, we propose to build a roof deck over the front section of the building. In order to gain access to this roof deck, we want to extend the interior stairway up into a small access structure on the roof. The total additional floor space represented by the stairs and access structure is 130 SF, less than 2 percent of the existing space. (Although the permit office had included a proposed enclosed pavilion on the roof in its definition of the variances needed, we have since calculated that this pavilion would be defined as a roof structure, and couldn't be habitable. so we have deleted it from the plan.)

The property does not provide for usable outdoor open space for the residence other than the parking, and has not done so since it was built, long before the enactment of the zoning rules. The proposed roof deck would relieve this practical difficulty by providing usable open space on the roof. The deck would be bounded by the existing parapet on the front of the house, a short extension of it to a chimney on the west side, and a fence at the rear of the deck, in the middle of the house. The only additional structure visible from the street would be the short extension of the parapet on the west side. The structure would have no impact whatever on surrounding properties; it would be invisible from the street or from the alley in the rear. It is consistent with the rules and the Comprehensive Plan in that it:

- Meets housing objectives by making this structure more viable as a residence
- Meets transportation objectives by maintaining adequate off-street parking
- Meets design objectives by providing outdoor space without altering a building that enhances the streetscape and is compatible with its surroundings
- Meets historic preservation objectives by providing open space without altering the appearance of a historic structure

The proposal, with the pavilion, was approved enthusiastically by the Dupont Circle Conservancy, and the staff of the historic preservation office has recommended approval to the board. ANC (Advisory Neighborhood Council) 2B approved the proposal unanimously at its May meeting, and I understand that the Office of Planning is proposing approval of the variances. We know of no opposition from anyone.

If the Board can approve this, we would like to get the decision as soon as possible. This has been a long process: we applied for a permit last September 6, but we didn't get the letter from DCRA permitting us to apply for a variance until late February, so we are really eager to start, if the variances can be granted.

Our architects, Robert Froman and Michael Fox, are here to answer technical questions.

Letter to D.C. Board of Zoning Adjustments. Used with the permission of Arthur H. Fawcett, Jr.

EGRESS

[LSC94 A-8-2.3.2]

Demonstrating compliance with egress requirements in graphic form is useful for more than reducing the time a project spends in the permit process at city hall. Egress diagrams can be generated at several points in the development of a design and with varying degrees of specificity to explain conceptual design considerations to a project's senior designer or the owner, to check compliance at the level of schematic design, or to demonstrate compliance just before submitting for a building permit.

When architects prepare preliminary egress diagrams and use them to check basic compliance before making schematic design client presentations, they can catch any noncompliant design features when they can still be changed relatively painlessly. Once owners approve conceptual designs, it can be difficult and embarrassing for their architects to explain that those designs must be altered, sometimes radically, to meet egress provisions of the code. It's hard to explain that prime square footage will need to be reallocated for an additional exit, or that core efficiencies will have to be grossly reduced in order to shorten exit distances. The documentation process is fairly straightforward and does not have to be onerous.

Examine the Program

Determine Use Classification(s)

This is not a question of getting permission, as it was with the zoning analysis. Project teams need only confirm the nomenclature that corresponds with their intended program.

Determine Program Relationships

Adjacency relationships, stacking relationships, minimum workable floor plate, and so forth.

Examine the Design Concept

Make Basic Design Decisions

Construction classification
Extent of sprinklering, if any
Number of floors
Size of fire areas

Examine the Exiting

Calculate Egress Occupancy

Calculate occupancy per programmed space (or calculate programmed space per egress occupancy).

Calculate aggregate occupancy per floor.

Diagram the Layout of Individual Spaces

Determine circulation within space: aisles.

Determine maximum distances allowed from most remote corners of programmed spaces to doors leading to exit access corridors.

Exit Access Corridors

Check maximum allowable length of dead-end (single choice, or "common use") corridors.

Determine widths of corridors and doors.

Exits

SHOW EXIT COUNT AND DISTANCES

Show the required number of exits per floor for those classifications.

Show the minimum distances required between those exits. Required distance refers only to the most remote two exits. The other exits could be anywhere.

Show that remote corners of the plan don't exceed maximum travel distances. Draw circles to show radii where appropriate.

Show that dead-end corridors are within allowed maximums. Radius lines help here, too.

SHOW COMPLIANCE WITH REQUIRED DIMENSIONS

Determine and show spacing of railings on a wide stair, landing widths, corridor widths, etc.

Document All Incentives and Options Used in the Design

Is it different because it includes sprinklers?

Is it different due to an unexpected construction classification?

Is it different due to conformance with requirements for high-rise buildings, even if it isn't one?

Highlighting where "tricks of the trade" have been used:

- Scissor stairs
- Exiting through the lobby (50 percent maximum)
- Mezzanine allowances and restrictions
- Circumventing common closure requirements

HVAC AND PLUMBING WORK

In addition to showing all work, calculations are required. Many jurisdictions look for:

- Fire protection calculations, including pipe schedule calculations and hydraulic calculations, with the latter based on pressure drops, flow rates, and fixture counts. These are often delegated to fire suppression subcontractors by specifications.
- Energy budgets.
- Ventilation schedules.

Check in state and local ordinances for documentation requirements.

Negotiations are based on the calculations, in marked contrast to negotiations over planning, architectural, preservation, or accessibility issues. That difference allows for a lot of optimization, so long as numbers can demonstrate equivalency. For example, a storm line installed with a reverse slope might prove to be acceptable, as is using calculations to show spare capacity in the line. Other project disciplines can have a much tougher time demonstrating compliance precisely because calculations would often be meaningless.

LIGHTING AND POWER WORK

Many jurisdictions look for energy budgets and lighting tables. Check in state and local ordinances for documentation requirements.

STRUCTURAL WORK

Like mechanical work, structural design can be demonstrated with calculations. Codes require the following:

- Design loads
- Live loads, broken down as uniform, concentrated, and impact loads for all floor plans, with roof live loads documented as well.
- Ground snow load (P_g) for all projects, with values over 10 PSF additionally listing the flat roof snow load, the snow exposure factor, and the snow importance factor.
- Wind loads for all projects, including the basic wind speed, the wind importance factor and building category, wind exposure factor, and wind design pressure.

OCCUPATIONAL SAFETY

Some diagramming is required by OSHA and by BOCA96 1603.

WETLANDS AND FLOODPLAINS

Based on Executive Order 11988, "Floodplain Management" (May 24, 1987), federal agencies must evaluate action taken in a floodplain. The Floodplain and Wetlands Assessment is prepared pursuant to the DOE "Compliance with Floodplain/Wetlands Environmental Review Requirements" [10 CFR, 1022, 1979]. There must be a floodplain assessment (a document sent to the public) for any action inside a floodplain.[51]

HAZARDOUS SUBSTANCES

Background

There are two distinct types of reports structured by published rules.

- Environmental impact: These evaluate the effects proposed developments are expected to have on ecosystems, particularly as defined by watersheds. Rules for such studies are mostly found in federal and state codes.
- Assessment: These evaluate the environmental hazards present in existing properties. Rules for such studies are mostly found in standards published by ASTM (See pp. 91–92.).

Documentation Required by Governments

Project teams must call the government with jurisdiction over the project site to get their documentation requirements. While most simply look for the documentation required by model codes and other regulations, they may want to see specific diagrams or certifications. Bear in mind that approvals may need to be obtained from agencies and organizations other than those of the local government, including Design Review Boards, the EPA, and state OSHA and historic preservation offices.

Documentation Required by Private Entities

Background

Documentation is often required not by the EPA or other state or federal agency but by projects' lenders.

Banks know the potential for risk related to environmental impact and use rules to help manage this risk. In effect they are saying "Want a construction loan or mortgage? Get tested."

These reports also help when pursuing claims against neighboring properties whose hazardous materials are migrating, causing liability to others. They can also help establish an "innocent landowner defense" for property owners with respect to CERCLA liability (meaning that they tried).

ASTM publishes four "standard practice" documents explaining procedures for "Transaction Screening," "Phase I and II" environmental site assessments, and "Phase III" remediation.

The Recommendations

TRANSACTION SCREENING

This is a cursory and very preliminary review and takes no more than an hour or two using a checklist [ASTM E1528].

PHASE I ASSESSMENTS

Phase I assessments [ASTM E1527] involve a review of available information but no physical testing or sampling. They have four components:

- A review of available records of proposed properties and others within a "minimum search distance" established by the standard. This is intended to uncover known information about the way the site is and has been used, any violations previously recorded, and the like. Sources of information include government records and aerial photographs.
- A site visit for purposes of direct observation. Evaluators look for direct and indirect evidence of hazardous wastes.
- Interviews with property managers and occupants.
- Preparation of a report.

PHASE II ASSESSMENTS

When Phase I assessments suggest contamination, Phase II assessments are done. They involve physical testing and sampling for radon, lead, asbestos, and other contaminants suspected in the Phase I report. Testing and sampling are conducted at the property, with results confirmed through laboratory analysis. When contamination is confirmed, Phase II assessments also recommend corrective measures.

PHASE III REMEDIATION

This is the actual work of removing, encapsulating, or otherwise eliminating the dangers posed by hazardous materials.

The Report

The ASTM Standard Practices include recommended tables of contents for reporting the results of assessments, going so far as to suggest documentation appropriate for optional appendixes.

HISTORIC PRESERVATION

Inclusion in the National Register of Historic Places

Existing buildings must be surveyed, and a set of drawings prepared for the HABS/HAER (Historic American Buildings Survey/Historic American Engineering Record) initiative. Other requirements vary by local jurisdiction, with each having its own historic designation requirements.

Getting Approval for Rehabilitation Designs

When a historic review committee must review work before construction starts, project approvals can take a while. Bear in mind the multiple levels of review many projects must undergo, from local, state, regional, and federal authorities, and allow time in the development schedule. Some direction may be found in William J. Murtagh's *Keeping Time: The History and Theory of Preservation in America* (New York: John Wiley & Sons, 1997).

For directions on the actual preparation of a report, the General Service Administration (GSA) publishes a Historic Building Plan (HBP), which suggests a format for getting the job done.

ACCESSIBILITY

Documentation

- **Interior:** Indicate turning circles in each required accessible toilet and the distribution of accessible stalls and areas of refuge.
- **Parking:** Indicate the count and position of accessible stalls.
- **Site plan:** Indicate curb cuts and grade accesses to buildings.

APPENDIX
Citations

THIS APPENDIX CONTAINS a series of matrices citing related sections from many of the applicable rules. It will be particularly useful to readers, structured as it is around the Table of Contents of this book. Readers are encouraged to get the background on particular issues from the text and then check this Appendix for specific rules to help them with planning or design decisions. Where the text lists one citation, the parallel citations from other sets of rules will be found here. This will be particularly helpful with references to the forthcoming *International Building Code.* Each matrix spans two facing pages, reading approximately as if the facing pages were aligned left and right; the first column lists each issue's location in this book.

Readers will note that many boxes are left unfilled. The matrices, and the blank spaces they contain, can be used by readers as a place to keep their own notes and code findings. The author welcomes comments and additional citations from readers.

For issues related to urban land use planning, listings are limited to check marks, indicating that the topic is probably addressed in local rules—specific citations are not possible because there are no national model ordinances or even predominantly accepted formats for them.

Subject Locator and Citator

	Location in this book	As described in the rules	Planning: Zon'g Ord.	Planning: Subdv Regs.	Planning: C C & R's	Accessibility: ADAAG	Accessibility: ANSI A117.1	Accessibility: UFAS	Sect'y of Int Stds	OSHA (29 CFR 1910)	Other U.S. Code	Other Code of Fed Reg's	Loc & State Codes	Other Rules: Name	Other Rules: Citation	Catalogs: Master-format	Catalogs: Uni-format
1	**People**																
	Context						Chpt 1										
	Participants																
	Investment Team															01	
	Design Team	Licensing Reqmt					Chpt 2						√			01	01
	Construction Team	Licensing Reqmt											√				
	Users Group	Building Departments	√	√	√	36.600							√			01	01
		Powers and Duties of the Building Official	√	√	√								√			01	01
	Trends in Code Review																
2	**Rules**																
	The Concept																
	Basic Approaches																
	Basic Types																
	Planning Rules																
	Building Rules											16: Conservation	√			01	01
	Civil Rights Rules																
		Accessibility Rules				√	√										
		Fair Housing Rules															
	Advocacy Rules											29: Labor				01	01
	Material and Workmanship Rules	References									15-47 15-50					01	01
	Applicability																

Subject Locator and Citator

Model Building Codes										Other Codes		
IBC Affiliate Building Codes							Specialty Codes		CABO	NFPA		NBCC
IBC	BOCA		SBC		UBC		Name	Citation	(Residential)	Life Safety	Other	(Canadian)
5/97 Draft	Pre '93	Post '93	Pre '93	Post '93	Pre '93	Post '93			All	NFPA 101	Stds	All
	108	114	103, 501	104, 3503					R-109	---		
	109			3503					R-109			
103	110	104	101	102					R-104	---		
104		105	102, Chpt 3	103, 3503	202					1-4.3		
					6001-03							2.2.1;2,3,1
		3500	Chpt 3	3501-03					R-108, R-113			2.6.4

Subject Locator and Citator

Location in this book	As described in the rules	Planning: Zon'g Ord.	Planning: Subdv Regs.	Planning: C C & R's	Accessibility: ADAAG	Accessibility: ANSI A117.1	Accessibility: UFAS	Sect'y of Int Stds	OSHA (29 CFR 1910)	Other U.S. Code	Other Code of Fed Reg's	Loc & State Codes	Other Rules: Name	Other Rules: Citation	Catalogs: Master-format	Catalogs: Uni-format
Do Rules Apply?	Existing Buildings: General	√	√		36.400	2.2.3		√		16-1.A					01	01
	Existing Buildings: Unsafe														01	01
Which Rules Apply?	Code Applicability	√	√	√	36.102, A4.1.1	1.1.1, 1.1.2				16-30 16-59a		√			01	01
Which Sections Apply?	Validity-Severability	√	√	√	36.103			√							01	01
Using Rules																
Definitions	General	√	√	√	36.104, A3.4, A3.5	Chpt 3									01	01
	Occupancy Classification														01	01
	Constructn Classif														01	01
	Fire Resistance							√				√			01	01
	Fire Protection & Prevention											√				
Getting Approvals																
	Administration	√	√	√		2.2.1				41-1					01	01
	Approvals	√	√	√	36.600	2.2.5									01	01
	General, Definitions	√	√	√		Chpt 3									01	01
	Orders to Stop Work														01	01
	Violations and Penalties	√	√	√	36.501 to 36.505										01	01
	Workmanship														01	01
	Permits	√	√	√											01	01

Subject Locator and Citator

Model Building Codes										Other Codes		
IBC Affiliate Building Codes							Specialty Codes		CABO	NFPA		NBCC
IBC	BOCA		SBC		UBC		Name	Citation	(Residential)	Life Safety	Other	(Canadian)
5/97 Draft	Pre '93	Post '93	Pre '93	Post '93	Pre '93	Post '93			All	NFPA 101	Stds	All
201; 3401-7		3401	101, 402, 1201	3401-02	104				R-116	1-4.1, 1-4.4 to 1-4.6, odd-numbered Chapters 9 to 19 and 23 to 27		
	120	119	102	103	203					---		
102	101	102	101	101	103-104				R-103	1-4		2.1-2.2.3; 2.1.4
	102	103	106	109	303					---		
201	200; 201	201	Chpt 2	201-02	Chapter 4				R-118, S26-000-01, S26-118	3		1.1.3,1.1.4,9.1, 9.1.1,9.2, 9.2,1
301,02	300;301	301, 302	401	301, 302	501				R-103	4-1.1		3.1.1; 3.1.2
601-03	401	601-06	601-07	601-610, Table 600	1701-17							
701,02,04; 416		701, 702	608, 1001, 1003	701, 702						6		2.3.3;
101-15	107	101	101	101	101					1-1 to 1-3		2.6.5
	107	106;1704		---	305,306				R-110	---		2.6.5
1701-11	200; 1300	1701, 1702	Chpt 2, Chpt 25	201-02, 1701, 1702					R-118	1701, 1702		1.1.3,1.1.4
115	118	117	102, 107	103.3	202				R-117	---		
114	117	116	107	110	205		BOCA Prop Mgt	PM-106	R-106	---		
110	116	115		---						---		2.6.4
105	111; 112; 2704; 2803	107 - 108	501	104, App B	301, 5110, 7003				R-110	---		

Subject Locator and Citator

	Location in this book	As described in the rules	Planning			Regulations, Standards, and Guidelines										Catalogs	
			Zon'g Ord.	Subdv Regs.	C C & R's	Accessibility ADAAG	ANSI A117.1	UFAS	Sect'y of Int Stds	OSHA (29 CFR 1910)	Other U.S. Code	Other Code of Fed Reg's	Loc & State Codes	Other Rules Name	Citation	Master-format	Uni-format
		Inspections														01	01
		Required Testing														01	01
		Certificates	√	√	√											01	01
	Optimization																
		Adjustments and Appeals				36.506, 36.606					41-4 41-9					01	01
		Planning															
		Building															
		Accessibility															
		Historic Preserv															
3	**Use**																
	Intended Site																
	Public Needs																
	Regional Conditions	Dead, Live, Snow, Wind, Earthquake									15-47					N/A	
	Site Conditions																
	Intended Purpose																
	Sources of Permission	Enterprise Zones															
		Fire Districts	√													01	01
	Degrees of Permission																
4	**Position**																
	Dealing w/Prop Lines																
	Encroachments	Construction in the Public Right of Way											√			01	01
		Canopies & Awnings									42-86					13	

Subject Locator and Citator

Model Building Codes										Other Codes		
IBC Affiliate Building Codes							Specialty Codes		CABO	NFPA		NBCC
IBC	BOCA		SBC		UBC		Name	Citation	(Residential)	Life Safety	Other	(Canadian)
5/97 Draft	Pre '93	Post '93	Pre '93	Post '93	Pre '93	Post '93			All	NFPA 101	Stds	All
109	115; 2502; 2702; 3003	105.4; 113; 1705; 2704; 2804; 3004; 3303	103, Chpt 25	105, 1709	305				R-105, R-112, H-115	---		
703; 1701-11; 1903	903; 1202, 03; 1300 01, 03, 06; 2602; 2702	1701,10; 2704,2804; 3004	104, Chpt 25	107, 1703, 1706-1708	107				M-1711, P-2014, P-2208, P-2412	LSCH Suppl 5		2.5.3
111	119; 2603; 2704; 2803	118; 2904; 3005	103	106	308				R-115	1-6		2.7.1
113	124; 3022	121	105	108	204				R-105			
									R-116			
1605-09, 13; 2404,04	1102, 03, 04, 11-14; 1204	1601-10,13; 1710; 1805	Chpt 12	1603 - 1607, 1704-05, 1912-13, 2115, 2213, 2312, App A	Chpt 23, Div 1,2,3							4.1-4.1.10; 5.2; 9.4-9.4.4
706	1014	---	Chpt 30, App F	505, App F								3.1.2
3201, 02	50607,10; 3006	3201-05; 3304, 08	Chpt 22	3201-07, 3303-09,3311-12	Chpt 44				R-111, R-201-02			8.2.5; 8.2.6
3105	510	3105			4501, 06							

Subject Locator and Citator

	Location in this book	As described in the rules	Planning: Zon'g Ord.	Planning: Subdv Regs.	Planning: C C & R's	Accessibility: ADAAG	Accessibility: ANSI A117.1	Accessibility: UFAS	Sect'y of Int Stds	OSHA (29 CFR 1910)	Other U.S. Code	Other Code of Fed Reg's	Loc & State Codes	Other Rules: Name	Other Rules: Citation	Catalogs: Master-format	Catalogs: Uni-format
	Cones of Vision																
	Dealing w/Adj. Bldgs																
		Buildings Sharing a Site		√												01	01
	Dealing with Site Features																
	Floodplains	Flood-resistant Construction	√								42-50		√			01	01
	Wetlands																
	Soil Conditions												√				
	Rights of Passage																
	Archaeological Finds																
	Other Site Features																
	Position for Pedestrian Access																
	Position for Vehicular Access																
	Roads						4.7						√				
	Parking	Accessible Parking		√	√		4.6				15-49 42-15A		√	ANSI CFR	4.6 A4.6.3	01	01
	Loading						4.6.3										
5	**Massing**																
	Context																
		General, Definitions	√						√							01	01
	Establishing Size & Height																
	Estab. Total Size, Size per Floor, Total Height	Height and Area, Compartmentation	√				4.1, 4.2		√							01	01
	Area & Height	Height and Area Reqmts	√	√			4.1, 4.2		√							01	01
		Height and Area Modifications	√	√			4.1, 4.2		√							01	01
	Estab. Shape																
	Architectural Style																

Subject Locator and Citator

Model Building Codes										Other Codes		
IBC Affiliate Building Codes							Specialty Codes		CABO	NFPA		NBCC
IBC	BOCA		SBC		UBC		Name	Citation	(Residential)	Life Safety	Other	(Canadian)
5/97 Draft	Pre '93	Post '93	Pre '93	Post '93	Pre '93	Post '93			All	NFPA 101	Stds	All
503,08, 3402	502	---	610	504					R-201-02			
1611,12	2102	3107		3107	2390-96							
					7013							
1103-05,10,11		407, 408, 1105		1104	3107, 3108				R-209			4.4.2; 4.4
501,02	500.01	501, 502		501, 502					R-201-02			2.1.6
503	502;503	503		503, Table 500	407, 409				R-204-05, R-207	6-2.2		2.1.6; 9.5-9.5.4
Table 503	Table 602	Table 503	Table 400	503, Table 500	Tables 5C & 5D				R-204-05	---		3.71; 9.5-9.54
504, 506, 507	502,03	504, 506, 507		610								3.7-3.7.5

Subject Locator and Citator

	Location in this book	As described in the rules	Planning: Zon'g Ord.	Planning: Subdv Regs.	Planning: C C & R's	Accessibility: ADAAG	Accessibility: ANSI A117.1	Accessibility: UFAS	Sect'y of Int Stds	OSHA (29 CFR 1910)	Other U.S. Cod	Other Code of Fed Reg's	Loc & State Codes	Other Rules: Name	Other Rules: Citation	Catalogs: Master-format	Catalogs: Uni-format
	Projections	Trims, Balconies, & Bays							√		15-49 42-15A					01	01
	Roof Design	General, Definitions							√		15-49 42-15A					07	
		Roof Storm Runoff									15-49 42-15A		√			07	
		Parapets									15-47 15-50 15-63 36-21 40-6 40-12					07	
		Other Rooftop Structures									15-47 15-49 15-50 15-63 36-21 40-6 40-12 42-15A					07	
		Roof Cooling Towers									15-47 15-49 15-50 15-63 36-21 40-6 40-12 42-15A					07	
		Roof Penthouses									15-47 15-49 15-50 15-63 36-21 40-6 40-12 42-15A					07	
		Roof Tanks														07	
	Seismic Concerns																
6	**Enclosure**																
	Context	General, Definitions							√							07, 08	
	Wall & Window Design																

Subject Locator and Citator

Model Building Codes										Other Codes		
IBC Affiliate Building Codes							Specialty Codes		CABO	NFPA		NBCC
IBC	BOCA		SBC		UBC		Name	Citation	(Residential)	Life Safety	Other	(Canadian)
5/97 Draft	Pre '93	Post '93	Pre '93	Post '93	Pre '93	Post '93			All	NFPA 101	Stds	All
508	507			1404	4504-06, 1711							
1501-2	2300-06	1501, 1502		1501, 1502	2305, 3203-04, 3601-02				R-701, R-801, 26.701			3.1.15; 5.1.4, 5.1.5; 9.19-2; 9.23.13,9.23.15;
1508	2304	1510;1511; 3105		1508								5.6.1.5.6.2; 9.26.18
1511	927			1507	1710							
1511;1614; 3006	927			1506	3602							
508; 1511; 2801	927			1505	3601							
508; 1511; 2801	927			1503	3601							
508; 1511; 2801				1504	3601							3.1.14
1401,02	2100,01,04	1401, 1402		1401, 1402	2301-05							5.3, 5.4, 5.5, 5.6, 5.7, 5.8; 9.27.2, 9.28.1
					2309				R-208			

Subject Locator and Citator

Location in this book	As described in the rules	Planning			Regulations, Standards, and Guidelines										Catalogs	
					Accessibility								Other Rules			
		Zon'g Ord.	Subdv Regs.	C C & R's	ADAAG	ANSI A117.1	UFAS	Sect'y of Int Stds	OSHA (29 CFR 1910)	Other U.S. Cod	Other Code of Fed Reg's	Loc & State Codes	Name	Citation	Master-format	Uni-format
Wall Assemblies	Fire Resistance									15-49; 15-50					07, 08	
Wall Materials	Performance Reqmts									15-49; 15-50					07, 08	
	Veneered Exterior Walls							√		15-49; 15-50					07, 08	
Windows						4.12										
Balconies & Railings	Glass: Assemblies—Railings									15-47; 15-50; 15-63; 36-21;40-12					08	
	Means of Egress									15-49; 42-15A						
Exterior Circulation	Balcony Egress									15-49; 42-15A						
Wall & Window Details																
Wall & Window Details	Installation					4.12									07, 08	
	Signs														13	
Special Construction	Window Cleaning														13	
Roofs																
	Roofing							√		15-47; 15-50; 15-63; 36-21;40-12					07	
Foundations																
	Definitions							√		15-47; 15-50; 15-63; 36-21;40-12					03, 04	Sub-structure
	Excavations											√			02	

Subject Locator and Citator

Model Building Codes										Other Codes		
IBC Affiliate Building Codes							Specialty Codes		CABO	NFPA		NBCC
IBC	BOCA		SBC		UBC		Name	Citation	(Residential)	Life Safety	Other	(Canadian)
5/97 Draft	Pre '93	Post '93	Pre '93	Post '93	Pre '93	Post '93			All	NFPA 101	Stds	All
704	906,20,22	705		704, Table 700								3.1., 3.1.7, 3.1.8, 3.1.10; 3.2
1403	904			709, Table 700					R-401-02			5.3, 5.3.1; 5.4, 5.4; 5.5, 5.5.1; 5.6-5.6.2
1405	2106-09	1403		1403	Chpt 30, Chpt 55				R-503, S26-503			5.3.1; 5.4.1; 5.6.1-2; 5.7.1; 5.8.1-2; 9.27-.13; 9.28-.6
					5401				R-411, S26-411			
2406	815,16,28	1022; 2406		1015					R-214			9.8.7
1003.13	828	1022		1015	408					5-2.2.4		9.8.7-8
1004.6	807	1006		1014	403					5-2.4.4		
1405; 2504	1410; 1604; 1704				5401-02, 5404-05				R-411-12			
1109; 3107		3102										
	713	3103										
1501-07,12	2300	1512		1509	3201-04, 3208-15				R-802-09, S26-705, S26-801, S26-807-09			9.26-9.26.18
1801	1200	1801, 1802		1801, 1802	2901-02, 2904				Chptr 3, S26-302-04			4.2-.4; 9.4.4; 9.35.4
1803	3007	3310		1803	2903, 7001-02, 7004-05, 7010				P2004			4.2.5; 8.2.4; 9.12-.4; 9.15-.6; 9.23.13,.15; 9.35.3

Subject Locator and Citator

Location in this book	As described in the rules	Planning: Zon'g Ord.	Planning: Subdv Regs.	Planning: C C & R's	Accessibility: ADAAG	Accessibility: ANSI A117.1	Accessibility: UFAS	Sect'y of Int Stds	OSHA (29 CFR 1910)	Other U.S. Cod	Other Code of Fed Reg's	Loc & State Codes	Other Rules: Name	Other Rules: Citation	Catalogs: Master-format	Catalogs: Uni-format
	Footings, Foundations									15-47; 15-50; 15-63; 36-21; 40-12					03, 04	
	Pile Materials									15-47; 15-50; 15-63; 36-21; 40-12					03, 04	
	Piles & Pile Loads									15-47; 15-50; 15-63; 36-21; 40-12					03, 04	
	Water- & Damp-proofing									15-47; 15-50; 15-63; 36-21; 40-12					07	
7 Strategies																
Egress																
The Scale of the Problem	Occupant Load and Capacity														N/A	
Conceptual Exiting	Occupancy—Specific Egress: Requirements					4.3.10				42-51		√			N/A	
Subdividing Space																
Use Group	Occupancy-Specific Egress: Concept					4.3.1-10		√		15-49; 42-15A					N/A	
	Assembly Occupancies				A4.32, A4.33, A5.0, A8.0, A10.0	4.31						√			N/A	N/A
	Business Occupancies				A4.34, A7.0						12: Bank'g; 15: Trade				N/A	N/A
	Educational Occupancies										20: Educ'n	√			N/A	N/A
	Factory-Industrial Occupancies														N/A	N/A

Subject Locator and Citator

Model Building Codes										Other Codes		
IBC Affiliate Building Codes							Specialty Codes		CABO	NFPA		NBCC
IBC	BOCA		SBC		UBC		Name	Citation	(Residential)	Life Safety	Other	(Canadian)
5/97 Draft	Pre '93	Post '93	Pre '93	Post '93	Pre '93	Post '93			All	NFPA 101	Stds	All
1805	1205-12,1222	1801;1805-1808; 1810-11; 1815-16		1804								4.2.6; 4.26; 4.28
	1213-21			1807 - 1813	2908							4.2.3
1808-10	1212-21	1813,16-24		1805, 1806	2909							4.2.4
1806	1224			1814	2911-20				R-306, S-26.306			5.8-.2; 9.13-.3; 9.14-.6
1003				1003	3302				R-210, R-708	5-3		3.1.2;3.8.2; 9.10.8
	300-01			1019 - 1027	3301, 3303				R-210			3.3.3; 3.1.16
1001-03	300-01; 809-10	301,021010; 1011		1018	3301-03					9-2.1		3.2.2; 9.9; 9.10.8
303;1011	302	303		304, 704.1.2.3						4-1.2, Chpt 8/9		3.3.2
304	303	304		305						4-1.8, Chpt 26/27		3.1.2
305	304	305		306						4-1.3, Chpt 10/11		3.1.2
306	305	306		307						4-1.9, Chpt 28		3.3.5

Subject Locator and Citator

Location in this book	As described in the rules	Planning			Regulations, Standards, and Guidelines										Catalogs	
		Zon'g Ord.	Subdv Regs.	C C & R's	Accessibility: ADAAG	Accessibility: ANSI A117.1	Accessibility: UFAS	Sect'y of Int Stds	OSHA (29 CFR 1910)	Other U.S. Code	Other Code of Fed Reg's	Loc & State Codes	Other Rules: Name	Other Rules: Citation	Master-format	Uni-format
	Hazardous Occupancies									15-53; 42-23; 42-92; 42-104; 42-108		√			N/A	N/A
	Institutional Occupancies														N/A	N/A
	Mercantile Occupancies				A7.0										N/A	N/A
	Residential Occupancies				A9.0	4.32									N/A	N/A
	Storage Occupancies					4.23									N/A	N/A
	Utility and Miscellaneous Occupancies									42-23,71,73,74,77,81,84,91,92, 104					N/A	N/A
Single & Non-separated	Incidental Uses														N/A	N/A
Mixed Use Groups	Mixed Occupancies														N/A	N/A
Reinforcing Cultural Identity																
Alternative Design Approaches																
	High-rise (Busin & Resid)					4.32									N/A	N/A
	Covered Mall Buildings														N/A	N/A
	Mobile Units									42-70					N/A	N/A
	Underground Buildings														N/A	N/A
Special Occupancies	General, Definitions									15-53; 42-23,92,108					N/A	N/A
	Airport Control Towers														N/A	N/A
	Amusement Buildings											√			N/A	N/A
	Application of Flammable Finishes											√			N/A	N/A

Subject Locator and Citator

Model Building Codes										Other Codes		
IBC Affiliate Building Codes							Specialty Codes		CABO	NFPA		NBCC
IBC	BOCA		SBC		UBC		Name	Citation	(Residential)	Life Safety	Other	(Canadian)
5/97 Draft	Pre '93	Post '93	Pre '93	Post '93	Pre '93	Post '93			All	NFPA 101	Stds	All
307	306	307		308						4-2, 5-11		3.1.2
308	307	308		309						4-1.4, 4-1.5		3.3.5
309	308	309		310						4-1.7, Chpt 24/25		3.1.2
310	309	310		311, 410, App C					R-210-11, R-213, R-708	4-1.6, Chpt 16 to 23		3.3.4
311	310	311		312						4-1.10, Chpt 29		3.1.2
312	311	312		---						---		3.1.3; 3.6-.2; 3.6.4-.5
		302.1.1		704, 704.1.2						App 4-1.7		3.1.3
313	313	313		303, 704						4-1.11		9.10.8
403	602	403		412						30-8		3.2.6
402	601	402		413	Chpt 56					24-4.4, 25-4.4		3.1.2
	621	420		1611, App H								
405		405		415						30-7		3.1.2
401	600	401, 402		401, 402								
412	616	414		---	711-716							3.1.2
411	612,15	413		403								3.1.2
416	622	419		---								3.1.4

Subject Locator and Citator

Location in this book	As described in the rules	Planning			Regulations, Standards, and Guidelines										Catalogs	
		Zon'g Ord.	Subdv Regs.	C C & R's	Accessibility: ADAAG	Accessibility: ANSI A117.1	Accessibility: UFAS	Sect'y of Int Stds	OSHA (29 CFR 1910)	Other U.S. Code	Other Code of Fed Reg's	Loc & State Codes	Other Rules: Name	Other Rules: Citation	Master-format	Uni-format
	Automotive Service Stations														N/A	N/A
	Bowling Alleys														N/A	N/A
	Buildings Making Hazardous Materials											√			N/A	N/A
	Buildings Using Hazardous Materials During Production									15-53; 42-23,92,108		√			N/A	N/A
	Buildings Using Hazardous Materials in General									15-53; 42-23, 92,108		√			N/A	N/A
	High Rack Storage														N/A	N/A
	Institutional: Restrained														N/A	N/A
	Institutional: Unrestrained incapacitated				A6.0						24				N/A	N/A
	Outdoor Processing Facilities														N/A	N/A
	Parking: Lots, Garages & Hangars				A4.6	4.6									N/A	N/A
	Swimming Pools											√			N/A	N/A
Providing for Systems																
Providing Structure	General, Definitions							√		15-47; 36-21; 40-12					N/A	
	Allowable Loads & Stresses							√							N/A	

Subject Locator and Citator

Model Building Codes										Other Codes		
IBC Affiliate Building Codes							Specialty Codes		CABO	NFPA		NBCC
IBC	BOCA		SBC		UBC		Name	Citation	(Residential)	Life Safety	Other	(Canadian)
5/97 Draft	Pre '93	Post '93	Pre '93	Post '93	Pre '93	Post '93			All	NFPA 101	Stds	All
406	609,19	408		404								3.1.2
411		---		404								3.1.2
415	603,18.19	417		407								3.1.2
415	603,18.19	416		408								3.1.2
415	603,18.19	418		407						6-4.1		3.1.2
		---		411, SBC/FP 36	2706							3.1.2
408	611	410		409, 704	1011-23					Chpt 14/15		3.1.2
407	610	409		409						Chpt 12/13		3.1.2
		415		---								3.1.2
406, 412	607,08,09	406- 408		411, 610	701-710					29-6, 29-8, 30-6		9.35-.4
	625	421		---	1241-1243				R-114			3.1.2
1601-1604	1100	1601 - 1602		1601 - 1602					R-301, R-401, R-601, R-701, S26.201			4.0-.10; 5.2-.2
1605-15; 1804	1100-15	1603-14		1601-11, 1704								4.1-4.1.10; 5.2-.2

Subject Locator and Citator

	Location in this book	As described in the rules	Planning: Zon'g Ord.	Planning: Subdv Regs.	Planning: C C & R's	Accessibility: ADAAG	Accessibility: ANSI A117.1	Accessibility: UFAS	Sect'y of Int Stds	OSHA (29 CFR 1910)	Other U.S. Code	Other Code of Fed Reg's	Loc & State Codes	Other Rules: Name	Other Rules: Citation	Catalogs: Master-format	Catalogs: Uni-format
		Special & Combination Loads, Deflections							√							N/A	
		Structural Loads									42-70					N/A	
	Providing Electrical	General, Definitions							√		15-47,49,50,63;36-21		√			16	
		Elect: Emergency Systems									42-15A					16	
		Elect: Other					4.25					47: Telecomm				16	
		Elect: Standby Power Systems														16	
	Providing Mechanical	General, Definitions							√		15-47, 49, 50, 63					15	
		Mech Other														15	
	Providing Plumbing	General, Definitions							√				√			15	
		Plumbing Other														15	
		Plumbing Reference														15	
8	**Rooms**																
	Context																
	Rooms	General, Definitions					4.1		√		36-21; 40-6,12; 42-51	42: Health & Welfare					
	Meeting Basic Needs																

Subject Locator and Citator

Model Building Codes										Other Codes		
IBC Affiliate Building Codes							Specialty Codes		CABO	NFPA		NBCC
IBC	BOCA		SBC		UBC		Name	Citation	(Residential)	Life Safety	Other	(Canadian)
5/97 Draft	Pre '93	Post '93	Pre '93	Post '93	Pre '93	Post '93			All	NFPA 101	Stds	All
1605; 1804	1114	1613		1608 -1610, 1912, 2125, 2215, 2312								4.1.1; 4.4
				1611								
2701	2700,01; 3103	2701		2701-02, App E					Pt 6, S-26-6000, S-26-7000			9.34-.3; 9.10.17,.18
2702	2706,07	2706,07		905, 1109					Pt 6, S-26-6000, S-26-7000			9.34.3
	2703,05								Pt 6, S-26-6000, S-26-7000			
2703	2707	2707							Pt 6, S-26-6000, S-26-7000			9.34
2801	2500,01; 3104,05	2801		2801-02, App E					M-1001	7-2, 7-3		2.3.5; 6.1-.3; 6.2-.11; 9.33-.9
	2503,04,06-10			2803-11					Chpt 10-19, Refer to S-26-1000-1100.S-26-1300-1900			
2901	2800,01; 3105	2901		2901-02, App E					P-2001			7.1-.4; 9.31-.6
	2802,04-07								Chpt 20-25, Refer to S26-2000,S26-2200-2400			
1301	3100, 02, 06	1300		1300, App E	5300				S-26.2001, R-2207, 2309-10, 2314			
1201	700, 01	1201, 1202		1201, 1202		10-1.3			R-204, S26.215			3.3, 3.3.1; 9.5

Subject Locator and Citator

	Location in this book	As described in the rules	Planning: Zon'g Ord.	Planning: Subdv Regs.	Planning: C C & R's	Accessibility: ADAAG	Accessibility: ANSI A117.1	Accessibility: UFAS	Sect'y of Int Stds	OSHA (29 CFR 1910)	Other U.S. Code	Other Code of Fed Reg's	Loc & State Codes	Other Rules: Name	Other Rules: Citation	Catalogs: Master-format	Catalogs: Uni-format
	Proportion	Dimensions					4.2, 4.3, 4.30, 4.31.1, 4.32.1										
	Environment	Surrounding Materials					4.5		√		15-77,80						
		Lighting							√		15-56;36-21					16	
		Ventilation							√		15-56;36-21; 40-12;42-56					15	
		Noise							√								
		Temperature Control							√							15	
	Hygiene	Sanitation					4.15-.22, 4.24, 4.32.4, 4.32.6		√							15	
		Food Service					4.32.5					21: Food	√				
	Special Spaces																
	Atriums	Atriums							√							N/A	N/A
	Courts and Yards	Yards and Courts							√							N/A	N/A
	Mezzanines and Penthouses	Mezzanines							√							N/A	N/A
	Misc. Spaces	Computer Rooms							√							N/A	N/A
		Drying Rooms							√							N/A	N/A
		Projection Rooms							√							N/A	N/A
		Stages and Platforms							√							N/A	N/A
9	**Circulation**																
	Circulation	Egress: General, Definitions					4.3-.5, 4.8		√		40-6,12; 42-51					N/A	N/A
	General Circulation																
		Accessibility: General, Definitions					4.3, 4,8		√		40-6,12; 42-51					N/A	N/A

Subject Locator and Citator

Model Building Codes										Other Codes		
IBC Affiliate Building Codes							Specialty Codes		CABO	NFPA		NBCC
IBC	BOCA		SBC		UBC		Name	Citation	(Residential)	Life Safety	Other	(Canadian)
5/97 Draft	Pre '93	Post '93	Pre '93	Post '93	Pre '93	Post '93			All	NFPA 101	Stds	All
1207	708	1204							R-204			9.5
1209									R-203			
1204	703-05	1205-07		1203			BOCA Plumb	403	R-203, R-707			9.34-.2
1202	703, 06, 06, 09	1205,08,09		1203, 2893, 2810			BOCA Plumb	404	M-1001			2.3.5; 9.32-.3
1206	714	1214		708, 2603	Chpt 35							9.11-.2
1203				2801-06, 2809-11					R-203			9.33.2-.4; 9.25-.4
2901		2905		1204, 2901-02			BOCA Plumb	601	R-206, P-2001			9.31.4-.5
404	606	404		414						6-2.4.5 to 6-2.4.7		9.5.2
1205	508	1212										
505	605; 710-12	507			1717					6-2.5		3.2.8; 9.5.2
		---		417								
417	2505	2806		416								
409	613	411		403, 405	4001-07					8-3.2.2, 9-3.2.2		
410	615	412		403	3901-03					8-3.2.1		
1001-03; 1101,02	800-03	1001, 1002		1001, 1002	3301				R-210			9.8-9.8.6; 9.9-9.9.11
1101.02,11, 12	800-03	1101, 1102		1101, 1102	3101-02				R-211-13			9.8-.10; 9.9-.11

Subject Locator and Citator

Location in this book	As described in the rules	Planning: Zon'g Ord.	Planning: Subdv Regs.	Planning: C C & R's	Regulations, Standards, and Guidelines: Accessibility: ADAAG	Accessibility: ANSI A117.1	Accessibility: UFAS	Sect'y of Int Stds	OSHA (29 CFR 1910)	Other U.S. Code	Other Code of Fed Reg's	Loc & State Codes	Other Rules: Name	Other Rules: Citation	Catalogs: Master-format	Catalogs: Uni-format
	Airports														N/A	
	Bus Stops and Stations									40-6,12; 42-51					N/A	
	Existing Buildings							√		40-6,12; 42-51					N/A	
	Hotels									40-6,12; 42-51					N/A	
	Rail Stops and Stations									40-6,12; 42-51					N/A	
	Special Occupancy Reqmts for Accessibility					4.3, 4.5				40-6,12; 42-51						
Accommodating Infrastructure	Access to Unoccupied Spaces									40-6,12; 42-51					N/A	
Conveyances	Construction Documents														14	
	Conveyors														14	
	Emergency Signals					4.10.14									14	
	Existing Conveyances														14	
	General, Definitions					4.10.1, 4.11.1		√		15-47,63; 36-21					14	
	Machine Rooms									15-47,63; 36-21					14	
	Maintaining Conveyances									15-47,63; 36-21					14	
	Tests and Approvals									15-47,63; 36-21					14	
	Elevators					4.10				15-47,63; 36-21					14	
	Escalators and Walks									15-47,63; 36-21					14	
Egress Issues	Accessibility				A4.3	4.3				40-6,12; 42-51					N/A	N/A
Emergency: Occupants																
	Special Exit Requirements									40-6,12; 42-51		√			N/A	N/A

Subject Locator and Citator

Model Building Codes										Other Codes		
IBC Affiliate Building Codes							Specialty Codes		CABO	NFPA		NBCC
IBC	BOCA		SBC		UBC		Name	Citation	(Residential)	Life Safety	Other	(Canadian)
5/97 Draft	Pre '93	Post '93	Pre '93	Post '93	Pre '93	Post '93			All	NFPA 101	Stds	All
1120												
1118												
1110, 1117	512;804	1101		1110, 3403	3303				R-210	3-2, 9-1		
1113												
1119												
1107				1005, 1105	3301-03					3-2		
1208				1027								
	2600,01	3003										
3005		3010										
3003	2612	3009										3.5.4
	2605,06	3013										
3000	2600,01	3001,02		3001-02	5101-02, 5107-09					7-4		3.5-.2
3006	2610				5705							3.6.3
	2604	3012			5113							
	2602.03	3004 - 3005		105, 107	5110-11, 5114							
	2607,09	3006-08		3003	5103, 5106, 5112							3.5-.4
	820, 2617	3011		3104, 3106	3309, Chpt 51							
1003.14	800, 01	1002,07; 1102		1108	3104, 5112					5-5.4		9.8-.6
1010-12				1005, 3403								3.4
	825			1017	3301					5-1.7, 5-7.2		9.9.5

Subject Locator and Citator

		Planning			Regulations, Standards, and Guidelines										Catalogs	
					Accessibility								Other Rules			
Location in this book	As described in the rules	Zon'g Ord.	Subdv Regs.	C C & R's	ADAAG	ANSI A117.1	UFAS	Sect'y of Int Stds	OSHA (29 CFR 1910)	Other U.S. Code	Other Code of Fed Reg's	Loc & State Codes	Name	Citation	Master-format	Uni-format
Smoke and Fire Control	General, Definitions							√				√			15	
	Smoke Control Systems									15-47, 49,50; 36-21; 42-15A					15	
Suppression Systems	Automatic Sprinkler Extinguishing									15-47, 49,50; 36-21; 42-15A		√			15	
Suppression Systems	Portable Extinguishers									15-47, 49,50; 36-21; 42-15A		√			15	
	Automatic Non-sprinkler Extinguishing									15-47, 49,50; 36-21; 42-15A		√			15	
	Standpipe and Hose Systems									15-47, 49,50; 36-21; 42-15A		√			15	
	Water Supplies for Fire Fighting									15-47, 49,50; 36-21; 42-15A					15	
	Explosion Prevention									15-47, 49,50; 36-21; 42-15A		√			N/A	
Ventilation	Smoke and Heat Vents									15-47, 49,50; 36-21; 42-15A					15	
	Hoistway Venting									15-47, 49,50; 36-21; 40-6,12: 42-15A					14	
	Alarms									15-47, 49 50,60; 36-21;40-12; 42-15A		√			16	
	Emergency Alarms					4.26				15-47, 49 50,60; 36-21;40-12; 42-15A					16	
	Exit Illumination & Signs					4.28				15-47, 49 50,60; 36-21;40-12; 42-15A					16	

Subject Locator and Citator

Model Building Codes										Other Codes		
IBC Affiliate Building Codes							Specialty Codes		CABO	NFPA		NBCC
IBC	BOCA		SBC		UBC		Name	Citation	(Residential)	Life Safety	Other	(Canadian)
5/97 Draft	Pre '93	Post '93	Pre '93	Post '93	Pre '93	Post '93			All	NFPA 101	Stds	All
902	900,01; 1000,01, 19	901, 902		901, 902						7		3.1; 9.10-.4
909	1019	922							R-215			
903	1002-11	904,06-14		903						6-5.7, 7-7, LSCH Suppl 3		
906	1021	921										
904	1002,15-17	904,05										
905	1012,13	915		904						7-7		
912	1014	916,17										3.2.5
911												
910	930	923; 1209		2803					Chpt 14			9.33.10
3004	2608	3007			5104							
907,08	1017	918-20		905, 1109						5-2.1.5		3.2.4
908	1017	918-20		905, 1109								3.2.4;.9, .10, .17, .18
1003.11, 1003.12; 1010; 1109,16	823;2904	1023		1016, 1107, 3108	3313-3314					5-8 to 5-10		3.2.7;3.4.5; 9.34.3

Subject Locator and Citator

Location in this book	As described in the rules	Planning			Regulations, Standards, and Guidelines										Catalogs	
		Zon'g Ord.	Subdv Regs.	C C & R's	Accessibility			Sect'y of Int Stds	OSHA (29 CFR 1910)	Other U.S. Code	Other Code of Fed Reg's	Loc & State Codes	Other Rules		Master-format	Uni-format
					ADAAG	ANSI A117.1	UFAS						Name	Citation		
Egress Issues	Accessible Route					4.3				40-6,12; 42-51		√			N/A	
Internal Conditions	Ramps				A4.8	4.8				15-47, 49,50; 36-21; 40-6,12: 42-15A					N/A	N/A
Exits	Exit Number and Arrangement									15-47, 49,50; 36-21; 40-6,12: 42-15A					N/A	N/A
	Fire Escapes									15-47, 49,50; 36-21; 40-6,12: 42-15A		√			05	
	Stairway Protection and Construction				A4.9	4.9				15-47, 49,50; 36-21; 40-6,12: 42-15A		√			N/A	
	Horizontal Exits					4.14				15-47, 49,50; 36-21; 40-6,12: 42-15A		√			N/A	
Exit Discharges and Safe Places	Exit Discharge				A4.14	4.14				15-47, 49,50; 36-21; 40-6,12: 42-15A		√			N/A	
	Roof Access									15-47, 49,50; 36-21; 40-6,12: 42-15A					N/A	
Egress Issues	Accessible Entrances					4.14				15-47, 49,50; 36-21; 40-6,12: 42-15A					N/A	
Doors and Gates	Doors				A4.13	4.13-14, 4.3.9				15-47, 49,50; 36-21; 40-6,12: 42-15A					08	
Emergency: Fire Depts																

Subject Locator and Citator

Model Building Codes										Other Codes		
IBC Affiliate Building Codes							Specialty Codes		CABO	NFPA		NBCC
IBC	BOCA		SBC		UBC		Name	Citation	(Residential)	Life Safety	Other	(Canadian)
5/97 Draft	Pre '93	Post '93	Pre '93	Post '93	Pre '93	Post '93			All	NFPA 101	Stds	All
1104		1104		1103								
1004.5	810,15	1016		1013	3307				R-219	5-2.5		9.6.6
1005, 1007	806-09	1006,10,11		1004						5-4		3.4-.3
	821	1025		1011	3312					5-2.1		3.4.7
1004.4, 1008.3	805, 3005,16,18	1003,10, 14		1006 - 1007								
1008.4, 1008.6	814	1019		1009						5-2.4		9.9.8
1007.4, 1009	810,11,16, 19	1006,07,11,17, 220, 1110		1010								3.4.6
1007.2.2	811,17	1027		1108, 1110								9.19.2
1005,06	804,10	1007,10,17,18, 1106										
1004.2, 1004.3	810,12,13	717, 1017		1012	3304				R-211, R-412, S-26.412	5-2.8.3		9.6-.8

Subject Locator and Citator

	Location in this book	As described in the rules	Planning: Zon'g Ord.	Planning: Subdv Regs.	Planning: C C & R's	Accessibility: ADAAG	Accessibility: ANSI A117.1	Accessibility: UFAS	Sect'y of Int Stds	OSHA (29 CFR 1910)	Other U.S. Code	Other Code of Fed Reg's	Loc & State Codes	Other Rules: Name	Other Rules: Citation	Catalogs: Master-format	Catalogs: Uni-format
	Facilitating Fire Depts	Fire Department Access									15-47, 49,50, 63; 36-21; 40-6,12: 42-15A					N/A	N/A
		Fire Alarms and Detectors				A4.28	4.26				15-47, 49 50,60; 36-21;40-12; 42-15A		√			16	
		Fire Dept Access in Exterior Walls									15-47, 49,50, 63; 36-21; 40-6,12: 42-15A					N/A	N/A
10	**Detailing**																
	Walls & Floors																
	Rooms: Enclosing Partitions	Table of Assembly Ratings									15-47, 49,50, 63; 36-21; 40-6,12: 42-15A					N/A	N/A
	Fire Resistant Details	Combustibles in Rated Assemblies									15-47, 49,50, 63; 36-21; 40-6,12: 42-15A					N/A	
		Materials for Fire Resistance									15-47, 49,50, 63; 36-21; 40-6,12: 42-15A					N/A	
	Poles	Structural Members									15-47, 49,50, 63, 77, 80; 36-21; 40-6,12: 42-15A					N/A	
		Inspecting Spray Fireproofing									15-47, 49,50, 63; 36-21; 40-6,12: 42-15A					07	
	Planes: Vertical	Special Occupancy Separations									15-47, 49,50, 63, 77, 80; 36-21; 40-6,12: 42-15A					N/A	N/A

Subject Locator and Citator

Model Building Codes										Other Codes		
IBC Affiliate Building Codes							Specialty Codes		CABO	NFPA		NBCC
IBC	BOCA		SBC		UBC		Name	Citation	(Residential)	Life Safety	Other	(Canadian)
5/97 Draft	Pre '93	Post '93	Pre '93	Post '93	Pre '93	Post '93			All	NFPA 101	Stds	All
913	1014,17,18	918										9.10.19
907,08	1018	918-20		905, 1109						7-6, LSCH Suppl 2		3.2.4;.9, .10, .17, .18
	1014, 2100, 01	705		1405								3.2.5
Tables 601 (General), 602 (Exter. Walls), & 1006.2.1 (Corridors)	Table 902	Table 602		Table 600, Table 700						Table App 6-2.1, LSCH Table 6-1		App D
703,18	906,12,13, 14,24,27	709,13,16		706								3.1.4, 3.1.13-14; 3.1.9; 9.10.6
703,12,13,16,17; 1406	905,17,19, 20,22,26	717, 19-22		703								3.1.5
713	912	715,16		701								9.10.7
717				1709								9.10.7
		710,12; 1014		704								3.1.6; 3.2.8

Subject Locator and Citator

Location in this book	As described in the rules	Planning: Zon'g Ord.	Planning: Subdv Regs.	Planning: C C & R's	Accessibility: ADAAG	Accessibility: ANSI A117.1	Accessibility: UFAS	Sect'y of Int Stds	OSHA (29 CFR 1910)	Other U.S. Code	Other Code of Fed Reg's	Loc & State Codes	Other Rules: Name	Other Rules: Citation	Catalogs: Master-format	Catalogs: Uni-format
	Tenant Fire Separations									15-47, 49,50, 63, 77, 80; 36-21; 40-6,12: 42-15A					N/A	
	Combustible Materials									15-47; 15-49					07, 08	
	Exterior Walls							√		42-109A					07, 08	
Planes: Vertical	Building Separation Walls, Fire Walls									15-47, 49,50, 63, 77, 80; 36-21; 40-6,12: 42-15A					N/A	
Conveyances	Protecting Elevator Openings									15-47, 49,50, 63, 77, 80; 36-21; 40-6,12: 42-15A					14	
Fire Resistant Details	Protection of Floor Openings (Shafts)									15-47, 49,50, 63, 77, 80; 36-21; 40-6,12: 42-15A		√			N/A	
Planes: Vertical	Fire Separation Assemblies									15-47, 49,50, 63, 77, 80; 36-21; 40-6,12: 42-15A					N/A	
Planes: Vertical	Fire Partitions									15-47, 49,50, 63, 77, 80; 36-21; 40-6,12: 42-15A					N/A	
Conveyances	Hoistway Enclosures									15-47, 49,50, 63, 77, 80; 36-21; 40-6,12: 42-15A					14	
Planes: Vertical	Smoke Barriers									15-47, 49,50, 63, 77, 80; 36-21; 40-6,12: 42-15A					N/A	

Subject Locator and Citator

Model Building Codes										Other Codes		
IBC Affiliate Building Codes							Specialty Codes		CABO	NFPA		NBCC
IBC	BOCA		SBC		UBC		Name	Citation	(Residential)	Life Safety	Other	(Canadian)
5/97 Draft	Pre '93	Post '93	Pre '93	Post '93	Pre '93	Post '93			All	NFPA 101	Stds	All
705,06,08;1209 ;2504-06	908,10,15	707		704					R-218			9.10.9-.15
1406	924	1406		706								3.1.4
1404 ; 2512,13	2105-09								R-503			5.3.1; 5.4.1; 5.5.15.7.1; 5.8.1-2; 6.1; 9.27.6-.13; 9.28-.6; 9.35.4
705	910	707		704					R-218			3.1.10; 9.10.9,.11-.12
707,11,14	904	3008		705	1706							3.5.3; 9.10.5
707	915, 2604	710		705.2	1706					6-2.4		3.5.3; 9.10.5
705-08,10	908,10	709		704						6-2.3.2		3.1.9; 9.9,.11-.14
708	908,10	711		704					R-218			3.1.9-.11; 9.10.5,.9,.13

Subject Locator and Citator

		Planning			Regulations, Standards, and Guidelines										Catalogs	
Location in this book	As described in the rules	Zon'g Ord.	Subdv Regs.	C C & R's	Accessibility			Sect'y of Int Stds	OSHA (29 CFR 1910)	Other U.S. Code	Other Code of Fed Reg's	Loc & State Codes	Other Rules		Master-format	Uni-format
					ADAAG	ANSI A117.1	UFAS						Name	Citation		
Conveyances	Hoistway Enclosures									15-47, 49,50, 63, 77, 80; 36-21; 40-6, 12: 42-15A					14	
Planes: Vertical	Smoke Barriers									15-47, 49,50, 63, 77, 80; 36-21; 40-6, 12: 42-15A					N/A	
Structural Work	Concrete—Struct							√		42-86					03	
	Concrete—Seismic									42-86					03	
	Masonry							√		42-86					04	
	Steel							√		15-77,80; 42-86					05	
	Structural Aluminum									15-77,80; 42-86					05	
	Wood									42-86					06	
Planes: Horizontal	Floor and Roof Assemblies									15-47, 49,50, 63, 77, 80; 36-21; 40-6, 12: 42-15A					N/A	
Strategies for Partitions	Fire Resistance by Testing									15-47, 49,50, 63, 77, 80; 36-21; 40-6, 12: 42-15A					N/A	

Subject Locator and Citator

Model Building Codes										Other Codes		
IBC Affiliate Building Codes							Specialty Codes		CABO	NFPA		NBCC
IBC	BOCA		SBC		UBC		Name	Citation	(Residential)	Life Safety	Other	(Canadian)
5/97 Draft	Pre '93	Post '93	Pre '93	Post '93	Pre '93	Post '93			All	NFPA 101	Stds	All
3002	2608	3007								6-2.4.8 to 6-2.4.11		3.1; 3.2.1,.3;3.3-.1; 9.26-.18; 9.30-.6
709	911	712							R-217-18	6-3		3.1.12
1904,06,07,09, 11-15	1505-09	1905-09,10-13		1908	Chpt 26				R-603			4.3.3; 9,3.1; 9.15.4; 9.16.4
1910	1501	1903,04		1912 -1913, 3406	2625							4.1.9
2106-09	1100-15;1400	2105-12		2105 - 2109	Chpt 24				R-404-05, R-407-10, S26-404, S26-408,S26-410			4.3.2; 9.15.5, .6; 9.17.5; 9.20- .16
2004	1100-15; 1800	2205-10			Chpt 27							4.3.4; 9.17.3
2001-2	1100-15; 1900	2001 - 2002		2001 - 2003	Chpt 28				R-403, R-605, R-705, R-805, S26-403,S26-605, S26-705			4.3., .5; 9.3, .3
2309		2306		2312, 3406	Chpt 25				R-402, R-412-14, R-602, R-606-07, R-702-04, R-802,			4.1.9
710;803,04;1505,08,09		713, 714		701, 704	Chpt 32				R-601, R-603-07, R-701-05, S-26.601, S-26.701	6-2.3.1, 6-2.4		3.1.14
703	902,04,05	704		701, Table 700								

Subject Locator and Citator

		Planning			Regulations, Standards, and Guidelines										Catalogs	
					Accessibility								Other Rules			
Location in this book	As described in the rules	Zon'g Ord.	Subdv Regs.	C C & R's	ADAAG	ANSI A117.1	UFAS	Sect'y of Int Stds	OSHA (29 CFR 1910)	Other U.S. Code	Other Code of Fed Reg's	Loc & State Codes	Name	Citation	Master-format	Uni-format
	Fire Resistance by Calculation									15-47, 49,50, 63, 77, 80; 36-21; 40-6,12: 42-15A					N/A	
Openings In Walls & Floors																
Door, Shutters & Dampers	Protection of Wall Openings (doors, dampers)									15-47, 49,50, 63, 77, 80; 36-21; 40-6,12: 42-15A					N/A	
	Egress: Interior Finishes					4.3.9, 4.13.8, 4.27				15-47, 49,50, 63, 77, 80; 36-21; 40-6,12: 42-15A		√			09	
Plugged Gaps	Combustibles in Concealed Spaces									15-47, 49,50, 63, 77, 80; 36-21; 40-6,12: 42-15A		√			N/A	
	Fireblocking and Draftstopping									15-47, 49,50, 63, 77, 80; 36-21; 40-6,12: 42-15A					N/A	
	Thermal Insulation									15-47, 49,50, 63; 36-21; 40-6,12: 42-15A					07	
	Penetrations									15-47, 49,50, 63, 77, 80; 36-21; 40-6,12: 42-15A					N/A	
General Design Concerns																
Materials	Concrete: General, Definitions							√		15-47, 50,60; 36-21;40-12					03	

Subject Locator and Citator

Model Building Codes										Other Codes		
IBC Affiliate Building Codes							Specialty Codes		CABO	NFPA		NBCC
IBC	BOCA		SBC		UBC		Name	Citation	(Residential)	Life Safety	Other	(Canadian)
5/97 Draft	Pre '93	Post '93	Pre '93	Post '93	Pre '93	Post '93			All	NFPA 101	Stds	All
718	1000,01; 2301	721,22		709								
710,14,2405	907,09	706, 08, 14, 16, 18		705.1					S-26.412	6-2.3.3 to 6-2.3.5		3.1.8; 3.4.4
1001-12	920,23-25	709,11,13, 16,17,22,23; 1000-18		703, 706					R-502, S26.217, S26.502	5-1.4		3.1.5
715	924,29	720		707						6-2.6		3.1.4; 9.10.10
705,06,08	912	720,21, 23		705								9.10.13
717	928	723		708, 2603	1714							9.25-.4
711	914,16,17	714,17, 18		705						6-2.3.6		3.5.3; 9.10.5
1901,02	1500-09	1901, 1902		1901, 1902	2601-02				R-603			4.3-.3; 9.3,.1; 9.17-.2, .6

Subject Locator and Citator

		Planning			Regulations, Standards, and Guidelines										Catalogs	
					Accessibility								Other Rules			
Location in this book	As described in the rules	Zon'g Ord.	Subdv Regs.	C C & R's	ADAAG	ANSI A117.1	UFAS	Sect'y of Int Stds	OSHA (29 CFR 1910)	Other U.S. Code	Other Code of Fed Reg's	Loc & State Codes	Name	Citation	Master-format	Uni-format
	Concrete Assemblies: Forms, Embeds									15-47,50,60; 36-21					03	
	Concrete: GFRC, Gyp Conc									15-47, 50,60; 36-21;40-12					03	
	Concrete: Locations—On Grade, Parapets									15-47, 50,60; 36-21;40-12					03	
	Concrete: Plain									15-47, 50,60; 36-21;40-12					03	
	Masonry: Adobe							√		15-47, 50,60; 36-21;40-12					04	
	Masonry: Anchorage									15-47, 50,60; 36-21;40-12					04	
	Masonry: General							√		15-47, 50,60; 36-21;40-12					04	
	Masonry: General, Definitions							√		15-47, 50,60; 36-21;40-12					04	
	Masonry: Glass Block							√		15-47, 50,60; 36-21;40-12					04	
	Masonry: Locations—Chimneys, Fireplaces							√		15-47, 50,60; 36-21;40-12					04	
	Steel: Assemblies, Joists, Cables, Composites									15-47, 50,60, 77; 36-21;40-12					05	
	Steel: General, Definitions							√		15-47, 50,60, 77; 36-21;40-12					05	

Subject Locator and Citator

Model Building Codes										Other Codes		
IBC Affiliate Building Codes							Specialty Codes		CABO	NFPA		NBCC
IBC	BOCA		SBC		UBC		Name	Citation	(Residential)	Life Safety	Other	(Canadian)
5/97 Draft	Pre '93	Post '93	Pre '93	Post '93	Pre '93	Post '93			All	NFPA 101	Stds	All
		1909		1906, 1907								
				1910, 1914	2627							
				1909, 1911								9.16-.4
1909		1902, 04		1903 - 1905	2622							9.16.4
				2116, App G								
				2110, 2111					R-406			9.20.4, .5, .9, .11
2101	1400-22	2101		2104	2401-02, 2407				R-404, S-26.404			9.20,.1
2101,02	1400-08	2101 - 2103		2101 - 2103					R-404			4.3, .2; 9.17-.2, .5; 9.20-.16
2110	1420	2118		2112					R-901, R-903, S-26.1605			
2111	2400-02	2113-17		2113 - 2114, 2804-07	3701-04, 3707							9.21-.5; 9.22-.10; 9.33.10
2201,06,07	1801-03,05-08	2203, 05-09		2205 - 2206, 2211	2705, 2707							
2201,02	1800-04	2201, 2202		2201, 2202	2701-02							4.3.,4; 9.3.3; 9.17-.3

Subject Locator and Citator

Location in this book	As described in the rules	Planning			Regulations, Standards, and Guidelines										Catalogs	
					Accessibility								Other Rules			
		Zon'g Ord.	Subdv Regs.	C C & R's	ADAAG	ANSI A117.1	UFAS	Sect'y of Int Stds	OSHA (29 CFR 1910)	Other U.S. Code	Other Code of Fed Reg's	Loc & State Codes	Name	Citation	Master-format	Uni-format
	Steel: Locations—Parapets									15-47, 50,60, 77; 36-21;40-12					05	
	Steel: Standards, Rolled, Formed									15-47, 50,60, 77; 36-21;40-12					05	
	Steel: Welds, Bolts									15-47, 50,60, 77; 36-21;40-12					05	
	Gypsum: Assemblies									15-47, 50,60; 36-21;40-12					06	
	Gypsum: Board, Plaster, Stucco									15-47, 50,60; 36-21;40-12					06	
	Gypsum: General, Definitions							√		15-47, 50,60; 36-21;40-12					06	
	Gypsum: Locations—Shear Walls									15-47, 50,60; 36-21;40-12					06	
	Plastics: General, Definitions							√		15-47, 50,60; 36-21;40-12					06	
	Plastics: Special Approval									15-47, 50,60, 77; 36-21;40-12					06	
	Wood: Assemblies—Timber, framed							√		15-47, 50,60; 36-21;40-12					06	
	Wood: Components—Hangers, etc.									15-47, 50,60; 36-21;40-12					06	
	Wood: General, Definitions							√		15-47, 50,60; 36-21;40-12					06	

Subject Locator and Citator

Model Building Codes										Other Codes		
IBC Affiliate Building Codes							Specialty Codes		CABO	NFPA		NBCC
IBC	BOCA		SBC		UBC		Name	Citation	(Residential)	Life Safety	Other	(Canadian)
5/97 Draft	Pre '93	Post '93	Pre '93	Post '93	Pre '93	Post '93			All	NFPA 101	Stds	All
				2210								
2205				2203, 2204, 2209	2703-04							9.24-.3
2209,09				2207 - 2208	2708-09							
2509	1604	2504			4703-04							
2510-12	1601-03	2503, 2505, 2506			4707--12							3.1.13
2501,02	1600	2501		2502-03	4701-02							3.1.13
		2502		2504-06	4712							3.1.13
2601-02	2000	2601		2601-02	3007, 5201-02				R-216, R-220, S26-216, S26-220			
2605												
	3020	2304, 2305			2511-15							9.23 -.17
	1701-12	2312, 2313		2306	2510							9.23 .3, .5, .7
2301.02		2301, 2302		2301, 2302	2501-03, 2516-17				R-402,S26-402, 13-14, S26-602, 04, 06-07, S26-702, 04			4.3, .1; 9.3, .2; 9.17-.2, .4; 9.23 .29.23-.2

Subject Locator and Citator

Location in this book	As described in the rules	Planning			Regulations, Standards, and Guidelines										Catalogs	
		Zon'g Ord.	Subdv Regs.	C C & R's	Accessibility			Sect'y of Int Stds	OSHA (29 CFR 1910)	Other U.S. Code	Other Code of Fed Reg's	Loc & State Codes	Other Rules		Master-format	Uni-format
					ADAAG	ANSI A117.1	UFAS						Name	Citation		
	Wood: Locations									15-47, 50,60; 36-21;40-12					06	
	Wood: Natural and Engineered									15-47, 50,60; 36-21;40-12					06	
	Plastics: Foams									15-47, 50,60; 36-21;40-12					06	
	Glass: Assemblies—Skylights									15-47, 50,60; 36-21;40-12					08	
	Glass: General, Definitions							√		15-47, 50,60; 36-21;40-12					08	
	Glass: Locations—Floors & Walks									15-47, 50,60; 36-21;40-12					08	
	Glass: Locations—Sports Courts									15-47, 50,60; 36-21;40-12					08	
	Glass: Plain									15-47, 50,60; 36-21;40-12					08	
	Glass: Safety									15-47, 50,60; 36-21;40-12					08	
	Glass: Structural Design									15-47, 50,60; 36-21;40-12					08	
	Plastic Glazing									15-47, 50,60; 36-21;40-12					06	
Materials	Interior Finishes: Restrictions					4.3.6, 4.31.4		√							09	

Subject Locator and Citator

Model Building Codes										Other Codes		
IBC Affiliate Building Codes							Specialty Codes		CABO	NFPA		NBCC
IBC	BOCA		SBC		UBC		Name	Citation	(Residential)	Life Safety	Other	(Canadian)
5/97 Draft	Pre '93	Post '93	Pre '93	Post '93	Pre '93	Post '93			All	NFPA 101	Stds	All
	1700	2303		2307-11	2506-09, 4705-06, 4713				R-402,, R-413-14, R-602, R-604, R-606-07, R-702-04, R-802, Refer to S26-400,600,700			9.14.4; 9.16 -.3, .5; 9.17- .2; 9.23.8 - .16
2305-07	1705-08; 2302	2303, 2307 - 2311			2511-15				R-402, R-602, R-702, R-802, S-26.607, S-26-702			9.23.4
2603	2002	2603			1713				R-216			
2404	2204	2404		2707	3401-06							9.7-.7
2401,02	2200-02	2401		2401-03	3401-02, 3405, 5401-02				R-411, S-26.208, S 26-411			9.7-.1
2408	2206	2406			3406							
2407	2207	2407										
2402		2402		2404								
2405	2203	2405			5406							9.7.6
2403	1420	2118; 2403		2406								4.3,4.3.6
2607-2612	2001;2205; 2306	2604 - 2608		2604	5201-14							
		114		706, 803					R-502	6-5		

Subject Locator and Citator

	Location in this book	As described in the rules	Planning			Regulations, Standards, and Guidelines										Catalogs	
			Zon'g Ord.	Subdv Regs.	C C & R's	Accessibility: ADAAG	Accessibility: ANSI A117.1	Accessibility: UFAS	Sect'y of Int Stds	OSHA (29 CFR 1910)	Other U.S. Code	Other Code of Fed Reg's	Loc & State Codes	Other Rules: Name	Other Rules: Citation	Master-format	Uni-format
		Ceilings							√		42-63,63A					09	
		Decorations and Trim							√		42-63,63A					09	
		Floor Finishes					4.3.6, 4.5		√							09	
		Interior Finishes: General, Definitions							√		15-47, 50,60; 36-21;40-12					09	
		Plastic Veneer									15-47, 50,60; 36-21;40-12					06	
		Plastics: Interior Finishes									15-47, 50,60; 36-21;40-12					06	
		Wall Finishes							√							09	
	Fit	Features & Facilities									15-47, 50,60; 36-21;40-12						
		Telephones					4.29				15-47, 50,60; 36-21;40-12					16	
	Communication	Signage					4.28				15-47, 50,60; 36-21;40-12					16	
	Vermin Control	Rodent-Proofing											√				
	Special Construction	Special Construction—Definitions							√		42-86					13	
		Broadcast Towers									42-86					13	
		Demolition									42-86					01	
		Emergency Measures									15-49;42-15A, 86					01	
		Skybridges									42-86					13	
		Temporary Structures				36.407, A4.1.1(4)					42-86					13	
		Tent Structures									42-86, 109A					13	
11	**Method**															01	01

Subject Locator and Citator

Model Building Codes										Other Codes		
IBC Affiliate Building Codes							Specialty Codes		CABO	NFPA		NBCC
IBC	BOCA		SBC		UBC		Name	Citation	(Residential)	Life Safety	Other	(Canadian)
5/97 Draft	Pre '93	Post '93	Pre '93	Post '93	Pre '93	Post '93			All	NFPA 101	Stds	All
803	923,-25; 1600-04	801,04;2504		804, 2504					R-706, S-26.701			9.26-9.26.10
805	922	803,04										
804	1705	805,06							R-604			9.30-.6
801,02	1300	801, 802		801, 802					R-501, R-701			9.29.1
2606	2001	1405										
2604									R-502			
803	1600-04; 1704; 1900,01	803,04,07;2504-06		2504-06					R-502, S-26.502			9.29-.10
1108, 1114		1101, 1103		1106	511, 3105							
1115		1023; 1109;2609; 2706; 3009; 3102										
1109, 1116	2900-15			1016, 1107, 3108	3105(e)							3.45
1210	2103	1215		1205, 2304, 3405					R-310			
3101		3101		3101-02								
3108	623,24	3108, 3109										
3303	105;3007	110;3101		3301-02, 3314, App D						---		8.2.2
	121	120		---						---		3.4.6
3104		3106		3104								
107, 3101	509, 511, 626	109, 3104			104					---		
3102	604	3103		3103, 3105	5501-05							

Subject Locator and Citator

Location in this book	As described in the rules	Planning			Regulations, Standards, and Guidelines										Catalogs	
					Accessibility								Other Rules			
		Zon'g Ord.	Subdv Regs.	C C & R's	ADAAG	ANSI A117.1	UFAS	Sect'y of Int Stds	OSHA (29 CFR 1910)	Other U.S. Code	Other Code of Fed Reg's	Loc & State Codes	Name	Citation	Master-format	Uni-format
Analyzing Rules																
Suggested Procedures																
Sequence																
Recording the Design																
	Fire Resistance: Passive									42-86		√			01	01
Demonstrating Compliance																
Background	Administration									42-86					01	01
Zoning																
Egress																
Civil																
HVAC and Plumbing																
Lighting & Power																
Structure																
Occupational Safety																
Wetlands & Floodplains												√				
Hazardous Substances																
Historic Preservation																
Accessibility																

Subject Locator and Citator

Model Building Codes										Other Codes		
IBC Affiliate Building Codes							Specialty Codes		CABO	NFPA		NBCC
IBC	BOCA		SBC		UBC		Name	Citation	(Residential)	Life Safety	Other	(Canadian)
5/97 Draft	Pre '93	Post '93	Pre '93	Post '93	Pre '93	Post '93			All	NFPA 101	Stds	All
704		703										
106		107.5 - 107.7, 111		104, 106	302					---		
									M-1009, P-2014			

SUGGESTED READING

GUIDES

Handbooks

Quoting and expanding on a single code.

1. Goldberg, Alfred, P E, *Designer's Guide to the 1991 Uniform Building Code*, third edition. New York: John Wiley & Sons. 1993.
2. Parish, Scott, *BOCA Code Manual: A Compliance Guide for Architects, Builders and Design Profesionals.* McGraw-Hill. 1998.
3. *Building Code Commentary.* Birmingham, AL: Southern Building Code Congress International, 1997.
4. *One and Two Family Dwelling Commentary.* Falls Church, VA: International Code Council, 1997.
5. Coté, Ron, P E, ed., *Life Safety Code Handbook*, seventh edition. National Fire Protection Association. 1997.
6. Woodson, R. Dodge, *National Plumbing Code Handbook*, second edition. McGraw-Hill. 1998.
7. Rosenberg, Paul, *Guide to the 1993 National Electric Code.* MacMillan General Reference. 1993

Graphic Guides

Illustrating design alternatives to regulatory requirements.

1. Hopf, Peter S., *AIA, Designer's Guide to OSHA.* New York: McGraw-Hill. 1975.
2. Morton, W. Brown III, *The Secretary of the Interior's Standards for Rehabilitation with Illustrated Guidelines for Rehabilitating Historic Buildings*, National Park Service. 1992.
3. Kardon, Redwood, *Code Check: A Field Guide to Building a Safe House.* Newtown, CT: The Taunton Press, Inc. 1995, revised 1996.

PROFESSIONAL REFERENCES

General Issues

1. Greenstreet, Robert C., "Impact of Building Codes and Legislation on the Development of Tall Buildings," *Architronic* 5, no. 2 (1996).
2. Centuori, Jeanine, "Building Codes as Dress Codes for the Protective Clothing of Buildings," *Architronic* 1, no. 1.
3. McGowan, Maryrose, *Specifying Interiors: A Guide to Construction and FF&E for Commercial Interiors Projects.* New York: John Wiley & Sons. 1996.
4. Allen, Edward, *The Architect's Studio Companion*, second edition. New York: John Wiley & Sons. 1995.
5. Lassar, Terry Jill, *Carrots & Sticks: New Zoning Downtown.* Urban Land Institute, 1989.

About Process

Reviewing intent and method

1. Lerable, Charles A., *Preparing a Conventional Zoning Ordinance*, Planning Advisory Service Report Number 460. American Planning Association. December 1995.
2. *An Architect's Guide to Building Codes & Standards*, second edition. Washington, D.C.: The American Institute of Architects. 1990.

Case Studies

1. Carmody, John, and Ray Sterling, *Underground Space Design.* New York: Van Nostrand Reinhold. 1993. See Appendix D: "Life Safety Features of Three Underground Facilities."

NOTES

[1] Of course, these three project role designations are somewhat generic. Builders' roles may be filled by construction managers, and those acting as owners may lease rather than own their properties. In addition, there may not be three players in the project relationship. Design-build contracts may be interpreted as having only two; construction management projects may be seen as having four. However, the differences they represent are not significant to understanding the connection between governments and projects built by private groups.

[2] Stephens, Spencer, Esq., "A/Es With Cash Flow Blues May Soon Sing 'Lien On Me'," *Engineering Times,* National Society of Professional Engineers 19, no. 8 (August 1997).

[3] Quotes taken from Winston, Sherie, "Disabilities Lawsuit Dismissed Against Arena Designer," *Engineering News-Record,* McGraw Hill (July 29, 1996).

[4] For an excellent analysis of the communication and delegation factors involved in this case, see Marshall, R.D. et al., "Investigation of the Kansas City Hyatt Regency Walkways Collapse," *NBS Building Science Series #143,* U.S. Department of Commerce, National Bureau of Standards (May 1982).

[5] Excerpted from "Waterlines," the newsletter of the Indiana Department of Natural Resources, July–December 1995.

[6] Quoted in Schmertz, Mildred F., "Dictating Design," *Architecture* (February 1993): pp. 33–34.

[7] Lobell, John, *Between Silence and Light: Spirit in the Architecture of Louis I. Kahn.* Boston: Shambhala Press, 1985, p. 26.

[8] Reprinted with permission of the American Planning Association from Lerable, Charles A., *Preparing a Conventional Zoning Ordinance,* Planning Advisory Service Report No. 460. (Chicago: American Planning Association, December 1995.)

[9] Information for this case was generously supplied by Steve Jacobson, director of outreach for the Reston Association. Quoted development guidelines are excerpted from *Design Guidelines (Initial Construction),* Reston Association, Reston, 2/13/96. Quoted covenant rules are excerpted from *Governing Documents of the Reston Association,* Reston Association, Reston, 1987.

[10] Anthony O'Neill, National Fire Protection Association, Arlington, Virginia, office.

[11] Summarized from "Elderly Housing Upheld in City," The *News-Times, Danbury,* Connecticut, January 24, 1996.

[12] For an extended review of the history of preservation, see Keune, Russell, ed., *The Historic Preservation Yearbook,* First Edition 1984–85, Baltimore: Adler & Adler, 1984.

[13] Russek, Karl "Trends in Environmental Cleanup Costs," *The Corridor Real Estate Journal* 9, no. 3 (January 24–30, 1997).

[14] Reverend Alexander A. DiLalla, OFM, Professor of Biblical Studies, The Catholic University of America.

[15] For additional exploration of the financial benefits of historic preservation, see Rypkema, Donovan, "The Economics of Rehabilitation," the Information Series of the National Trust for Historic Preservation, Washington, D.C.

[16] HABS/HAER is a combined effort to measure and draw existing projects of historic note, particularly if they are in danger of demolition or collapse. It stands for the Historic American Buildings Survey and the Historic American Engineering Record.

[17] Adapted from the Sequence of Inspections chart on the cover page of Kardon, Redwood, *Code Check.* The Taunton Press, 1995.

[18] Reprinted with permission of the American Planning Association from Getzels, Judith, and Martin Jaffe, *Zoning Bonuses in Central Cities,* Planning Advisory Service Report No. 410. Chicago: American Planning Association, September 1988, p. 2.

[19] Summarized, with extensive quotation from Franke, Caitlin, "Facility for Aged Loses Bid to Grow," *Baltimore Sun,* September 4, 1997, p. 1B.

[20] Based in part on an interview with George Calomiris, an architect, building manager, and developer who has had many years of experience with historic preservation projects and community groups in the Washington, D.C., real estate market.

[21] Summarized from account in Levinson, Nancy, "Give and Takings: When Do Land Use Rules Trample Property Rights," *Architectural Record,* October 1994.

[22] For an excellent discussion of the implications of multiuse versus limited- or single-use districts, see Calthorpe, Peter, *The Next American Metropolis: Ecology, Community, and the American Dream.* New York: Princeton Architectural Press, 1993. Also, see Van der Ryn, Sim, and Peter Calthorpe, *Sustainable Communities: A New Design Synthesis for Cities, Suburbs, and Towns,* San Francisco: Sierra Club Books, 1986. Another thoughtful book is Wolf, Peter M., *Land in America, Its Value, Use, and Control.* New York: Pantheon Books, 1981.

[23] From Document B141-1997, *Standard Form of Agreement Between Owner and Architect in Multi-Part Format,* American Insti-tute of Architects, 1997.

[24] Excerpted from "Environmental Compliance Manual 15.0, Floodplains and Wetlands," Lawrence Livermore National Laboratories. Approval date: November 15, 1994. Revision date: June 1996. www.llnl.gov/es_and_h/ecm/ chapter_15/ chap15.html.

This document was prepared by an agency of the United States government. Neither the United States government nor the University of California nor any of their employees, makes any warranty, express or implied, or assumes any legal liability or responsibility for the accuracy, completeness, or usefulness of any information, apparatus, product, or process disclosed, or represents that its use would not infringe privately owned rights. Reference herein to any specific commercial products, process or service by trade name, trademark, manufacturer, or otherwise, does not necessarily constitute or imply its endorsement, recommendation, or favoring by the United States government or the University of California. The views and opinions of authors expressed herein do not necessarily state or reflect those of the United States government or the University of California, and shall not be used for advertising or product endorsement purposes.

[25] Quoted are Tom Kelsch, an EPA scientist, and Clifford Rader, the EPA's chief wetlands enforcement officer, in Lehman, Jane H., "Banking on Wetlands," *Builder,* March 1994, p. 57.

[26] Lehman, p. 59, quoting Scott Middleton, an analyst with the Urban Land Institute.

[27] Schreiber, Stephen, "Anatomy of an Infill Project," *Design/Builder,* September 1995, pp. 13–15.

[28] Adapted from Lassar, Terry Jill, "Great Expectations: The Limits of Incentive Zoning," *Urban Land,* May 1990, pp. 12–15.

[29] Based in part on interviews with Thomas Rohrbaugh, PE, of U.R.S. Greiner.

[30] For further information on energy budgeting, see "A Baseline for Energy Design," *Progressive Architecture,* April 1982, pp. 110–115, and follow-up articles in June, August, October, and December 1982, and February, March, and April 1983.

[31] Coté, Ron, PE, editor, *Life Safety Code Handbook,* sixth edition, National Fire Protection Association, 1994, p. 101.

[32] Coté, p. 33.

[33] Coté, p. 793.

[34] This is a term used in the IBC. The BOCA code refers to "building areas" as "buildings."

[35] This is another IBC term. BOCA refers to "building separation walls" as "fire walls."

[36] *Compartmentation* carries two distinctly different definitions, depending on the code. As used here, it refers to the separation of a project into building areas. In some of the building and fire prevention codes, it refers to the separation of a project into fire areas by fire barriers. That definition is more specialized and is not discussed elsewhere in this book [IBC405.4].

[37] Illustration 1-8 from Karlen, Mark, and Kate Ruggeri, *Site Planning Basics.* New York: Van Nostrand Reinhold, 1993.

[38] For additional exploration of the financial benefits of historic preservation, see Rypkema, Donovan, "The Economics of Rehabilitation," *The Information Series of The National Trust for Historic Preservation,* Washington, D.C.

[39] "Indoor Air Quality Basics for Schools," U.S. Environmental Protection Agency, EPA-402-F-96-004, October 1996.

[40] Quotes are taken from, and descriptions of SBS and BRI are derived from, "Indoor Air Facts No. 4 (revised)," Environmental Protection Agency, April 1991.

[41] This quote and parts of the ensuing section are excerpted from "Indoor Air Quality Basics for Schools—Fact Sheet," U.S. Environmental Protection Agency, EPA-402-F-96-004, October 1996.

[42] Kira, Alexander, *The Bathroom.* New York: Bantam Books, 1976. This is an incredibly thorough and thoughtful exploration of the design possibilities associated with personal hygiene. It is well worth reading, as much for its methodology as for its insights.

[43] For an in-depth exploration of these issues, see Rolf Jensen & Associates, Inc., "Atrium Fire Safety, A Guide to Fire Protection Requirements for the Atrium Portions of Buildings." Deerfield, Ill. 1982, 1985.

[44] Coté, pp. 188–189.

[45] Coté, p. 188.

[46] Bunker, Merton W., Jr., and Wayne D. Moore, eds., *The National Fire Alarm Code Handbook,* second edition. Quincy, Mass.: The National Fire Protection Association, 1997 p. 249.

[47] Coté, p. 122.

[48] For more, see Dorris, Virginia K., "Engineering Fire Safety," *Architecture,* February 1994, pp. 111–115.

[49] Coté, p. 71.

[50] Barnard, K. and F. Johnson, "Ergonomic Design for a More Comfortable Bottom Line," *Corridor Real Estate Journal* 9, no. 12, (3/28/97): A17.

[51] Excerpted from "Environmental Compliance Manual, 15.0 Floodplains and Wetlands," Lawrence Livermore National Laboratories. Approval date: November 15, 1994. Revision date: June 1996. www.llnl.gov/es_and_h/ecm/ chapter_15/ chap15.html.

INDEX